KU-539-762

MANAGING THE ENGINEERING DESIGN FUNCTION

RAYMOND J. BRONIKOWSKI, P. E.

Manager
Product Planning and Ventures
RTE Corporation
Waukesha, Wisconsin

Van Nostrand Reinhold Series in Managerial Skills in
Engineering and Science

 VAN NOSTRAND REINHOLD COMPANY
————————————————————— New York

Copyright © 1986 by Van Nostrand Reinhold Company Inc.

Library of Congress Catalog Card Number: 85-3202
ISBN: 0-442-21440-5

All rights reserved. No part of this work covered by the copyright hereon may be reproduced or used in any form or by any means—graphic, electronic, or mechanical, including photocopying, recording, taping, or information storage and retrieval systems—without permission of the publisher.

Manufactured in the United States of America

Published by Van Nostrand Reinhold Company Inc.
135 West 50th Street
New York, New York 10020

Van Nostrand Reinhold Company Limited
Molly Millars Lane
Wokingham, Berkshire RG11 2PY, England

Van Nost. and Reinhold
480 Latrobe Street
Melbourne, Victoria 3000, Australia

Macmillan of Canada
Division of Gage Publishing Limited
164 Commander Boulevard
Agincourt, Ontario MIS 3C7, Canada

15   14   13   12   11   10   9   8   7   6   5   4   3   2   1

Library of Congress Cataloging in Publication Data

Bronikowski, R. J.
  Managing the engineering design function.

  (Van Nostrand Reinhold series in managerial skills in engineering and science)
  Bibliography: p.
  Includes index.
  1. Engineering design—Management.   I. Title.
II. Series: Van Nostrand Reinhold series in managerial skill development in engineering and science.
TA174.B75   1985   620'.00425'068   85-3202
ISBN 0-442-21440-5

D
620. 0042
BRO

MANAGING THE ENGINEERING DESIGN FUNCTION

To Carol
for loving support
and encouragement
when I needed it most

Van Nostrand Reinhold Series in Managerial Skills
in Engineering and Science

Michael K. Badawy, Editor-in-Chief
Virginia Polytechnic Institute
and State University

Developing Managerial Skills in Engineers and Scientists: Succeeding as a Technical Manager, by M. K. Badawy

Modern Management Techniques in Engineering and R & D, by J. Balderston, P. Birnbaum, R. Goodman, and M. Stahl

Improving Office Operations: A Primer for Professionals, by Jack Balderston

Managing the Engineering Design Function, by Raymond J. Bronikowski

Series Introduction

Managing the Engineering Design function is the fourth volume in the Van Nostrand Reinhold Series in Managerial Skills in Engineering and Science. The series will embody concise and practical treatments of specific topics within the broad area of engineering and R&D management. The primary aim of the series is to provide a set of principles, concepts, tools, and practical techniques for those wishing to enhance their managerial skills and potential.

The series will provide both practitioners and students with the information they must know and the skills they must acquire in order to sharpen their managerial performance and advance their careers. Authors contributing to the series are carefully selected for their experience and expertise. While series books will vary in subject matter as well as approach, one major feature will be common to all volumes: a blend of practical applications and hands-on techniques supported by sound research and relevant theory.

The target audience for the series includes engineers and scientists making the transition to management, technical managers and supervisors, upper-level executives and directors of engineering and R&D, corporate technical development managers and executives, continuing management education specialists, and students in technical management programs and related fields.

We hope that this dynamic series will help readers to become better managers and to lead most rewarding professional careers.

M. K. BADAWY
Series Editor

Preface

The engineer who becomes manager of an engineering design department faces a multiple responsibility. First of all, he is responsible for the technical progress of his company or the division he is part of. He does this by means of the products that are conceived, designed, tested, and approved by him through his engineering design groups. His second responsibility is for these products, which must meet user requirements and be safe, and must not give rise to product liability lawsuits from use or misuse in service.

To meet these requirements, his engineering staff must be competent and creative. The engineering design manager must see to this important managerial responsibility by both staff selection and staff training, at all levels, from expert professional to incoming support person.

Beyond these engineering-related, substantially technical responsibilities, the engineering design manager must also be a competent manager of people, budgets, and time. He must understand his staff and learn what motivates them, how they respond to challenges and stress. He must delegate effectively in order to multiply his capabilities through the development of his staff members. At the same time, he must live and work with budget constraints, time constraints, and shifting priorities, as all managers must. While the engineering design manager is both engineer and manager of engineers, he must not allow his skill as an engineer to interfere with his being an effective engineering manager.

Because of the complex, interrelated responsibilities of an engineering design manager, this book focuses on managing a design engineering department, blending both the theoretical and practical aspects of the job. It is addressed to engineers, scientists, technical managers, design engineering managers, engineers, engineering students—who either want to become better managers, or who are thinking about engineering management as a career goal. And, it may provide insights for those wondering whether the job of engineering design manager is suited to them.

This book was written to provide a framework for an overview and study of the functions and responsibilities of an engineering design manager in a company environment. It blends the experience of the author and his past and present associates with the writings of many experts whose insights and views help to illuminate particular facets of design engineering management problem areas.

The book covers a broad range of subjects, grouped into three parts. The first part, Chapters 1 through 6, covers the design function, project selection, design objectives, the design process, and control, showing how they relate to overall company objectives and strategies. Chapter 6 deals with the design engineering department, covering the more familiar types of organizations and how these departments function. Part II, Chapters 7 through 12, deals with the design manager doing his job as technical leader and supervisor in such a way that projects and non-projects are done effectively. It includes discussions on the design manager, design experimentation, reviews, communications, and publications. Chapter 11, on meetings, is intentionally short, as all well-run meetings should also be.

The third part, Chapters 13 through 16, deals with the important areas of creativity and innovation, patents, and products liability, as well as an assessment of the major factors that comprise the everyday, heavy workload of the design engineering manager. The chapters covering patents and products liability were reviewed by practicing attorneys in both fields, for accuracy and content. They are not intended to be a substitute for a legal opinion on any issue, but rather for information and background in the areas covered.

Throughout the book, there are more references to ''he'' than to ''she.'' The use of the male pronoun rather than the female is not meant to reflect any sexism or discrimination. Rather, it reflects a problem that many other authors have, as well, with using the English language.

Writing this book has been possible because of the help of many people. The engineering, marketing, manufacturing, and general managers I have known and worked with have contributed much to what I have learned about engineering design management in a business environment. I thank them collectively.

On the personal side, my deep appreciation goes to my wife, to whom I've dedicated this book, for her love, encouragement, and understanding. My thanks to Veronica Bronikowski, my daughter-in-law, for typing the book's first draft into my word processor. Thanks also to John Majewski, who drew most of the illustrations for the book. However, while thanking others, let me assert that any mistakes in the book belong to me, the author, not to anyone else.

I hope this book provides each reader with one or more insights into how to do his or her job as an engineering design manager better, more effectively. If this occurs, I have achieved my goal in writing it.

R. J. BRONIKOWSKI

Contents

MANAGING THE ENGINEERING DESIGN FUNCTION

Part I
THE ENGINEERING
DESIGN FUNCTION

1
The Design Function— Bridge between Dreams and Drawings

COMPANY OBJECTIVES AND THEIR RELATION TO DESIGN

Every company expects not only to survive, but to grow stronger in its marketplace. Its basis for growth is providing needed goods or services at a reasonable price that yields an acceptable profit. Its management must make decisions about what must be done each year to place the company in the best competitive position despite drastic changes that likely will occur. How it chooses to do this depends on the industry it is part of and where that industry is going.

If a company is in an industry where growth is rapid and technology is expanding, it may have the objective of capturing market share and becoming a dominant factor in one or more segments of its market. If it is a large company, it may wish to dominate an entire technology area. In a growth situation, the numbers of people both in the design areas and in the product engineering areas are relatively high. Timing is of utmost importance, more so than cost. Products must be designed well and hit the market promptly.

On the other hand, if the industry growth is slow (relatively flat), it may decide to enlarge the market share by becoming the low-cost producer. In a "flat" situation, a company's objectives would be to accentuate the role of product engineering, rather than design. Continuing the major-effort projects would be focused on cost reduction and value analysis. They would concentrate on making products less costly to manufacture and easier to assemble, so that internal costs would be reduced to a minimum consistent with the manufacturing processes available, or to a capital outlay justifiable for new equipment to reduce costs and maintain product quality.

In a flat, no-growth situation, a company may alternatively decide that it has better profit opportunities in other business areas and go into a "harvest" mode. In a harvest mode, the company makes few if any changes in the product, but uses it as a vehicle for improved profit. Prices are increased, and market share is allowed to decline. This is a difficult area for engineering managers to feel comfortable in because most are interested in growth and expansion. But harvest mode is a reality that some must face.

Overall, a company's objectives are constrained by at least two factors: market and money. Money at certain times in a company's history can be difficult to obtain, when the economy is tight, or interest rates are high. Even if the market is expanding, the company may not be able to afford the expenditures necessary to grow with the market, or ahead of it. Or, it may not wish to grow. Some highly successful companies plan not to grow too large. They are of a size to be very effective in their market, and worry about growing so large that their management loses touch with the details of its business and market.

Clearly, no set of objectives is right for all companies. Each must decide its objectives, and the strategies to meet them, based on the industry, its growth characteristics, and the company's own size. Depending on the direction it chooses, these strategies will have a major effect on its engineering design functions.

COMPANY STRATEGIES AFFECTING DESIGN

The technical strategies a company uses are at least four. They are:

- First in the field
- Fast follow
- High-volume, low-cost producer
- Customer needs producer

The particular strategy used by a company will be determined both by the company's management and the market it serves and by the products it manufactures. A small company can use only one of them, if it wishes to prosper. A large company can use several of them, particularly if it is a multidivision firm. Each division could employ a different strategy. But it is difficult to try to employ more than one of these strategies within a division. If it is not done well, the financial results can be disastrous.

Each strategy has its own facets to differentiate it from the others, and each poses different challenges for the design group and its engineering manager.

First in the Field

This strategy captures the imagination and energies of a company's management and its marketing and engineering staffs. There is an exhilarating feeling about being the leader, in first place, plowing the first furrow, setting the pace. The resulting enthusiasm permeates the organization, enhancing commitment to company objectives.

When a new discovery has been made by the development group or a potentially new and significant product need is uncovered by marketing, the desire to be first is like a strong magnet, pulling everything together and focusing everyone on the needs to be filled.

This first in the field strategy impels many companies to greatness. Such a pioneer company usually needs the financial resources of a Du Pont or a General Electric because the price tag associated with being the leader is high. But small companies can become leaders as well if they have that strong objective in mind and develop the products that lead the competition.

This strategy fosters strong interactions between marketing, engineering, and manufacturing. To be first in the field means to "beat the competition," who may or may not be as close to the product as one's own company. But this threat encourages the key groups within a company or division to work quickly, to work closely with one another. Being first means being cooperative and interfacing well. These are pluses.

The manufacturing engineer must be thinking about his fixtures and tooling as the design engineer is developing his parts. The marketeer must be planning the promotion, the ads, and the customer proposals. A recent article in *Advertising Age* promoted the concept "Write the ad before you design the product."* It fits the concept of first in the field, even though it was intended to focus on the need for marketing to develop adequate product specifications as an aid to product design.

There are several minuses to being first. If a design needs to be done in the shortest possible time, it must be the first workable design. It may not be the best design, one that would solve all problems. It probably will not be the least costly design because value engineering will not have been seriously attempted because of lack of time. Internally, it will have higher than usual development and product costs because time is of greater importance than budget. More people, more material, more waste are all a natural consequence of being first.

Priorities, planning, and close scheduling are also keys to making the promise of first in the field a reality. Everything must come together at the correct time, done correctly, or the project will fail.

Follow-through after the product is in the hands of customers is as important as all the foregoing, for a company to stay first in the field. If the product needs improvements to satisfy customers or to bring in new ones, response from the product group must again be rapid, timely. It is important to respond to these needs in order to "grab" market share, and hold it with a new product.

Lack of follow-through to changed customer needs is one major reason

---

* Reprinted with permission from the December 17, 1979 issue of *Advertising Age*. Copyright 1979 by Crain Communications Inc.

why some companies start off as first in the field and introduce new products, but do not maintain the lead. Someone else starts second and becomes the leader.

Fast Follow

A second-place start is frequently an advantage. This is particularly true if the market is expanding rapidly with the acceptance of a new product. While the first in the field strategy may be driven either by technology, by discovery, or by marketing's finding an unfilled customer need, the fast follow strategy is marketing-driven.

Fast follow has one major advantage over first in the field; the product design concept, no matter how new or novel, is an accomplished fact—it is doable. Knowing that something can be done is much more confidence-building than believing it can be done, and working hard to make it happen.

Fast follow strategies are similar to the concept of ''innovative imitation,'' described by educator and business writer Theodore Levitt.* He says that supporting a strategy of imitation starts imitative activities early, shortly after the introduction of a competitor's product. This demonstrates in an organization that while innovation is valued, so is creative imitation. Levitt believes it makes sense to have just as clear and carefully developed a method of planning innovative imitation as planning innovation itself.

The key to success in the fast follow strategy is not merely to ''follow the leader'' but to ''leapfrog the leader,'' and move ahead. The innovative aspect must be in performance or some other customer benefit. The benefit can be a reduction in price, but price reduction is a two-edged sword.

If price is reduced by the incoming competitor, the first in the field will be forced to match the reduced price, and the higher profit will be quickly lost. A seesaw pricing battle may be waged in the short term. It is doubtful that anyone benefits in the long term from price cutting. Both may struggle along at near break-even pricing levels. One of the competitors may decide to drop the product line, take its losses, and channel its resources to other, more profitable areas. Then, prices will rise if only one producer remains.

However, price reduction by the first in the field is a potent weapon if accomplished before the second competitor gets into the market. A low price reduces the projected rate of return on investment on the fast follower. It may cause him to rethink his position and possibly drop the project. So, price is an interesting but not always superior tactic to be considered by the fast follower.

Improved performance is a better inducement to attract customers for a

---

* Reprinted with permission from *Marketing for Business Growth* by T. Levitt. Copyright 1974 by McGraw-Hill Inc.

fast follower than price reduction, and a more practical one. There often are shortcomings in the first product in the field. The device marketed meets the perceived customer needs, but as more and more customers begin to use the device, this changes. New, different customer needs arise, or shortcomings in the equipment are found by users.

If the fast follow company has its marketing and sales people talking to customers, these weaknesses and shortcomings will surface. This information, compiled and analyzed, provides valuable insights as to what performance or operating improvements are needed. If the most important of these can be incorporated in the product being developed, the fast follow company has some key advantages in its new product.

Another factor favoring a performance or operating improvement lies in the natural reluctance of the first in the field manufacturer to make significant changes in his product soon after introducing it to the market. He should be willing to listen, and should be receptive to market intelligence. However, management may insist that the sales force sell the existing product, and criticize them for wanting to sell a better one instead. Thus, the sales group may be accused of inadequacy, rather than of determining the true reactions of the customer.

So, the benefit of improved product plus the reluctance of the first in the field to make rapid changes offer the fast follower an opportunity to be a strong contender, and potentially to take over as product leader.

From a technical standpoint, the fast follow company uses less R & D but more development and product engineers. It must be an innovative engineering group, capable of responding to market inputs, must be concerned with cost and value, and must interact with marketing and manufacturing as much as the first in the field.

Being a fast follower can be just as exciting as being first in the field because the proud feeling of passing the leader is almost as rewarding as being the pacesetter. By being an innovative imitator, the fast follower company may find itself first in the field.

One major concern a fast follower must address in that of patents. If the leader has tied up the basic idea with patents, the innovative imitation must be sufficiently novel to avoid any patents that have issued to the leader. This is easier said than done. It currently takes over two years for a typical U.S. patent to issue. And until it issues, all of the documents and correspondence are confidential. The fast follower has little if any idea of potential patent coverage during the time it is developing an innovative improvement.

A later chapter will discuss patents and their business implications in greater detail. However, for the fast follower, ignorance of the patent situation is a major problem to address. Patent licensing and cross licensing are one way in which businesses cooperatively cope with patents owned by another. However, that can be an expensive approach.

High-Volume, Low-Cost Producer

Despite the aura and excitement of being first in the field, or the innovative fast follower who tries to leapfrog over the leader, there is much to be said for another type of engineering strategy. By becoming a low-cost producer of high-volume products, a company can be successful and profitable as well.

Not all industries see rapid technological change; not all must move at great speed in order not to be left behind. It is in slower-moving business segments that the low-cost design strategy is effective. It is not nearly so glamorous or exciting as the first two strategies, but is certainly is just as important.

If the industry is reasonably mature and the market stable and changing rather slowly, a low-priced product of good quality will sell well. In this engineering (and business) strategy, the approach is to get the best value out of older, established products.

The design group is the heart of the operation, since an R & D function is not needed. With the product performance criteria and customer needs known, the major objectives are to provide production with a better design, one that is easy to build and assemble, looks better, handles more easily, and at the same time, costs less to manufacture. It is not an easy task to build something better and less expensive than the competition, but that is the challenge for this engineering group.

Since the market is stable and time is important rather than urgent, the design criterion for something better than the competition will include a substantial effort in value engineering. Value engineering is functional design the first time, whereas value analysis is functional redesign of a product being manufactured.

Value engineering will be detailed in Chapter 9, but in an overview the following principles need to be recognized. Engineering for best value means taking all unnecessary costs out of a product without affecting its performance. It is not cost reduction or cost cutting but creatively taking out unnecessary costs. It uses a technique called functional analysis.

The functions of a device are those attributes that either make it work or make it salable. After the functions are defined, the cost of providing each function is analyzed and compared with other alternatives. Value analysis works on reducing the cost of providing a given function, rather than the cost of a given part.

Applied value engineering usually results in mechanisms with fewer parts, easier assembly, substantially lowered costs, and improved reliability. Similarity of engineering design does not mean the design group copied the original design. Rather, with value engineering applied, a better product at lower cost is brought to market.

Quality is another attribute that a low-cost producer can bring to the market. The word quality has many definitions. Phillip Crosby, in his provocative book *Quality Is Free,** defines quality as "conformance to requirements." This definition, rather than "without any defects" or "a degree of excellence," is appropriate to measure product quality. Using this definition avoids the ambiguities of "good quality," "poor quality," or "marginal quality." If a product conforms to the requirements of the user, it is a quality product. If it does not conform, it is not a quality product. The subject of product quality will be expanded upon later, in Chapter 9.

All products developed should conform to the user requirements. The "low-cost" engineering design group has the best opportunity to build quality into a product because they have observed products in the market, studied customer requirements, and reviewed deficiencies of existing products in meeting these requirements. With that array of information, they are able to design with product quality as a major criterion.

The objectives of a low-cost strategy can be well paraphrased in the words of Mr. Jack Tramiel, chief executive of Commodore International, a personal computer firm. In early January 1982 a *Wall Street Journal* article described the home computer that his company had just introduced. They believed it would have a major market impact. Yet it was bringing nothing new. Instead it was capable of interfacing with most of the programs of existing home computers—those who lead the field and the fast followers. Mr. Tramiel said, "We don't want to invent anything, we just want to make sure we have the best."†

Low-cost engineering has been looked down on by many, as copying, noninnovative, mundane. However, it could be better compared to the fabled race between the tortoise and the hare. The hare jumped off to an enormous lead and was certain to win the race. He was sidetracked by other diversions but was confident of winning because of his superior speed. The tortoise entered the race and was no speed demon. He very deliberately took one step after another, and stayed in the race. Suddenly the tortoise was approaching the finish line, and the hare found himself too far behind to win, even with a fantastic burst of speed at the finish. The fable has one implication that applies to this engineering design strategy, "You don't have to be the leader in order to be a winner."

Customer Needs Producer

This engineering strategy is different from the others because it involves meeting a customer's requirements in a very specific way. For the most part, it covers larger systems or networks in a customer's plant or operation. Ex-

---

* *Quality Is Free* by P. Crosby. McGraw-Hill, Inc. 1979.
† Reprinted by permission of *The Wall Street Journal.* © Dow Jones & Company, Inc. 1982. From "New Home Computer That Can Emulate Others Threatens to Shake Up the Industry" by R. Shaffer, in the January 12, 1982 edition.

amples are a communications system for a large company and a process control system for a chemical processing plant. It may also include manufacturers who build specialty products on order for customers.

In this engineering area, the technical salesman is a key factor, acting as a bridge between the customer and his company's application engineering and manufacturing capability. The sales representative will often request engineering assistance so that a design engineer may accompany him after the preliminary contact has been made with a prospective customer.

Whether this is done or not, the sales engineer must be able to understand the customer's real needs better than the customer. This is true because once the system is in operation, the customer will better understand his real needs and will appreciate the "extras" that the sales engineer convinced him were necessities to his system operation.

Within the company, some level of research is important to determine changes in technology that will affect the systems and network being designed. Over the years, the progression from mechanical relays to solid state switches, from stepping switches, cams, and mechanical, pin-type activators to programmable microprocessors, has simplified and revolutionized many of the communication or process control system designs. In the next few years, it will substantially change all of them.

The design group has two basic functions and should probably be two semi-separate groups. One group will be responsible for designing the system, the other for equipment and installation. Designing a system to perform as intended is extremely complex. It involves understanding, logic, and imagination. To an extent it also involves equipment, which is the bridge between the two groups. However, the design group deals in an important way with developing interacting networks that, particularly in process control, must monitor, change, and report the status of a variety of inputs and outputs.

The equipment and installation group has the responsibility for selecting the equipment, the actual hardware, and for being certain that it will function in the customer's environment, whether it be the desert heat and sand of the Middle East, the high humidity of the tropics, or the intense cold of northern Canada or Alaska. Beyond that, this group must determine how and where the monitoring and control devices will be installed, considering the needs for access to maintain and repair various items in the system. This responsibility often requires the same amount of logic and imagination as the system design.

The effectiveness of this customer engineering strategy depends on the interactions of these two groups because of the important interrelated roles they bring to each project. The value the design brings to a customer depends on the effectiveness of the total system—how well the total system performs. In other words, is the concept effective in controlling the customer's process? Of secondary importance is the particular equipment selected. Meeting the customer's total need both in performance and time is paramount, not the brand name of any of the items in the system.

Strategy Combinations

Can a company have a combination of first in the field, fast follower (equal or better), and low-cost, high-volume, customer engineering strategies, all at the same time? This was touched on earlier.

If it is a large company or division with multiple product lines, the answer is probably yes. If a small company, probably no. The one exception is customer needs engineering. That is such a specialized area that it would be inappropriate if not foolhardy for such a company to try to be anything different.

However, in the other three areas, a large division may well have all three strategies in place, and all of them going on at the same time. It is not difficult to imagine a company that is first in its field in several areas. Once its product has been on the market a year or more, it is quite likely that a fast follow company will introduce a competitive model. Since the original company's costs are high, its profits may be cut by the decreased sales and the competition. Its best strategy is to begin a low-cost, high-volume redesign of the product to be in a good long-term position on the production–cost experience curve. Similarly, the fast follow company needs to be in a good position to counter these efforts by the original company, or by a third company that sees a large market developing.

The smaller company that stays first in the field does so by bringing new and effective designs out quickly. Sometimes the best strategy is to stick with the first in the field strategy, bringing more and more new products to the market and making a profit in the short run. Such a company's staff works well in its present function and may do poorly if it tries to mix strategies.

On the other hand, the large division will probably have a variety of products, with some that fit each strategy. This can be either troublesome or advantageous. It will be troublesome if the same staff that concentrates on speed and timeliness without equal concern for cost or detail must try to develop a low-cost, high-volume design on a device that a competitor brought to the market earlier.

It is equally true that the low-cost, high-volume group that is vitally concerned about reducing cost 10% even if it takes an extra 90 days, will find it a difficult task to bring out a first in the field design where timeliness is of much greater significance than a 10% cost reduction.

What is the answer? There are at least two answers: doing what one does best and maintaining separate design groups, depending on the size of the company and the market. Again, the small company or division is best served by concentrating on what it can do best—design, marketing, or manufacturing—and the markets it serves. A small company can be a leader, a fast follower, or a low-cost, high-volume producer, but it will falter if it attempts more than one strategy.

A large division can employ two or even three of the strategies, but only if it is willing to pay the price of greater staff, some duplication, and separation of functions, so that individual contributors are not trying to be all things to all products.

COMPANY POSITION AND TECHNICAL POLICIES ON DESIGN

One of the continuing problems faced by a company making a significantly lower profit than its competitors is that it will have difficulty surviving in the long term. It will find its products undercut in price in one area and surpassed in performance or features in another.

If it meets the challenge by increasing staff, profits will drop further, and it may have difficulty borrowing money for the capital tooling required to bring the product improvements on stream. On the other hand, it can improve the P&L statement by reducing the engineering budget. However, the long term will have been sacrificed for short-term survival.

Beyond this simplistic viewpoint, the industry and the company's perceived position in that industry have a great effect on its technical policies. Some companies have a minor position in their industry and intend for it to grow to be a major one. Some companies do not want to grow appreciably, but rather wish to operate in a small, well-differentiated, and profitable market segment. Still others want to maintain their present position and improve their profitability.

Many of the policies made are governed by the major industry the company serves. Energy-related companies see an expanding market, with tremendous growth potential. So do electronics and computer-related areas, but the proliferation of companies and burgeoning technologies make the area one of survival of the fittest. Automotives and steel tend toward slow growth, related to the general economy, growth of the gross national product, and, importantly, foreign competition. Electric power utilities perceive a slow growth in the next few years, as do the companies providing equipment for their systems. However, within a given industry some companies will grow exponentially, others will grow only slightly, and some will decline and go out of business. It is quite possible to have a fast-growing company in a no-growth industry, just as it is possible to have a company shrink in an expanding growth industry.

The policies of a company, its financial stability and borrowing capability, the strength of its R&D, marketing, and manufacturing arms, and the capabilities of its design group are all factors in what a company accomplishes in its own industry and marketplace. Since this book focuses on the design aspect, the design function's important role will be stressed.

Policy statements are only broad generalizations of where a company plans to function. From that policy come major goals and objectives that should be obtained via a short-range or a long-range plan. Some of the important policy statements that lead to meaningful goals and objectives for such plans are these:

- The technology areas where the company intends to operate.
- New markets to be entered.
- The target rate of expansion for the company and its major business segments.

From these general statements, quantifiable objectives can be developed and approved. Once approved, they can be further developed into strategies and tactics that can bring them to reality.

STRATEGIC PLANS AND TACTICAL DESIGN ACTIONS

In the hierarchy of concepts for a company's growth and progress, policies lead to specific goals. Each goal has several measurable objectives, in both the time and the effort required to reach that goal.

To attain each objective there must be one or more strategies, or methods of attack to accomplish all facets of the objective. Each strategy is composed of a number of tactical actions that fulfill the requirements of each strategy. Each action may well be a completely planned sequence of events.

While this may sound complicated, it is not. It is a sequence of planning actions wherein the whole must equal the sum of the parts. In other words, all of the tactical actions must in total comprise the strategy. All of the strategies, when carried out, meet the objective, and so on.

Consider this example: A company's goal is to improve its long-term profitability from near zero to plus 10%. In order to do this, several objectives must be met. One could be to increase profit margins on existing products by 15% or more. Strategies to achieve that objective include:

- Increase prices.
- Find new markets where a higher price can be obtained for the existing product.
- Identify a market segment or niche that is not well served by the competition.

The tactical actions to find new markets could include a market research study to determine what different markets might utilize the product.

Although this example is not quantitative, it illustrates the concept that each tactical action is part of a larger strategy to meet specific objectives toward goals that issue from a company's policies.

INTERFACES ESSENTIAL TO DESIGN

An interface is a boundary between two materials, sometimes more than two. If two solids interface, the interface is clear-cut and well defined, on the macro level, if not on the micro. If the interface is between liquids of different densities, the interface is usually a distinct layer (oil on the surface of water) after any motion or turbulence has subsided. If the two materials have similar densities, the interface will be diffuse and difficult to define; and similarly if gases are involved.

The interface is the location where the action between the two materials takes place. If one material is a copper bar and the other is hydrochloric acid, the bulk of the copper and the acid is initially unaffected, with all the bubbling and copper–chlorine interaction occurring at the interface between the two.

What does this have to do with design? Quite a bit. Design is not an entity unto itself; it interfaces with a number of other areas. Key areas vital to effective design are R&D, marketing (the customer), manufacturing, and finance. There are many other parts to a company—personnel, accounting, maintenance, purchasing, material control, for example. These areas are all equally important, but for the sake of brevity, only the four key ones will be touched upon here. In another volume in this series, interfaces will be dealt with in greater depth.

Design with R&D

Many new products get their start in a research/development program where the dream of fulfilling a need is thought about, mulled over. It finally emerges in some type of working model that demonstrates feasibility. The device developed may or may not resemble a finished product, but it has taken a large amount of imagination and perspiration to get the concept to the point where it can be transferred to design.

The device has usually taken longer to develop than was originally planned, and the time remaining to take it into prototype and production may be tight. If R&D has not been in close contact with design, one predictable result will be the NIH (Not Invented Here) reaction. Design will look for all the problems, seeing few of the advantages and all of the disadvantages. It may even request additional time to redesign the device completely in order to create an acceptable unit.

The interface between design and R&D is different from the other design interfaces because, to an extent, it is competitive. Often, design believes that R&D has not been particularly innovative, and that it could have come up with an equally good device in less time; or it believes that R&D has been too innovative and spent too little time on development, leaving it all up to design.

To be effective, the interface between R&D and design must be informal, cooperative, and interacting. It must not be a structured firm interface, but must be relatively flexible. Since both groups have a major stake in the success of the device, involvement by design in the later stage of R&D is necessary, before it is complete. This is not always easy because an R&D group may feel design is dictating what it wants to R&D—and that may be true. However, interaction before technology transfer is a key success factor.

The technical success of an R&D project transferred to design depends on a high level of cross-communication and involvement. The transfer of the project from R&D to design should be smooth, but it should not be final. When design takes over responsibility, interaction should continue. Design should advise the R&D group of its progress, and not just give a recap of problems and shortcomings.

Design with Marketing

Not all products begin in R&D; some start in marketing, and flow to design engineering for product development. Whichever direction the product starts from, the requirement is the same; the interface between design and marketing is the key to commercial success.

Marketing is the key interface with customers. It learns and translates their diverse requirements into interrelated common needs that can be expressed in design. For that reason, in many companies, marketing represents the customer, and the design/marketing interface is also the design/customer interface. In order to make this interface effective, the tentative specifications from marketing must be clear and as concise as possible. If they are not, the ambiguities need to be clarified.

However, as the particular projects evolve during the design phase, customer requirements may change. Sometimes this occurs because customers get a better perspective of their true needs; sometimes additional prospective customer needs are factored in. It is important that these changes be factored into the project plan, and any effects on time, budget, or performance be evaluated. Sometimes the change can be incorporated readily; sometimes only with great difficulty and expense.

The interface between design and marketing must be kept fluid and informal. Formality is required in setting the specs and making modifications to them, but the day-to-day interactions must be informal to be effective.

Design with Manufacturing

The design interface with manufacturing is the key to bringing a successful design into low-cost, reliable production. However, in some companies a design is tested to completion, a set of production drawings is issued, copies are given to manufacturing, and then design waits for a response from manufacturing. Requests for dimensional revisions or material changes are, for the most part, denied, on the basis that too much testing would have to be repeated. Manufacturing would be under considerable time pressure to complete its work and get the design into production, to meet a commitment or a bidding deadline that marketing required.

In such an environment, a manufacturing group rapidly develops an NIH reflex of its own. Any errors in design, parts that do not quite fit, or tolerances that are inappropriate are seized as opportunities to return the product to design and have "them" do it right. At that point, some higher-level executive is forced to act as a Solomon and to decide who must do what. This "decree" makes the interactions between manufacturing and design so formal that they become nearly unworkable.

None of this is necessary. Sometimes it happens because the design engineer, young and inexperienced, does not know enough about the plant's manufacturing operations and is either too shy or too proud to ask questions. Such an attitude is stupid, not ignorant, because it is mutually destructive. Other times it happens because the design engineer is arrogant and believes that his design will perform only if it is made exactly as he designed it. He is unwilling to listen to any suggestions for change because only he knows the intricacies of the design. This attitude is equally troublesome and potentially divisive.

It is important for the design engineering manager to find out how effectively his interfacers are performing. He can do this by contacts with his peers, in the division or the company, with whom his staff members interface.

Interfacing is partly natural, partly learned. One simple guideline a manager can use to evaluate the effectiveness of an interfacer is to find out if that individual works toward mutual benefit (win–win), or tries to manipulate others to gain his own views without concern for the other group (win–lose).

The design interface with manufacturing, in some respects, is the same as the interface with R&D, but with the roles reversed. Design has completed a product design, and manufacturing is expected to convert it into its own—a manufacturable product. The same problems are likely to occur; if the interface is formal, and initiated only when the design is completed, look for problems in implementing, as mentioned before.

If the interface is started early enough in design, manufacturing engineers are aware of what is happening and have input into the shape and manufacturing requirements of parts. Two good results occur. First, the final design will be well understood, and manufacturing will cooperate willingly in doing its portion well. Second, if tight tolerances or finicky inspections are needed, manufacturing can plan for it in advance and will support these necessary requirements.

One method of fostering improved interfaces with design is the task force approach, with manufacturing engineering membership. Another method is to assign manufacturing engineers to design teams, so that design and manufacturing work together from early design to full-scale manufacturing. The extent to which these powerful interface methods can be employed depends on the size, complexity, and timing of the design project—its value to the company. It also depends on the size of the company and the resources it can afford to dedicate to the project. In either case, it is sheer folly to ignore the key interface between design and manufacturing. The "product" of the design group is words and paper, and samples, not the finished product. Only the manufacturing group can convert the paper to a finished product.

Design with Finance

In any design project, funding is involved. Before a major effort is started, return on investment estimates, project costs, expenses, and capital must all be quantified. As the project progresses, expenses must be monitored, unusual or unplanned requests considered, and the project budget balanced.

Doing this effectively requires a working interface with finance. That may mean the company comptroller in a small company, or a division accountant in a larger one. The financial function is an important one because the chief executive and his managers are judged on their financial performance, usually on a quarterly basis. Although one can argue that this is not the best way to run a business for long-term progress, one cannot argue against the reality of it.

Since the finance function is so important to top management evaluation, it is important to recognize this reality and develop working interfaces with the people involved. If an approval requires certain financial estimates, find out how the financial people need the information in order to get the project approved, and help them get accurate realistic estimates. Do not do it one way and then complain because that is not the way accounting wants it done. This is only common sense, but it is not always used.

Engineers have a tendency to call the finance people "bean counters" and "purse string holders." However, those people have problems and constraints of their own. Since they have "money control," the prudent manager makes sure his staff works effectively with finance, for a mutually beneficial result.

Other Interfaces

In the course of everyday project and product design effort, effective interfaces must be maintained with diverse other groups in the company: purchasing, quality control, inspection, production control, tool room, maintenance, and personnel, to name the most prominent ones. Each of these groups consists of people, and good human relations will foster effective interfaces without the design manager's losing sight of his major objectives.

DESIGN TIMING

The success of an engineering manager and of his design group depends on the timing with which product projects are brought to the marketplace. Timing a product's entry depends, first of all, on having the product designed and built by the required time. But in a broader sense, timing depends on the company's design strategy. If it leads the field, the company needs a sixth sense about what products are needed. It then makes its product available about the time the need is felt by a number of major customers.

Timing for the second in line, the innovative imitator, is more critical. It must bring the product out as customer acceptance begins, after the leader has interested major customers. It must come out as buying habits are beginning to change and before they have crystallized. Since some customers insist on second sources for many items, the fast follow company has a built-in advantage, provided the second product is there quickly and meets customer needs.

Timing for the low-cost, high-volume firm is important but perhaps not so critical as in the two previous examples. If the industry or technology is in a state of rapid technological change (as for small computers in the early 1980s), this firm can step in, powerfully, and command part of the market; but it must be cautious, as continuing change may make its product obsolete. If the industry is slower-moving, there are both opportunities and problems associated with timing. Again the firm must enter the market as demand is growing and becoming strongly established. This market growth enables it to sell the product at a high profit level and perhaps to wrest the commanding market share away from either the leader or his imitator.

If a company decides to maintain its lead by "jumping on the experience curve," it is difficult to establish good competitive timing. The experience curve is based on an analysis of major industries by the Boston Consulting Group. Their studies showed that in major industries—automotive and electronics, for example—the businesses that were long-term leaders consistently reduced their costs without "cheapening their product." Analytically, each time the total production of the product doubled, costs were reduced about 20%. Firms that did not do this eventually lost business and dropped out of the market, or had a drastically reduced share of it.

The experience curve is a powerful lever for establishing and maintaining market position. The first in the field may or may not utilize it because of its continuing "obsession" with new products. However, the company that uses this strategy will make the competition more difficult for other market entrants, regardless of timing.

Recognizing these factors, one realizes the critical importance of timing to the success of a design group and its manager. Timing in its broad sense must simultaneously encompass design, manufacture, marketing, and promotion, all as part of the same project. It is of little value to a company to have a design tested and built if the sales and marketing group has not completed its necessary assignments of sales training and information, advertising, promotion, and other important tasks needed to get the product into the customer's hands.

DESIGN PLANNING

Planning of a design must factor in four important facets: projects, priorities, performance, people. These can be considered the four P's of design, different from the four P's of marketing (product, price, place, promotion). These four factors will be developed in detail in the following chapters.

Design planning is substatially more than enumerating all the projects, ranking them according to priorities, agreeing to the performance parameters, and assigning projects to various people on the staff. In reality, design planning is a continuous process, sometimes a hectic juggling act. Projects must be initiated and kept going at a reasonable level, consistent with priorities that shift and change. This week's number-one-priority project may be replaced next week by one of greater urgency. Such rapid changes are not only unreasonable, but it is unrealistic to expect to function effectively under this degree of variability. The manager's ability to support the vital projects, resist the unnecessary ones, and be able to differentiate between the two, marks the difference between a mediocre manager and a superior one.

In addition to variable priorities, another planning problem is the change of performance requirements as the device being designed moves from experimental phase to final prototypes. This type of change cannot always be resisted because customers needs may have changed, or new potential customers with different needs have been providing marketing input.

The final variable, which tends to be considered a constant, is people. People—design staff—are a resource that must be used on projects. The estimated time to complete project tasks can be calculated, but the degree to which the project challenges the skills, self-motivation, and personal drives of the engineer assigned the project cannot be. However, these influences determine whether the project will be done effectively and on time, and to a much greater extent than one might anticipate.

Yet, it is within this changing spectrum of requirements, needs, and resources, this arena of changing values, that the manager of engineering design must function effectively. The chapters that follow describe major areas in which the manager functions, developing and improving his own expertise and learning to cope with the diverse requirements of his job.

SUMMARY

A company's design objectives are influenced by the business that it is in, market growth, and the newness or maturity of its products. It may employ strategies that range from being first in the field to being a producer of customer-designed products. Each strategy requires a different functioning design group, from strong emphasis on new, innovative products, to cost reduction, to specialty design. A large company may employ more than one strategy, depending on its product mix, but a small company usually cannot afford more than one.

The design group must effectively interface with other groups in the company because it is the bridge between the dream or product concept and final drawings and specifications that define the end product. These groups include R&D, marketing, manufacturing, finance, and several others. The purpose of interfacing is to time correctly the entry of a new or improved product to meet market needs, or competitive actions. Effective planning on the part of the engineering design manager and his key subordinates is needed to make interfacing effective.

2
Selecting Engineering
Design Projects

INTRODUCTION

Successful designs meet performance, schedule, and budget targets, often in that order. Most important, they meet customer needs, and are profitable for the company as well. This chapter deals with the key aspect of any design engineering strategy, the screening and selection of the most promising projects.

SELECTING PROFITABLE PROJECTS

No company begins a project with the expectation that it will be a loser. Every project is initiated with the perception that it will become a marketplace winner. Not only will it provide significant benefits for customers; it will also bowl over the competition, and provide significant gains for the company.

If this optimistic prediction were realized most of the time, more companies would be profitable and financially healthy. However, not all projects turn out as expected. Very few of them are better than expected, and only a few are on target. Most do not meet expectations; some fall significantly below them, and others are near disasters.

Sometimes the cause is a product that has too high a selling price because its costs are excessive. Sometimes it is timed too late; other competitors have become established in a relatively limited-size market. It may be timed too early; a solution is developed to a problem that does not yet exist or does not exist to the extent anticipated. There are many reasons why a product fails in the marketplace.

Product failures can be reduced. If a serious effort is made at the time when a project is begun to determine its probable value to the company, the success average will increase. Selecting profitable projects is neither science nor art, but it has elements of both. It replaces guesswork with thinking and preliminary calculations, but it does not rule out the role of intuition, hunches, or strong management support of a particular project.

EVALUATION GRIDS AND SCREENS

There usually is no shortage of potential products to be considered for further development and production. The problem is one of preselection or screening of many projects to determine which are most promising. Screening of some type is a necessity because no company has adequate resources to undertake every product project that is generated.

The general rule for choosing "the" product idea is to select the *right product at the right time.* For a given company, this means the product:

- Fills a need that is presently not served, or is served inadequately.
- Serves a current market where demand exceeds supply.
- Has a differentiating advantage over existing products such as improved features, better performance, or lower price.

Many proposed projects meet these generalized objectives to a greater or lesser extent; but in order to separate the better projects from the poorer ones, it is necessary to develop a methodology for screening. It must be a selective simple method, yet one that provides high assurance that a potentially valuable project has not been rejected and a poor one accepted.

One method, developed by the consulting group of Schello and associates, asks three basic questions about the product, the market, and the company itself. The questions are:

- Is it real?
- Can we win?
- Is is worth it?

Thus, for a proposed product one asks: Is it realistic to believe that this product can be successfully designed and built? Can the company market it, on the basis of new or unique features or benefits? Is it worth it, in the sense that the price obtainable will so exceed costs as to be sufficiently profitable? Is it worth diverting the needed technical talent into this design/development/manufacturing program, instead of a different one?

The approach is sound and usable. Depending on the amount of effort expended at the screening stage, this type of analysis could be done first as a screening method, then repeated in greater depth as part of a feasibility study on those projects that survive the screening process.

Other Screening Methods

Other types of screens are used for a preliminary evaluation if a relatively large number of potential product projects are being considered. The total requirements of all the projects probably exceed the capabilities of the design group to design them, the manufacturing group to tool up and manufacture them, and the financial resources of the company to pay the bills; but a priority ranking allows a company to match opportunities to resources.

Figure 2–1 is an example of a screen that can be filled out in a relatively short time using a highlighting yellow marker. All of the criteria are listed on the left, and the range of expectations is listed in five columns, from excellent to poor. In addition, there is a weighting factor, which in this case doubles the emphasis on gross profit, potential market size, fifth year market share, and time to introduction of the product. A total of 16 factors are considered. In this evaluation, a perfect product would receive a 100 rating, a poor product 20. Preselection screening will typically show products ranging from 40 to about 80. A product rating in the nineties would be rare; 70 or more would indicate a product worth further consideration.

A company in a different market will have different ideas of how to construct and use a screening chart. Return on investment criteria may be much lower, market share criteria higher; or the ranges may be acceptable, but product leadership may be rated more strongly, as well as technical ability to develop and produce the product against cost and timing criteria.

Alternatively, an entirely different screen may be used, with more attention paid to details. Figure 2–2 is an example of such a screen. It contains 27 technical screening items and 30 market and business items. Weightings vary from −3 to +3, so that a super product (perfect) has a chance at a 159 rating, and a "dog" would rate −171. A good product would rate at least a 70, and a marginal one about zero.

The form also provides a left-hand column for checking off key areas for more significant review or emphasis. Note that it includes an estimate of confidence in the market data and credibility of the benefits and liabilities. It also has a weighting factor for the "horsepower" of the project, assigning a "BAD" or −3 to an R&D-sponsored project, a "BEST" or +3 if two corporate vice-presidents champion it.

Both forms are designed to be completed in about 10 to 15 minutes, assuming that a reasonable amount of data is available. If too many blanks cannot be filled in, the value of the screen is much diminished.

The screens should be completed by several people working independently. Marketing, R&D, design engineering, and manufacturing managers should each prepare them, as should the planning manager. Then they compare results, looking for major differences of viewpoint. After that the group

CONSIDERATIONS	WT. FACTOR	EXCELLENT 5
EST. GROSS PROFIT	2	50% and over/yr
EST. IROI	1	90% and over
CURRENT ANNUAL AVAILABLE BUSINESS	2	$10 million +
MARKET POTENTIAL 5 YR	1	In growth stage. Increasing sales & demand at an increasing rate
EST MARKET SHARE 1 YR	1	25% and over
EST MARKET SHARE 5 YR	2	50% and over
STABILITY	1	Product resistant to economic change
DEGREE OF COMPETITION	1	No competitive products
PRODUCT LEADERSHIP	1	Fills a need not currently satisfied. Is original
CUSTOMER ACCEPTANCE	1	Readily accepted
INFLUENCE ON OTHER PRODUCTS	1	Complements and reinforces an otherwise incomplete line
MANUFACTURING CONTENT	1	Completely manufactured in-house
PATENT POSITION	1	Impregnable. Exclusive license or rights
SALES FORCE QUALIFICATION	1	Qualified sales force available
TIME TO INTRODUCTION	2	Less than 6 mo.
TECHNICAL ABILITY TO DEVELOP AND PRODUCE	1	Present technical know-how available and qualified

Fig. 2-1. Simple product screen.

ABOVE AVERAGE 4	AVERAGE 3	BELOW AVERAGE 2	POOR 1
49–30%/yr.	20–20%/yr	19–10%/yr.	9% and less/yr
89–75%	74–60%	59–49%	39% and less
$10–7.5 million	$7.5–$5 million	$5–$2.5 million	Less than $2.5 million
Reaching maturity Increasing sales but at a decreasing rate	Turning from maturity to saturation. Leveling of sales	Declining sales & profits	Demand, sales & profits declining at an increasing rate
24–15%	14–10%	9–5%	4% and less
49–30%	29–20%	19–10%	9% and less
Some resistance to economic change and out of phase	Sensitive to economic change—but out of phase	Sensitive to economic change and in phase	Highly sensitive to economic change and in phase
Only slight competition from alternative	Several competitors to different extents	Many competitors	Firmly entrenched competition
Improvement over existing competition	Some individual appeal, but basically a copy	Barely distinguished from competitors	Copy with no advantages, possibly some disadvantages
Slight resistance	Moderate resistance	Appreciable customer education needed	Extensive customer education need
Easily fits current line, but not necessary	Fits current line, but may compete with it	Competes with, and may decrease sales of current line	Endangers or replaces an otherwise successful line
Partially mfg'd, assembled & packaged in-house	Assembled and packaged in-house	Packaged in-house	No manufacturing content
Some resistance to infringement, few firms with similar patents	Probably not patentable; however, product difficult to duplicate	Not patentable, can be copied	Product may infringe on other patents
Sales force has basic know-how. Minor product orientation required	Sales force has basic know-how, requires product and application education	Sales force requires extensive product and application education	Existing sales force inadequate to handle market and/or product
6–12 mo.	12–24 mo.	24–36 mo.	over 36 mo.
Most technical know-how available	Some know-how available	Extensive technical support required	Ability to develop and produce with present technology questionable

FIG. 2-1. (*Cont.*)

TECHNICAL SCREENING	BAD	POOR	OK	GOOD	BEST
LEGALITY	Illegal	Publicly??		Unrest	Known Legal
SAFETY	Decrease	Slight Decrease		Slight Improvement	Improvement
CHARTER	Conflict	Questions			
CONFIDENCE IN SUCCESS	<40%	40-60%		60-80%	>80%
EXCLUSIVITY					
COMPANY'S PATENT POSITION	Open or None	Unknown		Limited Pat.	Patentable
TECHNOLOGICAL ADVANTAGE	No Advantages	Slight Advantage		Good Advantage	Breakthrough
COMPETITION: NOW	Strong Competition	Moderate Competition		Non Price	None
LATER	Strong Competition	Moderate Competition		Non Price	None
CUSTOMER BENEFITS					
PERFORMANCE	Decrease	Minor Decrease		Some Improvement	Obvious Improvement
EASE OF USAGE	Instructions	Demonstrate		Familiar	Easier
CONVENIENCE/COMFORT	Decrease	Minor Decrease		Some Improvement	Obvious Improvement
ECONOMY OF OPERATION	Decrease	Minor Decrease		Some Improvement	Obvious Improvement
RELIABILITY/DURABILITY	Decrease	Minor Decrease		Some Improvement	Obvious Improvement
EASE OF SERVICE	Decrease	Minor Decrease		Some Improvement	Obvious Improvement
CREDIBILITY OF: BENEFITS	Little	So-So		Good	Very High
LIABILITIES	Very High	Good		So-So	Little
OTHER COMPANY BENEFITS					
PROD COST VS COMP: NOW	Disadvantage	Slight Disadvantage		Slight Advantage	Advantage
LATER	Disadvantage	Slight Disadvantage		Slight Advantage	Advantage
EQUIPMENT & PLANT	New Plant	New Equipment		Some Equipment	Tooling Only
MANUFACTURING PROCESS	All New	Some New		Familiar	Routine
FIXED INVEST VS COMPETITION	Disadvantage	Slight Disadvantage		Slight Advantage	Advantage
EQUIPMENT & METHODS	>6 years old	3-6 years old		New	Latest
VERT. INTEGRATION: MATURE MKT	Little/None	Some		Good	Significant
GROWTH MKT	Significant	Good		Some	Little/None
COST OF WARRANTY	Increase	Slight Increase		Slight Decrease	Decrease
INTERNAL FOLLOW-ON	None	Need Men		Budgeted	
CHAMPION	R&D	Mkt/Eng		Profit Center	2 Corp VPs
OVERALL TECHNICAL RANK	-3 -2	-1		+1	+2 +3

Fig. 2-2. Detailed product screen.

MARKET & BUSINESS SCREENING	BAD	POOR	OK	GOOD	BEST
MARKET DATA					
MARKET GROWTH RATE	Decline	0-5		>10%	>20%
POTENTIAL - WORLD WIDE RTE	<$5 Million	$5-15 Million		$15-30 Million	>$30 Million
RTE	<$3 Million	$3-5 Million		$5-10 Million	>$10 Million
RTE EST SALES/YR	<$1 Million	$1-3 Million		$3-5 Million	>$5 Million
TIME TO ACHIEVE SALES	>4 Years	3-4 Years		2-3 Years	< 2 Years
PRICE ($ /Units)	Declining	Stable		Rising	Rising Rapidly
CONFIDENCE IN ABOVE	Good <80%	Very good 80-90%		High 90-95%	Very High >95%
EXCLUSIVITY					
MARKET SHARE NOW	<15%	15-25%		25-35%	>35%
LATER	<15%	15-25%		25-35%	>35%
NEXT 2 COMPETITORS VS RTE	Twice RTE	Greater than RTE		Less than RTE	Half of RTE
PRODUCT LEADERSHIP	Disadvantage	Slight Disadvantage		Advantage	Advantage
PRODUCT "IMPORTANCE" TO CUST	Critical	"A" Item		"B" Item	Minor
CUSTOMER BENEFITS					
APPEARANCE	Not Styled	Conventional		Good	Wow
CONSPICUOUSNESS	None	Identifiable		When in use	Easily Seen
MARKET DIFFERENTIATION	None	Minor		Can Advertise	Good Ad Copy
SOCIAL VALUE	Anti-social	Questionable		Some Improvement	Obvious Improvement
SERVICE AVAILABILITY	Factory Only	Training		Parts & Instructions	Available
PRICE VS COMPETITION	Disadvantage	Slight Disadvantage		Slight Advantage	Advantage
OTHER COMPANY BENEFITS					
DISTRIBUTION	Other New	New OEM Dist		Need Men	Suitable
PRESENT PRODUCT SALES	Replaces	Some Replace		Compliments	Increased
NEW PRODUCT LIFE	<3 Years	3-5 Years		5-7 Years	>7 Years
SIMILARITY TO PRESENT	All New Prod	Mostly New		Fits Line OK	Fits Well
CUSTOMERS	All Old	Mostly Old		Mostly New	All New
CHANGE IN MKT SHARE: MATURE	Increase	Slight Increase		Slight Increase	Decrease
GROWTH	Decrease	Slight Decrease		Slight Increase	Increase
MKTG EFFORT VS COMPETITION	Disadvantage	Slight Disadvantage		Advantage	Significant Advantage
R&D TO MARKETING RATIO	One Emphasized	Unbalanced		Balanced	
ESTIMATED ASSET RETURN	10-20%	20-30%		30-40%	40%
FORM OF REWARD: MATURE	Share	Share		Profit	Profit
GROWTH	Profit	Profit		Share	Share
OVERALL MARKET RANK	-3 -2	-1		+1	+2 +3

needs to review the differences and resolve the ones that can be resolved. Any that are not resolved need to be highlighted so that when the project proposal is discussed with the general manager, the differing views can be reconciled and factored into the evaluation.

Responsibilities for Screening

The question of who should be responsible for reviewing, summarizing, and reporting results of screening sequences is important, but not as important as recognizing that it should be an individual rather than a group of people. A collective responsibility means no responsibility for getting the job done. Stated more simply, if something is everyone's responsibility, then in reality, it is no one's.

The person who should be responsible may be the R&D project manager, the design project manager, or the product planning manager. Each of these individuals has a keen interest in a variety of projects, and in their thorough evaluation, before proceeding with or dropping any.

While input to the screening process comes from many areas—marketing, finance, production, and others—it may be desirable to have the responsibility at a top management level. However, the top manager may not have the time to sort and sift. Therefore either a technical area (R&D or design) or marketing should be responsible: marketing, because marketing imputs are essential for product evaluation; technical, because the products selected determine the expenditures of technical efforts, which represent the output of a limited technical resource—people.

Use of the screen has a major advantage: it forces the participants to think about the proposed product from several points of view. A well-thought-out screen covers all of the major technical, financial, and customer-oriented aspects of the problem. Some of the areas would not necessarily be considered in a simplistic product evaluation format.

The weakness of the screen approach is that the data can be biased in one direction by a proponent of the project, and in the opposite direction by an opponent. If several people screen the project and then compare results, these biases can be brought into the open and dealt with.

Screens are initially used as a method of separating potentially strong products from weak ones. After a portfolio of projects has been selected, the final composite screens can be kept and used, with periodic updating as the project moves forward and new marketing, technical, or financial data are developed.

In most cases, a company has a portfolio of projects under way, some nearing completion, others not. Addition of new projects obviously must be done with recognition of commitments already made and resources already allocated.

However, the screening method can be reapplied at this period of evaluation, to make sure no major changes have occurred in the market, the technology, the company's manufacturing and financial capability, the costs of the project, or the product. It is a type of zero base budgeting of ongoing projects to make sure none of them is carried on purely by inertia, sunk costs, or someone's pet idea. Thus, a screen has value over the life of a product's development as long as it is periodically updated.

MARKETING ESTIMATES

After project screening and evaluation have been completed, the list of proposed projects has been reduced to probably half its original length, or less. The remaining list is attractive, but it is likely still to represent a greater commitment of resources than the plant or division can handle. So, it remains necessary to do a further analysis of the key areas of marketing, manufacturing, and finance, in order to decide which projects should be undertaken.

Marketing inputs are essential to evaluate the worth of the project. In evaluating the screening method, marketing may have made some WAGs (Wide Angle Guesses) to estimate probable market size and market share. But now, in the business analysis phase, marketing expends effort to assess the likely market, the strength of competition, and selling prices that are current, as well as those that may result if competitors drop prices when the new product is released. Product cost targets also must be evaluated. Even though marketing does not always price on the basis of cost, expected cost is a major factor in the company's ability to keep the product viable in the face of unfriendly, untimely competitive price cutting.

Selling Price

Selling price is determined, in the final analysis, by the market. However, for the company that is first in its field, there is no market to compare against. In this case, marketing must evaluate the price for which products of similar features and benefits are sold in the same market. If no such product exists, the best guesstimate of price, based on the value of the functions provided, is the best number available.

For the company starting second, the innovating imitator, a price has been established. However, it may not be a stable one. The entry of a second supplier invariably forces price down, sometimes by the first supplier, who wants to prevent the second company from gaining a toehold on the market.

Sometimes the second supplier offers the product at a lower price for exactly that same reason. In still other cases, customers who now have two suppliers to choose from will try to get them to bid against each other and lower the price to them.

Thus, the estimate of market price is a constant only for a brief period. Sometimes it depends on product cost, but most of the time it depends on the market, and how it accepts the new product being offered it.

Volume

Price and volume go together. In the classical supply–demand examples, price and volume tend in opposite directions—high price meaning low volume, low price high volume. If that were always true, then the relationship would be:

$$Price \times Volume = Constant\ dollars$$

Here price and volume are inverse functions. In order to double the volume, halve the price, for example; or if price doubles, the volume is only half. Such cases rarely occur. However, there is a type of inverse relationship between price and volume, particularly if a large market is available and customers want the product.

In the general case of bringing a relatively new product into the market, the initial sales volume frequently is low, while customers try out the product to see how it meets their needs. During this period of introduction, marketing effort needs to be strong, but the effort seems slow to pay off. As the product gains acceptance, volume increases more rapidly as the life cycle moves into the growth stage.

Marketing must project these changing volumes on a year-by-year basis, to enable the product plan to be viable and reasonably accurate. The volume estimates will probably be wrong because reality is always different from what is predicted. The new product may be well accepted, and customers may purchase it in larger quantities than planned; or it may be slowly accepted, and no one will appear to be interested. Marketing volume estimates need to be reasonable, but they will never be correct. For that reason, frequent reassessment of the market and revision of volume estimates is important.

Timing

Timing of product introduction is another key element in product planning. But at the beginning of a project, timing of product introduction is partly subjective, unless a competitor has introduced a similar product. Then, timing has a sense of urgency.

Timing has several important ingredients, including the economic, technical factors and the political climate when the product reaches the market. In the case of a first in the field, there is no competition to provide clues. One danger in introduction is that the product is too early for the market.

As an example, in the early 1970s a company developed a system of accessories by which gas, electric, and water meters could be read automatically. Using telephone circuits late at night, a central billing computer would interrogate each customer's equipment on a monthly basis, read the meters, compare them with the previous month's readings, and compute the bills. Technically the idea was successful, and trial installations met every criterion set. However, the major utilities—gas, electric, and telephone—were not willing to support the concept with large-scale installations, and played politics to delay and defer all efforts to move ahead. Timing in the real world was not consistent with product timing because both economic and political factors were involved. The product system failed because it was at least ten years before its time.

Today, electric utilities are utilizing load management to defer capital equipment purchases and their computers are turning off water heaters and major appliances in homes for brief periods to shave peak demand. This is almost the inverse of reading meters. It appears that automatic meter reading is still another decade away.

Timing of a totally new concept can be completely unforecastable. However, marketing needs to do forecasting of the future as effectively as possible, under conditions of uncertainty and risk.

Timing is also important when a product improvement or new product is introduced in an existing market. If large users typically place blanket orders for their major needs early in the year, products must be qualified and meet specs, probably in early fall, before bid inquiries go out. This is analogous to producing items for Christmas purchases. Such products must be introduced in early spring so that store buyers can react and place orders in time for fall delivery, for the Christmas sales.

Only marketing can evaluate the timing requirements that need to be met in order to introduce the product at the best possible time and gain the customer impact desired. This timing must include all time elements: the engineering time estimates for development and approval and the time for manufacturing to tool up the product and produce the first pilot production units for approval tests, as well as time for preparations for higher volume production. These requirements will be discussed later.

Cost Targets

Cost targets need to be set by marketing and engineering with input from manufacturing. Price is necessarily set by the market, either by competition already there or by prices paid for somewhat similar items. However, it is of no value to develop a device whose costs are so high that it cannot be sold for a profit adequate to provide the needed return on the investments allocated to the project.

Because of this requirement, reasonably accurate costs must be estimated, based on anticipated volume. This can best be accomplished by setting a maximum cost target and having both engineering and manufacturing jointly prepare a cost estimate. Although a preliminary estimate on a design that has not been built cannot be precise, it can and must be reasonably correct.

Costs tend to rise, rather than drop, as design and development take place. This phenomenon is caused by changing requirements or by underestimation of the complexity involved. By recognizing that costs tend to rise, marketing can predict that if the estimated cost starts out above the maximum cost target, it will get higher, not lower. If the cost estimates come in well below the target, there is reason to believe the final cost will be close to target.

Thus, in gathering information for project selection, marketing estimates of selling price and volume, timing, and cost are necessary to make an effective estimate of the project. Another area of importance is the capability of manufacturing to produce the device economically.

MANUFACTURING ASSESSMENT

It is a major effort to design a key new product. Sometimes it is an even greater effort to manufacture it economically. For this reason it is important that manufacturing be involved during early screening and evaluation of projects.

While there may be no blueprints or prototype models to aid in the determination of cost and facilities capabilities, there will be sketches available. Also there may be photos, and perhaps a sample of a competitive device that is being considered by the innovative imitator.

The importance of cost awareness was mentioned earlier. Of equal importance in evaluation is an estimate of capital required for tooling and for the facilities required to produce the device at the projected volume levels.

Floor space requirements, temperature and environmental controls, air flow, special fixtures, and handling equipment must be evaluated also. The impact of the proposed new product on existing products must be considered, and vice versa. In some cases, existing tooling can be used on a product-line-expansion product, provided the volume levels do not exceed total capacity.

In other cases, a new product may require an entire new facility or a major addition to an existing one. Building layouts must be made and estimates developed from contractors or architects. The complexity of evaluating a new, major undertaking can be enormous, and significant expense may be necessary in order to arrive at a cost estimate for facilities expansion that is reasonable enough to provide data for a yes or no decision.

Most cases will not require this extensive a study of facilities, especially when the proposed product has some similarities to an existing line. However, if the projected volume exceeds the capacity level, a point is reached where additional floor space is required. This entails repositioning of existing manufacturing facilities or building of additional areas. If the company is doing an increasing amount of international business, the question also arises of whether the facility should be built overseas or not.

While estimates of capital tooling can be developed, the uncertainty of a new product's rate of growth makes another consideration worthwhile. Should the first six months or first year of production be built using expensed or short-run tooling rather than capital tooling? In some cases, say in a molding or forging operation, it is impossible to fabricate the parts without adequate tonnage available.

However, when individual parts are smaller, of lighter gauge and relatively uncomplicated contours, the use of short-run tooling that is expensed rather than capitalized has some distinct advantages. The time required to tool up is substantially shorter, and time frequently is a key commitment. Also, the final drawings may have errors that are discovered only when parts are fabricated. This often means reworking the tooling and sometimes scrapping it completely. With expensed short-run tooling, the same problem can occur, but the costs of reworking or refabricating the tooling are less. Also, if a design change is mandated by a changed requirement or a customer problem, the costs and time to change are less.

On the other hand, product cost will usually be higher. One reason to use capitalized tooling is to get maximum production per unit time and lowest possible part cost. The fact that expensed tooling is simpler to use usually means that more hand operations are required, increasing the cost.

The trade-off is time reduction and flexibility against higher cost with short-run tooling. With capitalized tooling there is a longer time required to complete tools and less flexibility, but lower-cost parts.

If there is uncertainty about the initial demand and growth rate of product acceptance, expensed tooling is an option for manufacturing to consider. Also, if there are capital constraints or rationing, it is often a necessary one.

RETURN ON INVESTMENT CALCULATIONS

In the evaluation of a proposed project, particularly one that promises to bring something new or vastly improved to the marketplace, there is a high level of interest and excitement, and a great push to get it started. The features of the product, probable pricing, size of the market, and growth potential sometimes are so intriguing that stars get in the eyes of marketing and planning, who are tempted to throw caution to the winds and plunge in. The results seldom are as great as predicted; more often they are near disasters because there was no advance financial planning.

The purpose of a preliminary set of ROIs (Return on Investments) is to create a pause to ponder, important in the consideration of any new product. The key questions to be answered are implicit in the words "return on investment." Over the expected life of the product, what will the returns (profits) be compared to the investments needed? Second, are they sufficient to meet or exceed the returns on other products already in the product line? Third, is the projected return on assets adequate for the risks involved in this particular project?

Before we perform a sample ROI calculation, returns and investments need to be qualified.

Returns

Profits, pretax or after tax, are a result of selling a number of products at a price greater than the total of all costs involved. Return is the total of the unit profit times the number of units sold. This is relatively straightforward, but it implies a good awareness of the market price and how it may change after the introduction and phase-in of the new product. It also requires a good idea of the product cost and how it will change as the product is manufactured in volume. The accuracy of the numbers generated is less than absolute, but "good" numbers are essential for a reasonably accurate estimate of future returns.

Investments

R&D and design engineering costs are considered expenses and not investments. They of course are investments of a limited resource—people and brains—but for the ROI calculations they are usually not computed. The primary investments are the capital tooling and facilities and the investments in inventories—raw material, work-in-process, and finished goods—plus the money tied up in accounts receivable. These are finished goods that have been purchased by the customer but have not been paid for.

Initially there is a careful review of R&D and engineering expenses, developmental tooling, and the money used for market research; but these expenses are minor compared to the size of investments needed to take a product to the marketplace and keep it there. Dr. Simon Ramo, in his book on managing technological organizations, states that "for every million dollars in research, development and marketing research that yield a manufacturable and salable product, perhaps some ten million dollars will be placed at risk, before the corporation makes substantial sales."* In other words, for every dollar of pre-production expense, there are probably ten dollars risked in getting the product into early production!

Considering such a ratio, the expenses of development pale by comparison to those needed to take a product into production, by an order of magnitude. This is an important reason for performing return on investment calculations, and doing them several times during the course of a product development program, as more of the input numbers become better known.

Every company has a target minimum return on investment. It is a pretax figure that relates the expected return on the new product to that of investing the same amount of money in other areas, such as Treasury bills, plus an additional amount for the risk of having no return at all. Pretax ROI targets can typically be 35 to 45%.

A point to remember: ROI is related to, but different from, percent profit on sale. ROI is the profit per dollar of investment, not profit per dollar of sales. With a low investment, high ROI's can be generated even with low profit margins on sales.

ROI calculations are of two types, those that use a constant dollar value throughout and those that use discounted cash flow. The latter recognizes that a dollar received a year later is valued less than a dollar received today. Both methods have value, but they have the same weakness, which will be discussed later.

Using the ROI Forms

Figure 2–3 illustrates a simple format for return on investment calculations, using constant dollars rather than discounted cash value. The estimates of sales, costs, and projected changes in each, plus capital and other investments, are entered as data on the form.

In the example, sales were estimated to rise smoothly from 5,000 the first year to 25,000 the fifth. Cost would stay at $50 and selling price at $100. The break-even or total cost was $75, providing 25% pretax profit on sales. In this projection, product costs and selling prices were assumed constant, the economic dip in 1981–82 was not anticipated, and all the data were entered in constant dollars. Development costs by engineering were $75,000, while tooling and other start-up capital expenses totaled $200,000. Completing the form is a simple arithmetic exercise which can be done quickly by a hand calculator or by computer.†

* Reprinted with permission from *The Management of Innovative Technological Corporations* by S. Ramo. Copyright © 1980 by John Wiley & Sons, Inc.
† Reprinted with permission from the July 6, 1978 issue of *Machine Design*.

NEW PRODUCT APPRAISAL

PRODUCT: Multiple Options Purging System (MOPS)

Selling Price $100
Standard Cost   50
Total Cost   75
Profit   25

Multiplier on standard cost for average total investment   0.85 (M)

	1st Yr	2nd Yr	3rd Yr	4th Yr	5th Yr	TOTALS
Unit Sales	5,000	10,000	15,000	20,000	25,000	
Sales (× $1,000)	500	1,000	1,500	2,000	2,500	7,500 (A)
Standard Cost (× $1,000)	250	500	750	1,000	1,250	3,750 (B)
Profit (× $1,000)	125	250	375	500	625	1,875 (C)

Expenses

Engineering Expense	$60,000
Expensed Tooling	15,000
Total	75,000 (D)

Investments

Capital Tooling	$175,000
Other Capital Expenditures	25,000
Total Capital Investment	200,000 (E)
Other Investments	3,188,000 (MB)

ROI, Profit Calculations

$$\text{ROI} = (\frac{C}{MB + E}) \times 100 = (\frac{1{,}875}{0.85 \times 3{,}750 + 200}) \times 100 = 55.34\%$$

$$\text{Profit} = (C/A) \times 100 = \frac{1{,}875}{7{,}500} \times 100 = 25.0\%$$

Fig. 2-3. ROI calculation form. Reprinted with permission from the July 6, 1978 issue of *Machine Design*.

This company used a shortcut method of calculating investments. Over a historical period, its accounting group had established ratios between the total standard cost and the average total investment. Investment included plant and facilities, all inventories (raw material, work-in-process, finished goods), and the net of accounts receivable less accounts payable. In this case, for every $1.00 in standard cost, $0.85 was required in investments to support the product. This is an aggregate number, which assumes that the new product will not have substantially different requirements in this regard from those of all the other products currently manufactured.

To calculate the ROI, one first calculates the total sales, costs, and profits (A, B, and C, respectively). Then, the ROI is computed from profit dollars (C) divided by the investments ($MB + E$). The 55.3% ROI is an excellent return on investment, as is the 25% profit on sales (C/A). Both of these are pretax calculations. The engineering development expenses rarely impact the ROI significantly. Had engineering been considered an investment, the ROI would have dropped to 54.1%; if it were considered an expense deducted from profits, ROI would be 53.1%. Neither approach would have had a significant impact, which is one reason why development costs are usually not used in ROI calculations.

Two other factors are determined: the number of years to recover (1) the costs of capital tooling and (2) engineering expense from profits. Since they total $275,000, the calculations predict that before the end of the second year, these costs would have been recovered, when a total of 11,000 units had been sold.

These are rather impressive projections. In many companies of medium size, a project that would reach a million dollars in sales the second year and keep on increasing would be approved with little further question.

SENSITIVITY AND RISK ANALYSIS

In the previous example, while everything looks great, several nagging questions arise. Will this optimistic scenario really occur? Will the constants stay constant, or are they variables? Are there any outside factors that might disrupt the predicted chain of events? Could a competitor introduce a similar model just as sales are beginning to pick up, and force prices down? Or, on the positive side, what if factory costs turn out to be lower than planned? Or what if volume is greater than predicted and prices do not erode? Any or all of these possibilities have some probability of occurring, although which ones and to what extent are not known.

In reality the ROI calculation is wrong, incorrect. There will be no single value of ROI that can be accurately predicted. The dilemma of incorrect ROI can be addressed by a simple technique called sensitivity analysis. By purposely varying the key factors over the extremes of expected values, changes in ROI can be determined and a better understanding of the possible future impacts can be obtained.

How to Do a Sensitivity Analysis

The key "constants" to be checked out as variables are possible changes in sales, cost, and investments. Before a sensitivity analysis is begun, the extent of possible change should be estimated by the appropriate groups— marketing, engineering, and manufacturing. A consensus on the extent of these changes should be reached.

For this example, assume that consensus was reached on a 10% variation in sales dollars, 10% variation in costs, and 50% variation in capital investment. All of these variations could be plus or minus from the benchmark point. Thus sales could vary from the original $7.5 million to a high of $8.25 million and a low of $6.75 million, and so on.

The change in sales, or in costs, could be due either to changes in units sold, or dollars per unit. In this simple analysis they will not be separated. A large change in the capital investment of $200,000 must be assumed because inventories and accounts receivable are a very large determinant of total investments during the five-year evaluation of the project.

To determine the effects of these variations, additional ROI calculations must be made. If we vary one item at a time and hold the others constant, the effects of each change by itself can be determined. Since there are three variables, six additional calculations are needed. However, what if more than one thing changes at the same time—sales go down and costs go up, for example? The one-at-a-time calculation scheme provides no answer.

A better approach to sensitivity analysis is to set up a series of calculations to check out the extreme values of the three major variables. Since there are two levels of each of the three variables, we can construct a model consisting of eight points, similar to an experimental design called a "2-cubed" factorial experiment. It can be visualized as a three-dimensional cube, with the eight evaluation points at the eight corners. The original ROI calculation is at the center of the cube.

ROI calculations are run with the variables set at their low and high levels in the following manner:

Run	Sales	Costs	Capital investment
A'	Low	Low	Low
B'	Low	Low	High
C'	Low	High	Low

Run	Sales	Costs	Capital investment
D'	Low	High	High
E'	High	Low	Low
F'	High	Low	High
G'	High	High	Low
H'	High	High	High

Before any ROI's are calculated, some of the probable outcomes can be predicted. The D' condition will be pessimistic because sales are low, while both costs and investments are high. Next, the E' condition will be quite optimistic because it is the opposite—high sales with low costs and investments. It is important to run all eight calculations because the actual returns cannot be predicted and may somewhat be different from what one would anticipate.

It is very easy to perform the constant dollar calculations. It takes less than 15 minutes with a pocket calculator and less than 5 minutes with a programmable one. Refer to Table 2-1 for the results.

The three-dimensional cube (Fig. 2-4) shows the outputs in a different form. Notice how a large variation in the tooling investments ($100,000 Low, $300,000 High) makes a relatively small change in ROI. This can be seen on the front and back faces of the cube, and comparing A' versus B', C' versus D', etc. It is only a three-point change in ROI, except in the extreme cases, E' versus F' and C' versus D'.

Points E' and F' represent the most favorable condition with an ROI in excess of 100%, while C' and D' represent the worst conditions, at about 15% ROI and 8% profit.

Given a choice, the product parameters should be pointed toward the top left corner of the cube, higher sales dollars with lower costs. The four points representing high E', F', H', and G' have an average ROI of 79.8% (the top surface of the cube). Those representing high costs are C', D' and G', H' (the latter two overlap both cases), whose average ROI is 35.4%.

Summarizing the surface responses, in tabular form below, it is apparent that with high costs, regardless of sales, low ROI develops. The other three are

Table 2-1

Characteristic	Cube surface	Average ROI
High sales	E'F'G'H'	79.8
Low sales	A'B'C'D'	35.1
High costs	C'D'G'H'	35.4
Low costs	A'B'E'F'	79.5

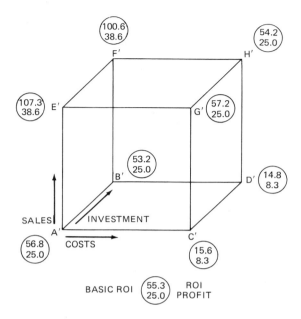

Fig. 2-4. Three-dimensional ROI plot.

straightforward: with low sales, low ROI, whether costs are low or high; but with high sales, ROI is high also.

If the minimum acceptable pretax ROI is 40%, a 10% increase in costs above estimated would wipe out the project. While the sales projections should, of course, be reviewed, it is vital that engineering cost estimates be restudied before the project proceeds. If the project looks promising, product costs and their reduction will be an area of major focus throughout project development.

Sensitivity analysis allows an early evaluation of what might occur in the uncertain future of a new product. It forces a deliberate second look at the estimates that were the basis of the original calculation. It also supports recognition that an ROI is an estimate of the future, not a future fact. By expanding the original single value of sales, costs, and capital tooling to an expected range of values, it expands the estimate from a one-point snapshot to a cubical volume of information. The actual range of ROI's is more likely to be within this cube than at any particular point.

Sometimes, there is no past history or data on actual product cost or sales variances from plan. Using 10% swings is a good starting point. However, it is not unusual for sales to be 20% off target because of competitive action, or for costs to vary as much owing to material shortages or changes in production quantities or part design and tolerances. The plus sides are usually optimistic. The minus sides, with high costs and/or low sales, occur more often than desired. Thus, it is important to study these areas carefully.

To recap, the weakness or sensitivity of a new product to changes in costs, sales, or capital tooling can be determined beforehand. If the product is still acceptable, there will be much greater awareness of the major areas of concern. The ROI cube is simple to use and provides considerably more information than changing one variable and holding the others constant, and the time required to perform eight calculations is not substantially greater than that needed for six.

Risk Analysis

All new product or product decisions are made under uncertainty. There is no way to be certain that a given scenario of the future will occur. The risk of being wrong is two-sided. First, there is the risk of being too pessimistic and missing an opportunity to capture a market segment at a profitable level with a product that meets customer needs in a unique way. Second, there is the risk of being too optimistic and tooling up and producing a product that does not meet customer needs, or grows slowly to a sales level much lower than that required or desired.

The cost of these errors is greatest when the projected sales and investments are greatest, and is of less significance (though important) when the financial needs of the product are less. Analysis of risk is thus a size- or cost-related requirement. There are several ways of analyzing risk. One is to use decision trees to assess the various possible outcomes, and then subjectively, or from experience, assign a probability of success. The product of the possible outcome times the probability of success becomes an expected value that can be compared to other decision choices.

Figure 2-5 is a simple decision tree. It shows that project A could lose $2 million dollars on a 20% probability basis, but there is 80% probability that it will have a return of $1,000,000. The probability value of the two are added algebraically, and the expected value of + $400,000 is compared to other projects being considered. If there are several possible outcomes, the probabilities must be adjusted so that they total 1.0.

The weakness of the decision-tree approach is the arbitrariness of assigning success probabilities based on history or intuition. The strength is that the building of the decision tree is a form of planning that forces one to study the possible outcomes, think of what might happen, and assess the seriousness of each consequence.

Another probabilistic approach in risk analysis is to develop probability

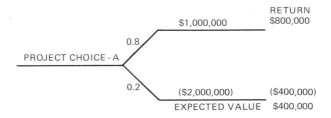

Fig. 2-5. Risk decision tree.

distribution curves of the major factors: cost, selling price, units sold, investments. Then a random selection of the variables is made, and the ROI calculated. This is repeated several hundred times by a computer, and a probability distribution of return on investment is developed.

This would be a cumbersome task, particularly in developing the data, but the task is made simpler by developing simple distribution curves. The value of such a probabilistic model is based on the value of the assets involved and uncertain outcomes. Risk is inherent in all ventures into the future.

The process begins with using a cumulative frequency distribution curve rather than the usual normal distribution curve. It is shown in Fig. 2-6.

For the curves to be utilized with minimum trauma, several estimates are required from each key manager. Using the previous sensitivity analysis example, where the standard cost was $50, the engineering manager or project leader would be asked these questions:

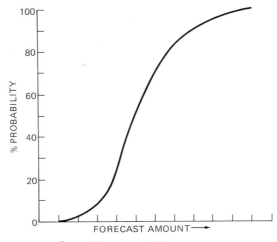

Fig. 2-6. Cumulative probability distribution curve.

1. What is the lowest cost that he expects to achieve on the product? He might review his original + 10% guess from sensitivity analysis and decide that there is a small chance that it could get as low as $44.
2. What is the expected cost? That still looks like the best number, and when pressed for a probability estimate, he might state that the probability that the cost would be $50 or below was at least 75%.
3. What is the highest cost? Assume that he reviewed his estimates and said that in the worst case the cost could reach $62, although that was not too likely.

The probability distribution on costs based on those inputs is plotted in Fig. 2-7.

Using the three points and their approximate probability, is cumulative cost distribution curve results. The expected or 50% value is about $46 because the project engineer is convinced that there it is a 75% probability that it will not exceed $50. There is some concern that he is "bluffing," but if his past history of cost estimating is accurate, the data must be accepted as presented.

In similar manner, the marketing manager develops probabilistic estimates of unit volume and dollar sales. He may estimate his five-year sales total at $7.5 million, and decide that his absolute top estimate is $9 million and the worst possible is $6 million, at $100 per unit. That curve is Fig. 2-8.

Alternatively he may speculate that an elastic price–volume relationship

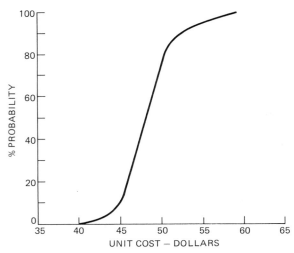

Fig. 2-7. Cost probability distribution.

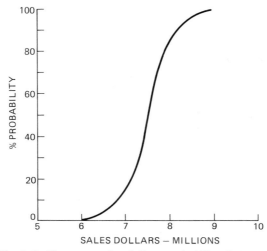

Fig. 2-8. Five-year sales dollars probability distribution.

might exist. In other words, a 10% price decrease might increase volume 30%, and a 10% price increase might decrease volume 30%, for example. This approach is more realistic but too complex for this review and will not be explored further.

Investments are also evaluated. The finance manager may decide he feels more comfortable in assuming that the need for the target $200,000 may occur only 25% of the time and that it could get as high as $350,000 and no lower than $150,000. This is based on his past experience on several recent projects. Figure 2-9 shows this. While we know from the sensitivity analysis that tooling capital investments do not impact the project ROI significantly, the new information must be considered for us to be sure.

With all of this information available and an interactive program in the computer, it is possible to compute an ROI probability distribution for this project. With the assumed distributions of cost, sales, and investments in the computer, a random number is selected for each factor, the corresponding value from the curve is placed in the formula, and when all are selected, the ROI is calculated.

This is a sample calculation. Selection of three random numbers resulted in 52, 81, 13. Using them as percentages with cost, sales, and investments, values from the probability curves were $48.50 cost, $7.9 million sales, and $188,000 capital investment. Using these data in the ROI formula, and ROI of 74% and a 31% profit were obtained. It is necessary to repeat this procedure several hundred times on a computer, retaining each of the ROI's in memory. Then a plot of the ROI distribution provides an expected 1 sigma and 2 sigma ranges.

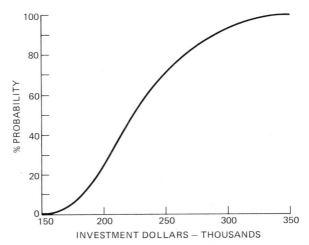

Fig. 2-9. Investment dollars probability distribution.

On an expensive, high-risk project, risk analysis is worthwhile. It forces the group responsible for the project to take a long look at it, try to peer into an uncertain future, and make some reasoned estimates as to the best and worst that can happen. With this information, the project will be either vigorously pursued or dropped.

Shortcomings of Sensitivity and Risk Analyses

The elegance of using estimates or educated guesses that provide ranges of ROI in one case and probability distributions in another has a hidden danger; that is, in considering the numbers as real instead of reasonable estimates, which may be in error. Yet numbers provide a sense of security.

On most projects the use of sensitivity analysis points out the areas of potential financial problems in a product proposal by determining which variables the returns are sensitive to. It is far superior to a single-value calculation that concludes "This *is* the ROI."

Risk analysis goes a step further by using probability distributions. It recognizes uncertainty in estimates and uses that factor to advantage, in determining an ROI distribution. It is of greatest value in a high-dollar-potential project requiring extensive engineering and marketing effort plus substantial capital.

Both methods provide a far superior approach to product planning and analysis, provided no one loses sight of the fact that all of the numbers are estimates—and any estimate is only an educated guess about the future.

PROJECT PRIORITIES AND PREFERENCES

Assigning priority status to a project is difficult. In one respect, no approved project has a low priority. It has dates that must be met if the project is to be completed on time and the new product introduced as planned.

Priority usually becomes a problem when testing is necessary. At that point, it is not unusual for the manager to receive test requests that *all* are top priority. Part of this is due to individual engineers' insistence that their own devices be tested immediately so that design work can continue as soon as the test results are analyzed. Thus, each project requires top priority.

However, such a situation must be carefully watched, as "priority push" by some engineers may impact the schedules of other engineers who are not so aggressive in pushing their projects through testing. It is important for the manager to have continuing contacts with the laboratories, to see how all of his projects are going, which are being held up and which are being scheduled.

Initial priority setting must take into account the need for the end product and the timing necessary to accomplish it. After the analysis phase has been completed, the result is a "portfolio" of potential products. Many of them are top priority, but obviously not all can be done simultaneously.

One way to arrive at relative priority is to rank the projects by expected ROI, or dollar return. There is nothing incorrect in this approach, but is is possible to have several projects clustered in various groups, without any agreement as to which has priority within any group.

Another way is to have each functional manager rank the projects in importance to the company, then compare the lists and negotiate conflicts. This is a reasonable approach that works if the project list is fairly short and a general consensus exists about the various projects and their importance to the company or division. While it is not likely, it still is possible for one manager to rank a project near the top and another to rank it near the bottom.

Project Preference Charts

Project preference charts provide a way out of the dilemma. They force the managers to make a priority decision on all projects by looking at two of them at a time and making a judgment as to which has priority over the other. Figure 2–10 is an example of a preference chart for ten projects.

It resembles a project screening form in that it allows rapid evaluation of a number of projects in a short time. Unlike screening, however, it provides no evaluation information, assuming that the manager has adequate knowledge about each of the projects.

Using the form is quite simple. Each of the projects is assigned a letter in alphabetical sequence, and the appropriate-size chart is used. A larger chart

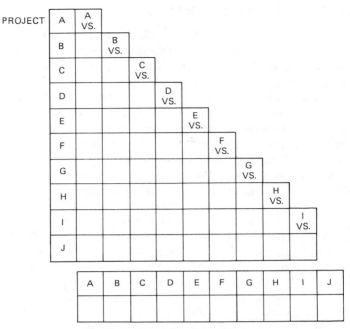

Fig. 2-10. Project preference chart.

can always be cut down to size by crossing out the unneeded portion. Projects are evaluated, not by looking at all of them but by making a decision on only one pair at a time. One asks the question, if only one project can be done, should it be Project A, or Project B? The appropriate choice is put in the box. Then is it Project A or Project C? Then A or D? This is continued through the entire list.

After all the preferences are made, one counts the number of times Project A is filled in, Project B, and so on, and the results are recorded for each below. In a list of 10 projects, 45 comparisons are made. It should take less than an hour to complete an evaluation of 15 or 20 projects. After that, the #1 priority preference determined is the project with the greatest number, #2 the second highest, and so forth.

In some cases there may be a tie, two projects receiving the same number of votes. There are two ways to approach it, assuming projects E and F each received the same number of votes. Recheck the choice made when only E and F were being considered, and rank them the same as the first time. The alternative is to consider both projects again and decide whether the original choice is still acceptable or should be revised. If the priority tie is very near the bottom, it does not matter greatly which way is chosen.

When several managers use a preference chart, significant disagreements may occur in selecting the top priority projects. Individuals may have strong feelings about certain projects because they represent their own ideas or utilize a segment of technology they believe to be important, or because the numbers say the project will be a big winner in the marketplace. The only solution is to arrive at a consensus, so that just one priority listing can be proposed to top management. If a consensus cannot be reached, then the reasons for differences need to be voiced.

The use of a preference chart makes it possible to set and adjust priorities when several projects have relatively equal perceived priority or profit potential so that it is difficult to decide between those projects competing for the same resources. Its greatest value lies in its use before priorities are set. Since it forces brief decisions about each project, it puts them all in perspective. It is an important aid in establishing priority of effort on all projects.

HOW MANY PROJECTS CAN BE HANDLED?

This question is more difficult to answer than it may appear at first glance. If one subtracts the number of estimated hours required for each project from the total number of people-hours available in a year, a project level will be reached when all of the available hours are used up. That would appear to be the maximum number of projects that could be handled.

However, that number is usually less than the number of projects that need to be done; and it does not consider that some projects will change their scope during the year, others will be raised or lowered in priority, some will encounter unexpected difficulties, while still others will encounter less difficulty than anticipated.

For all of these reasons, it is good practice to load each professional employee at 110% to 120% of the mathematical project load. If an engineering design department operates on a man-month basis, one can plan for 13 or 14 months of individual output per year. Or, if 1800 hours per man-year is the norm, allowing for vacations and holidays, each engineer can be loaded from 2000 to 2200 hours.

There may be some concern or some negative feedback about this approach, but most often it will work. Project estimates may be pessimistic or optimistic, but the effort needed invariably changes from the plan as the engineer gets into the project and makes things happen. With some projects encountering delays due to problems, vendor, or test lab schedules, other projects can be moved ahead, thus using available time to best advantage.

While there is no specific number of projects that can be predicted for a design group, because of the variations in size, scope, and resource requirements of a project, the guidelines that follow can be used as a starting point.

Top-priority projects should be few, one per engineer, certainly not more than two. In a large project more than one engineer will necessarily be assigned. The top projects should require 50% or more of his or her time, on the basis that 50% effort on the average could translate into almost 100% effort for an extended period and substantially less at other times.

In addition, each engineer should have two or three important projects in which timing is essential but the schedule is not so tight. These projects are frequently "stumpers" where the answers needed to bring the project to completion are somewhat obscure, and it will take time, persistence, and effort to conclude the project successfully. Depending on the complexity and demands of the major projects, there may be few of these secondary projects, or none.

Each engineer should also have one fill-in project, some problem that needs attention, has only a small payout, and cannot support much effort. These fill-in projects sometimes are training projects for young engineers, to give them full responsibility for the success or failure of a project. It is important that the reward for success here be much more than the punishment for failure.

On balance, each engineer should be somewhat overloaded, in the knowledge that estimating the effort required is more art than science. In most cases, a degree of overload will cause the engineer to become more effective rather than overburdened. While there is no optimum, it appears that from two to six projects represents a sensible project load for most engineers at any one time.

An important concern besides the number of projects is the occurrence of simultaneous key dates. If prototype approval dates, design reviews, or production releases have dates that coincide on two or more projects, the demands on the project engineer's time are extreme. It is nearly impossible to meet two major key dates at the same time. The possibilities of mix-ups and major errors due to overcommitment must be recognized and minimized.

Engineering Projects as Investments

Generally, the design engineering department can handle a fairly large or a fairly small number of projects. It depends on how many man-hours are needed to perform the design work effectively. However, the group of projects to be taken on by the engineering department must reflect the needs of the company and address the direction set by the long-range plan. In its broad context, projects are a portfolio of present investments with future returns.

Any investment portfolio is a combination of high- and low-risk items. With engineering projects, at least one major project should have high return, low risk, and a short time to payoff. A medium-return project may be a worthwhile alternative because this combination of low risk and quick, high return is unlikely. Several of the projects should be high-rate-of-return, medium-risk projects, which are the backbone of the department's projects. The risk may be either technical or marketing. Last of all, there should be a few long-range projects as insurance policies for the future. The future, especially uncertain because it is projected ahead, needs to be planned for.

Every investment portfolio needs to be examined periodically to make sure the investment strategy is correct and that each investment is still sound for the overall strategies, and that investment projections are still on target. The same holds true for engineering projects. If all answers are affirmative, no changes are needed; but if the external environment has changed, or if progress is better or worse than planned, it may be necessary to change the priority of effort on certain projects. Some ongoing projects must be stopped or rescheduled; some new ones must be initiated.

Shifts in Project Priorities

The engineering manager depends heavily on marketing for a correct assessment of changes in the environment, and hence in project emphasis. However, some marketing managers or marketing groups have a tendency to react too quickly. They may assess a small change as a major trend, and then in a few weeks or months realize they are wrong and change again.

This stop–start–stop approach when applied to engineering projects can be disastrous. Progress will stop abruptly; staff members will become demotivated and demoralized. When the project is restarted, a lot of wasted effort will be expended in retracing past steps and returning to speed. Any negative effects of this fluctuation can reflect badly on the design engineering manager, though he does not deserve it.

Coping with Priority Shifts

There is no simplistic way out of this dilemma, but here are some approaches that will reduce the frequency of such occurrences.

First, it is necessary to work closely with the marketing manager and make him aware that engineering effort has some resemblance to a speeding car. A car traveling at high velocity has difficulty stopping on a dime, or taking a corner smoothly. If stopped, it has difficulty getting up to speed again and making up for lost time. To avoid similar trouble, the engineering manager must resist any rapid changes in scope, direction, or speed, unless those changes are reasonably demonstrated to be trends rather than short-term anomalies.

Second, the impact of marketing trend changes on engineering projects, on dates and performance, must be reviewed before any changes are made.

Sometimes, the trade-off is not acceptable, and marketing would prefer the project be completed as originally planned.

Third, the engineering manager and his key subordinates must review and themselves assess how to cope with necessary market changes and become flexible enough to respond to changing conditions in an effective manner.

SUMMARY

The portfolio concept, which considers engineering projects as investments for future returns, allows a manager to take a more objective view of the whole process of project selection. Screening methods provide an initial separation between the better project candidates and poorer ones. With marketing, engineering, and manufacturing inputs, return on investment calculations provide a snapshot preview of the return on investment and whether or not a project is worthwhile.

The broader perspective of change in the real world, through sensitivity analysis and risk analysis, provides broader, better insights of what might actually happen if the product is brought to market.

Using these business analysis techniques, a portfolio of projects is developed for the near term and the long term. The use of preference charts helps the key managers focus on relative priorities. It allows differences in viewpoint to be recognized and dealt with. Priorities are key, of course, but projects of equal importance are otherwise difficult to sequence. While the number of projects depends entirely on the company's objectives, their variety and extent are the future direction of the business.

In one sense the engineering manager is the investment portfolio manager of his company. (The marketing and manufacturing managers are also investment managers, at a different point in time.) During the design and development phase of the projects, the engineering manager must look at his projects as investments, consuming resources and promising returns but not yet able to deliver. This approach allows better focus on results than is possible if the manager develops a paternal feeling toward "his" projects. He may tend to become protective of a project because of the technology involved and the people.

In the real world, it is difficult to remain purely objective and not have some emotional tie to major projects, since the emotional tie is also called "commitment." A balance of the two is necessary to remain in control, to know when to keep a project going despite opposition and also when to terminate a project, again despite opposition. These subjects will be dealt with again, in later chapters.

3
Design Objectives

INTRODUCTION

Before the design effort is begun, objectives must be set out and agreed to, and a project leader selected. If the project is to be structured along matrix lines, task force members must be selected from the organization and the project planned to meet the performance, time, and cost objectives. This chapter details some of the key factors that ensure that design objectives will be met during the course of the project, and helps to plan for checkpoints along the way, so that trouble spots are located early and corrected before they cause major delays.

KEY ROLE OF MARKETING

Design objectives are the required outputs of engineering effort that meet customer perceived needs. Since marketing provides the key interface between the customer and the product, it is marketing's role to establish what these needs are and translate them into performance requirements. Marketing must also have a broad-enough perspective to keep the company's interests in mind as various design projects are initiated and brought to market. The strategies employed by marketing have a significant impact on the profits and profitability of the company.

Over ten years ago, one of the planning groups at General Electric Company developed and published a landmark study in this area, called the "Profit Impact of Marketing Strategy" or PIMS. This study was based on evaluation of 800 "strategy experiments" in various companies using observable experiences. The group evaluated the competition faced, markets served, each company's cost structure and technology, and the profit results produced.

The study uncovered 55 variables that collectively accounted for 82% of the differences between high-, medium-, and low-profit occurrences. The variables had to be statistically significant, had to make sense to business managers, and had to be consistent with economic theories. Following are 11 of the key variables, and the effects they have on profitability.*

1. *Market share.* A business with less than a 15% market share of a product is rarely profitable, whereas with a 35% share it is almost always profitable. A linear relationship occurs between these percentages.
2. *Increased market share.* Increasing market share costs money. In the short term, profit drops because of the additional investments and expenses.
3. *Market growth rate.* The more rapid market growth is, the lower the cost (in expenses and investments) of acquiring a greater market share. The inverse is true in a flat or slow-growing market.
4. *Plant capacity utilized.* Utilizing present capacity for more business has a positive impact on profit to about 90% utilization. Beyond 90%, facilities become overcrowded and less efficient.
5. *Investment intensity.* This is the most important variable, one that has a sharp negative impact on profitability. Whether it is in inventory, receivables, plant, or equipment, the greater the investment per dollar of sales, the lower the percent ROI.
6. *Vertical integration.* In a stable, mature business, more vertical integration increases profit. The opposite is true if the market or technology changes rapidly.
7. *Price changes.* When prices are rising, profits rise. When they are falling, profits fall. When they stabilize, so do profits.
8. *Relative cost of product to customer.* This relates to the percentage of the customer's total expenditures. The higher the percentage for the products, the lower the profits.
9. *Ratio of R&D to marketing.* There are two ways to win. A high level of R&D and high level of marketing effort makes a product leader. A low level of R&D and low level of marketing makes a fast follower. Either way can be profitable, but profits do not increase by increased spending in only one of the two.
10. *Variable costs relative to competition.* If a company's variable costs are lower, there is considerable profit gain. This relates to the prime costs of production and distribution.
11. *Product leadership.* Product superiority and leadership are judged by customer criteria. The positive benefit is small, but an inferior product produces a severe profit loss.

If the business is in a growing market, the following three PIMS tactics are

------

* From the General Electric Company PIMS Study.

important. The objective of these tactics is to emerge quickly with a market share much greater than 15% and hopefully greater than 35%.

1. Take reward in market position and do it quickly.
2. Push manufacturing capacity faster than market growth to improve market position and discourage competition.
3. Avoid premature vertical integration.

This information is new, even though it is over ten years old. Unfortunately, many businesses do not utilize it—some because they do not accept it as true, some because it is contrary to present operating methods. These are not necessarily good reasons, but are realistic ones. Even the PIMS authors believe their study applies about 80 + % of the time. Thus, in some cases reluctance to use it may be appropriate. Since many companies and many marketing departments utilize the PIMS approach, one's competition may be following it. Therefore, it is important that an engineering manager be aware of these marketing strategies and why they are considered important. A company that ignores (or violates) these concepts may be relatively successful, but there is a high probability of being much more successful by following them.

PERFORMANCE AND COST OBJECTIVES

The development of specifications and objectives before the beginning of a project is vital to keep the project effort focused. One "nonstandard" approach was proposed recently by an article in *Advertising Age* magazine, titled "Write the Ad before You Design the Product."*

The author, head of an industrial design company, expresses his thesis clearly and simply. If during the conceptual stage of a product development program, marketing knows enough about the user's wants and needs to write a good advertisement, then the engineering group can begin the tasks of designing and building a product to answer those wants and needs.

Although an unusual approach, this concept has a strong undercurrent of logic. Unless significant user benefits have been perceived and are built into a product in the first place, a subsequent ad cannot develop them.

Writing a prototype ad is analogous to developing a product prototype on paper. The "paper" design and the hardware design must go hand in hand. Otherwise, if the marketing group does not know enough about customer needs to generate a prototype ad, engineering will not know enough about the end functions to develop a successful product.

---

* Reprinted with permission from the December 17, 1979 issue of *Advertising Age*. Copyright 1979 by Crain Communications Inc.

Performance Objectives

Meeting a customer's needs and wants can be a difficult process even if a successful prototype ad is developed. Needs sometimes are ambiguous factors such as improved reliability, better performance, lower price. When key features must be addressed and quantified, benefits such as "more" of this and "less" of that lead to difficulties in interpretation and implementation. Thus, there is a need in the conceptual stage of a project to develop a specific list of project objectives with as much quantification as possible.

This becomes a project/product specification in which the end point parameters and features are listed. When this specification is developed, it is important not to set up a single set of parameters that become *the* objectives. This has the unreality associated with a single value for ROI.

A better approach is to set up two columns in the specification, one labeled "Minimum requirements," the other labeled "Desired level." When the specification items have been tabulated at the two levels, the design group has a better idea of what the real requirements are, and what additions would be desirable if they could be attained.

By using the "Desired" column as the design targets and the "Minimum requirements" as the fall-back position, a better design can be generated within the existing time and budget constraints than could be obtained if only the minimum requirements were listed. It would be even more accurate if the only column were the desired end points, with no idea of how much negative tolerance could be taken without serious consequence. However, using a range of acceptable performance requirements gives some design latitude for performance trade-offs, as well.

Once the specification parameters have been prepared, often jointly by engineering and marketing, it is important to have them reviewed and approved by the engineering and marketing managers responsible for them. This review is a double check that the objectives of the project have been detailed adequately and clearly enough in the specification to define the end product performance.

Cost Objectives

Another important area to be clarified in the design specification is the target cost of the final product. This may appear contrary to the recognized dictum that the cost is not a basis for determining price, but rather that the market determines price. But this is all the more reason to be concerned about cost.

Each project, because of its ROI constraint, is expected to generate a given level of return, or profit. Knowing and committing to cost objectives recognizes that if the cost exceeds a given level, the price must increase proportionately in order for the desired return to be met.

Therefore enough work must be done in preliminary costing of the product to determine the likely cost and the highest expected cost. If either of these levels is too high, the project should be carefully reviewed. If there are areas where significant cost reduction or value engineering could bring the costs in line, these must be addressed. Otherwise, it may be necessary to cancel the project.

This should be done as early as possible. It is of no value and a considerable waste of time and talent to develop a product that meets its technical objectives but costs so much it cannot be sold profitably. For that single reason, cost objectives—maximum and desired—must be set early in the project, and monitored during the progress toward completion.

TIMING OF DESIGN COMPLETION

Time, the fourth dimension, has many implications and always is a critical design objective. Unfortunately, time is a one-directional parameter; it keeps on moving ahead regardless of the degree of progress or problems in the project.

Timing in project work is usually critical because various sequences of events by other support groups are initiated as portions of the design are completed. If the timing is off (and late is worse than early), these groups either sit relatively idle waiting for the completion, or may become involved in other important areas and not be able to respond as well to a late starting time. Both of these possibilities are undesirable, but the unavailability of a support group at a critical time is the worse of the two. This is just one of the many reasons why design timing is critical.

Another group severely impacted by engineering design delays is marketing. Much of the marketing effort, including promotion, advertising, and technical manual preparation, for example, depends on prompt completion of the engineering design effort.

Significant efforts must be made by engineering on a continuing basis to avoid most timing problems and to reduce their impact when timing delays are unavoidable. The key requirement is to complete a design project on schedule, so that the subsequent operations phase in and mesh with a minimum of waiting. This involves realistic scheduling and determined effort to cope with the unanticipated problems and delays that inevitably creep into a design program. This will be discussed in more depth in later chapters, and is the most important requirement.

The second requirement is to maintain open lines of communication with the design group and focus on keeping others informed of progress as well as any delays that are not resolvable. This is a two-edged sword and must be used with discretion. Too early a warning of delays can mean that a design engineer does not know how to solve a particular problem and chooses the delay route to avoid applying extra time and effort to solve it in a timely manner.

To avoid this kind of problem, managers should be the only persons "officially" allowed to delay a project, and then only after a considerable effort has been made to solve the problems first. Even then, a backup plan should be devised to make up the lost time by paralleling some series efforts in the project plan.

Another way to avoid costly time delays in a project is to allow enough time in the critical parts of the schedule to solve any problems that may surface and are not quickly resolved. All too often a schedule is so tight that no time has been allowed for any retest, or redesign. This can be a disaster.

If a project goes into a "crash" problem-solving mode, it either flies or fails. Intuitively, that drops success probability to fifty–fifty, a toss of the coin. Crashing is a successful method in some projects under some circumstances because the thrill of pulling an important project through, snatching it from defeat, makes all the effort worthwhile. However, to make it part of the normal operating routine is to set up too many project failures, instead of project successes.

A better approach to engineering design timing is to pre-study the project during the planning phase and determine which areas of the proposed design concept are the shakiest and at what points they are likely to cause difficult problems. These are the areas where additional design time should be planned into the project.

Because the timing of design completion is critical in many ways, it is important to plan ahead, reorganize where there may be delays, and allow for them. In some programs the possibility of delay occurs anyway. The negative effects can sometimes be reduced by timely communication between managers and taking corrective actions early.

Time is such a precious, one-directional asset in a project that its importance cannot be overemphasized. Time can be lost, wasted, or used with care. One key objective must be a plan to use time wisely, to do the project right the first time. As one pithy statement reminds, "There's never time to do it right but always time to do it twice." This is often true, but the timing of design completion is too important to plan for it inadequately.

PROJECT LEADER SELECTION

Each project needs a leader, a person responsibly in charge who knows what needs to be done and sees that it gets done. The best project leader will have a good track record on other projects, be committed to this new project, be a good planner, think well on his feet, solve problems effectively and creatively, relate well to other people, and have the confidence of his peers and the respect of his subordinates.

These superior people are hard to find and difficult to keep because they are very promotable. The manager who has one of these engineers on hand is fortunate. If he has more than one, he has been an excellent trainer and has brought people along in an environment where they have learned well.

In most cases, the selection of the project leader depends a great deal on the type and complexity of the project. To a lesser extent it depends on who is available, who has a project load this is light, or who can be shifted to take on a new assignment.

If there is a general rule, it is to appoint as project leader the person who has the most to gain if the project is successful. Sometimes that approach will provide surprising results. In one case the person may be an older, experienced engineer who does his job competently but has had no recent challenges to get him fired up. He may be in a rut and have no real incentive to do more than a competent job.

In another case it may be an engineer who "bombed out" on an earlier project and is not enjoying a good reputation. He may need a successful project performance to restore his own confidence and that of his peers.

In still another case, it may be a young engineer, a few years out of school, who has trained fast and is eager to prove his competency. A challenging, full-scale project may provide the right amount of seasoning to bring out hidden talent and find a rising star.

The choice of the best project leader may prove difficult. In many instances, the results of the choice will not be apparent until the project is well along, or nearly complete. Only then will the manager know whether his decision was a wise one.

The Team of Two

Another approach to choosing a project leader is to appoint a team of two. It has the advantage of having two minds working on problems—one designing while another tests, one the administrator and the other the activator. One creates new ideas; the other evaluates. This is rather like one composing the music, the other the words. In some specific areas a team of two can work, if their personalities and ambitions are mutually reinforcing.

However, most of the time using a team of two will lead to arguments and a struggle over who is the real leader. Other team members will tend to prefer one member to the other, because of either technical prowess or a better personality. The better leader may or may not emerge. The "winner" may just be the one with greater ambition or the more aggressive approach.

One way for the team of two to work is for one engineer to be older and more experienced, and charged with the training and development of a less experienced engineer. As the new engineer becomes more capable, the experienced engineer will gradually turn over more responsibility and authority to him. With the right blend of personalities, attitudes, and ambitions, the method can work.

One last way to combine the talents of two people in leading a project is to make the experienced engineer consultant to the project leader. This allows one engineer to be in charge of the project but to be able to obtain technical advice and counsel from the other. This method is more effective than the others because any project with dual leadership has difficulty in functioning properly any time that the two leaders disagree on a major issue.

Changing the Project Leader

What if the project falters because the project leader is not leading, has lost the way, or is incapable of doing the total job required? There are only two steps a manager can take: one is to counsel the project leader to find out whether he can do the job if he is given some direction or has the problem situation corrected; the other step is to replace the project leader if he is not working out.

This first step, counseling, is often the only one needed. The project leader may not realize how badly things are going. Often he will believe that things will straighten out given "a little more time." Missed schedule dates are, of course, the initial indication that events are not proceeding according to plan. The manager may be inclined to give the project leader a little more time to get the project back on target, but, before he does that, the project leader must have developed a plan of action to correct the situation and bring the project back on schedule. If that can be done to the manager's satisfaction, the project leader is generally in command of the project and has suffered only a temporary setback on it.

If the manager is not satisfied that the project will now meet its objectives, he has two choices: wait a little longer before making a change, in the expectation that performance will improve; or decide immediately to make a change. This is a matter of judgement, but the decision to change is usually the better one. The manager's discomfiture with an unsatisfactory project status will eventually be recognized by his own peers or, worse, by the next level of management—the director of engineering or, in a decentralized operation, the division general manager. Then his lack of decisive action will be looked upon unfavorably, and someone at the higher level may insist on immediate change.

A prompt decision is better for the foundering project leader. Sometimes a change in leadership is accepted with a sigh of relief by the former project leader, who has not wanted to admit openly that the problems were baffling him. If he can be reassigned immediately to another project of somewhat less complexity, the disappointment is eased.

The major reason to make a prompt change is for the good of the project. It is an old saying that "the project is the boss." In order to complete the project on time, within budget and meeting performance requirements, one must take only those actions that keep the project moving forward, including changing the project leader. Subordinates, supervisors, and the manager himself will benefit from this course of action.

One last observation on changing project leaders: it is important for the manager to know his project engineers and their reactions to problems; otherwise a wrong decision can be made because the manager misreads their messages. Some project leaders keep all problems to themselves, expecting that the team or the leader will solve the problems because that is expected. Therefore, even in the face of nearly insurmountable obstacles, they will say that everything is under control and the project is on schedule or nearly so.

Other project leaders bring nearly every problem to the manager's attention, either for help in solving it or to use him as a sounding board to bounce off suggestions or approaches. This type of project leader may appear to be in trouble all of the time, whether he really is or not.

Both types of project leaders can succeed or fail. A key to their success is the manager's understanding of them and what messages they are communicating, regardless of the words used. Once they are understood, the appropriate action can be decided.

A lot has been said about how to handle a project leader who may be in trouble, or failing; but little has been said about how to handle a project leader who is succeeding. The reason for this is that success is much easier to deal with than suspected failure.

If the project is moving on schedule with performance criteria met, and financially on target, the manager should let it alone. He must satisfy himself that the evaluations of progress are adequate and that key problems are under attack. Beyond that, the successful project leader needs some breathing room. He also needs an occasional word of compliment or praise to assure him that his accomplishments are noticed. The manager must avoid overcontrol of successful projects.

TASK FORCES—MATRIX OPERATION, GOAL CONGRUENCE

There has been a demonstrated increase in success by using the task force approach on complex design projects. The group functions effectively by living up to its naval namesake: a group of warships sharing a common mission and providing strength and support to the accomplishment of the mission by focusing its unique talents on a common goal.

In World War II and the Korean War a typical task force might have consisted of an aircraft carrier, a battleship, one or more cruisers, several destroyers, and supply ships that periodically replenished food, fuel, or ammunition. Command was usually headquartered on the carrier. Aircraft launch and recovery by the carrier were assisted by destroyers that protected the flanks and were positioned behind the carrier to pick up survivors of any aircraft that ditched on landing or takeoff. The awesome destructive power of the aircraft was embellished by the bombardment capability of the cannons in the battleship and cruisers.

As individual vessels, each had its characteristic strengths, weaknesses, and vulnerabilities. As a group of vessels, with a common mission and support assignments, the task force offered combined strength that towered over individual weaknesses.

Matrix Operation

In a less dramatic but equally important role to the company, the design task force has combined strengths that tower over any perceived individual weaknesses. Whether it consists of five people in a small company, or a dozen in a larger one, a task force has several common characteristics.

First, it is a matrix organization, one that cuts horizontally across the traditional vertical chains of command. The project leader usually is an engineer, with task force members including one or more additional engineers, plus representatives from marketing, manufacturing, finance, purchasing or materials, and perhaps legal and personnel, depending on the scope of the project.

Each of these individuals reports to his or her own manager and has regular assignments of a project nature that must be accomplished. Task force membership is an additional assignment. This extra assignment can create a conflict, particularly when there is a lack of goal congruence at the upper levels.

However, the matrix structure as such is not an unusual form. Most engineers, in their youth, worked for two or more bosses. In the home, for instance, there might have been several bosses—a father, a mother, an older brother or sister. Each of them made different demands that had to be satisfied. In school, there were several teachers, teaching assistants, coaches, each insisting on the student's full attention and effective results. In all of those areas, success required teamwork, ability to communicate with others, and an adeptness at coping successfully with conflicting requirements.

Membership in a matrix project team can be a growth experience if the project leader is effective and the group interacts effectively. There is an openness in the matrix concept—receptiveness to new ideas, sharing of information and problems—and an openness to uncovering mistakes and shortcomings. There is consensus decision making, as well as mutually designed courses of action for decision implementation and problem solving.

Communication and openness increase the possibilities of adversarial relationships, conflicts, and power struggles. Some members may question the strategies planned to achieve the goals, even the goals themselves. Others may attempt to communicate problems to their functional supervisor, in the expectation that higher-level influence may force a change of action by the project leader.

On the positive side, the matrix design is a powerful organizational mechanism for completing projects and for developing people at the same time. It can increase innovation and encourage entrepreneurial attitudes and actions on the part of its members. In an old, established bureaucracy it may have difficulty working at all. It will bring out organizational problems that are unseen in the conventional organization. On the negative side, it can stimulate power struggles and political influence attempts.

However, matrix management task forces can succeed dramatically, with top management support and some degree of patience when the arrangement is attempted for the first time.

Goal Congruence

Another major influence on the success of a project is goal congruence. If members of a task force do not all have the success of the project as their major goal, there will be serious problems. In a company environment where the standard chain of command structure is being altered by the introduction of task forces and matrix management, a perceived reduction of managerial authority often results.

If a particular manager feels threatened by having one or more of his key people working "too much" on a task force, he may react by reminding his staff members of who their boss "really is." He may also review their goals and make sure their routine, departmental goals are higher on the evaluation list than the goals of the task force. His rationale is that his department has continuing responsibilities to the plant, and those must come before any others.

Such actions severely undercut the authority of the project leader without reducing his project responsibility; the task force function is severely impaired. However, it is possible to avoid this intolerable situation if top management insists that certain key projects be given top priority in all departments that have members assigned to a task force.

Thus, goal congruence must be the rule. If it is not established and maintained by top management, project priority may be variable. Assignments will be missed, schedules will begin to slip, and the project will soon be in trouble. The best project leader will have great difficulty getting the necessary support and effort from other members of the team because they will notice the reduced effort by those whose functional supervisors perceive other priorities.

In some companies, goal congruence is well understood and applied. In others, only one major project is taken on at a time, so that goal congruence occurs automatically. In still other companies, the combination of multiple product lines and growth opportunities leads to the possibility of multiple goals by many task force members, especially those in purchasing, material control, manufacturing, engineering, and finance. These individuals often have herculean workloads, and additional project assignments are difficult to manage unless priorities and goals are set and adhered to. Goal congruence is a major factor in the success or failure of all these projects.

PROJECT PLANNING

In order to accomplish something of value, it is necessary to have a plan. The plan may be nothing more than a list of things to do written on a file card, or it may be an elaborate, computerized array of milestone tasks, dates, and resources required. In either case, a plan is a series of steps to get from the present to the future.

Before the various approaches to planning are reviewed, it is important to remember that the value of a project plan is that it is a thinking model of the actual effort that will go into the project. It is done at the beginning of the project and will predict the various events and outcomes with some degree of accuracy.

Because the future is different from what was planned, and it unfolds in a somewhat unpredictable manner, any plan is bound to have inaccuracies. Some events will take place as planned, whereas others will tend to be late; problems in achieving the desired end points will occur. External events may alter plans dramatically.

Though it is difficult to plan or schedule the future, it is the true goal of project planning to do so. In other words, one must achieve the performance, timing, and budget targets despite the unpredictable events of the future that would alter them as the project moves to completion.

The project plan is best done by the project leader, since he or she knows more than anyone else about the objectives and has a reasonably good idea of how to achieve them. The plan is, as mentioned, a thinking model of how to do the project. It includes the key events that must occur, the timing and costs necessary to get there, and the various tasks that must be completed.

One effective way to develop the project plan is to recognize that there are four major phases in most development projects: the conceptual or creative phase, the experimental phase, the prototyping phase, and the pilot run phase, followed by the release to production. A fifth phase is follow-up on early production. These phases will be described in Chapter 4.

If each of the phases is planned separately, on paper or on a computer screen, they can later be linked together in the total plan. Each is relatively independent of the others, though they are related.

It is necessary to list the beginning and end point of each phase. For example, the conceptual phase begins with the project assignment and ends with defining several design concepts for experimentation. The intermediate events include meetings with the task force, several creative thinking sessions, and one or more evaluation sessions, after which a number of design concepts evolve. A few of them, perhaps three, would be selected for the experimental phase.

Critical Path Mapping

One way to show the interrelationship of events is a critical path map (CPM). A CPM of the above portion, as well as the project plan approval, is shown in Fig. 3-1. This method of planning projects uses bubbles to indicate the com-

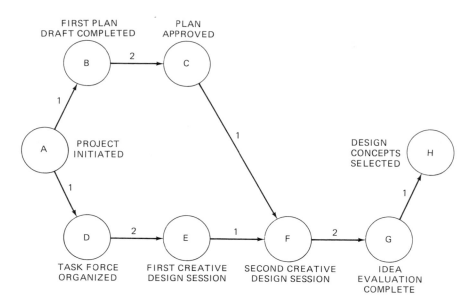

Fig. 3-1. CPM chart—conceptual phase.

pletion of the tasks culminating in a particular event. The length of time between events is sometimes listed as a single number, the expected time. Summing all of the times along a particular path provides a reasonable estimate of the time required to successfully complete the project or a portion of it. The critical path is the longest pathway in the map.

The conceptual phase shown in Fig. 3–1 would take seven weeks to complete. This phase is relatively uncomplicated, having only two partial branches. After one week the task force would be organized and the members given assignments. Two weeks later, the first creative thinking session for design concepts would have been completed, and one week later, the second one. Two weeks after that, the evaluation of ideas would be complete, and a week later, seven weeks after starting, the design concepts for the experimental phase would be selected. The second branch shows the plan development in parallel with task force organization and the first two creative thinking sessions. Since its time estimates are less than those of the other branch, it will not be the longer path through the conceptual phase.

Time estimation for these phases is developed initially by the project leader talking with various individuals to obtain their inputs on project timing. He may also use his own experience as a guide. Typically he would ask for three estimates, an average estimate, a pessimistic estimate, and an optimistic estimate. From these he would calculate an expected time from a formula often used in project estimating:

$$\text{Expected time} = \frac{\text{Optmistic} + 4\,(\text{average}) + \text{Pessimistic}}{6}$$

If the average time for a particular test were estimated at 2 weeks, the optimistic at 1 week, and the pessimistic at 5 weeks, the expected time (*TE*) would be:

$$TE = \frac{1 + 4(2) + 5}{6} = \frac{14}{6} = 2.33 \text{ weeks, use 2.4 weeks}$$

Other estimators may take a more pessimistic view and weight the pessimistic estimate heavier, yielding a somewhat longer expected time. However, consistency in method is necessary so that when the project is reviewed for comparisons between actual and estimated, the original basis for estimating can be reviewed and adjusted.

Of equal importance to estimating elapsed time is estimating the number and type of man-hours needed during the period. If 2.4 weeks of elapsed time is likely, how many people-hours are needed—96 hours at 100% time, or 24 hours at 25% time allocation? This applies equally for the time of engineers, technicians, and drafters involved, whoever is assigned project tasks.

Shortages of available time, the key resource, can occur all too often in a project. To avoid this, thorough estimation of tasks required and time to do them is important. On an ongoing basis, early and frequent reviews of manpower are also needed. Updating of the estimated versus actual time usage data will enable the project manager to stay on top of human resource needs and availability for his project.

PERT (Project Evaluation and Review Technique)

PERT diagrams are essentially the full project layout, wherein the critical path portion is the longest path through the interactions. Any delays along the critical path mean delays in the completion of the project. A PERT diagram is not illustrated here; examples can be found in most of the books that teach use of the technique. Generally they show an extremely complicated, interlaced series of events, which tend to appear insolvably complex. However, although complex, they are manageable and usable.

PERT diagrams can be managed manually on small projects, or by computer programs on large projects. After all of the events have been determined, and their elapsed times calculated, the interactions must be established. In other words, some events cannot be started until others have been completed. The PERT diagram can then be constructed, or the computer can do an equivalent construction, in order to determine the finish date of the project and the critical path.

Once this has been done, it is necessary to reconcile the PERT calculated finish date with the desired or required finish date earlier decided on by marketing or management. Often the project plan will extend beyond the target date.

Then, the value of determining the critical path is clear. When project timing is longer than allowed, the first area to be re-evaluated is the group of activities and events that comprise the critical path. These events need to be studied by the responsible individuals, and times reduced either by arranging some activities in parallel that were originally in series, or by looking for ways to pare down the amount of time estimated for those segments of the project.

After this has been done and timing is within limits, the PERT chart must again be reviewed to see if the original critical path is still critical—that is, the longest time—or if another one has developed. If a computer program is used to monitor the events and dates, any revisions to the PERT diagram will be factored in automatically, and if a new critical path evolves, it will be indicated by the computer. If a new critical path exists, it then is the area that must be monitored carefully as the project progresses.

PERT diagrams and critical path determinations are important in understanding the interrelationships of events. An intersection of several paths means that all of the events must be successfully completed before the succeeding events can be started. Relationships such as this are key to ensuring that loose ends are not forgotten or ignored before new activities are started.

GANTT Charts

A PERT diagram is somewhat difficult for reviewing management to follow at a glance, but a calendar chart with bars is relatively easy to follow, showing start and completion dates of key segments of a project. Such charts are called program calendars or GANTT charts. An example is shown in Fig. 3-2. These calendars are very useful for planning purposes and for indicating timing factors. For those reasons, GANTT charts are important management control devices. They also have value in planning for the support areas of marketing, manufacturing, advertising, and others, which must dovetail their efforts into the total project plan.

The GANTT or calendar chart falls short by being unable to show interrela-

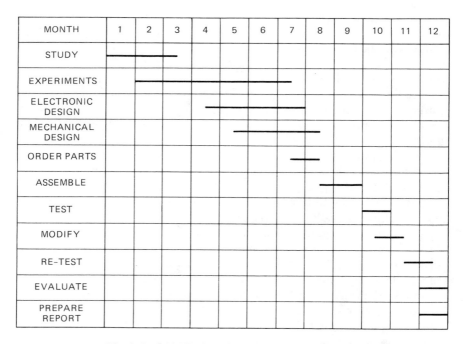

Fig. 3-2. GANTT chart for a development project.

tionships between key events of a project. The PERT diagram, which does show these interrelationships, does not necessarily show completion dates clearly enough to follow, unless one is familiar with the diagram and works with it regularly.

PERT–GANTT Combinations

One way to obtain the benefits of both is to combine PERT and GANTT, as in Fig. 3-3. This chart shows a PERT diagram stretched out on a calendar format. The diagram shows the interrelationship of key events and time-phases them onto a monthly or weekly calendar. The combined chart is a powerful tool because project engineers see both the events and the dates at a glance. Management observers can follow it without difficulty.

The combined calendar is more time-consuming to change than other charts, if major delays occur in a project. But that is a less serious problem if the diagram is developed on a computer. A further refinement of the diagram would include a row of squares below the diagram where the number of engineering, drafting, and technician man-hours planned for the month could be entered.

Value of Charting Methods

If a PERT or GANTT chart is developed for a project and then set aside, about half its value remains. That value is the thinking and planning that went into the development of the chart and gave the project engineer a good idea of the events that must be completed, as well as their interrelationships.

The remaining 50% of the value of the chart is in its utilization, in using and modifying the plan as the project progresses. Some will argue that the greater benefit derives from using the plan, and others will argue that the primary insights are associated with the planning process, which is a creative effort.

While opinions are divided as to which phase is more important, no one argues against the concept of project planning. So, comparisons of the value of developing versus using the plan are relatively academic. The author believes that 120% of value is achieved by both developing a plan and using it during the project—a synergistic effect.

Project planning using PERT or GANTT methods has a psychological value as well as a practical one. The sight of the manager or project leader reviewing the status of one's project frequently serves as a self-motivator, keeping the project's effort moving forward. There also is a second factor. In a long project, it is difficult to develop and maintain a sense of urgency, particularly in the project's early stages. A delay of two or three weeks in a two-year project seems of little consequence. However, a well-constructed PERT or GANTT chart, used on a regular basis, points out dramatically that a three-

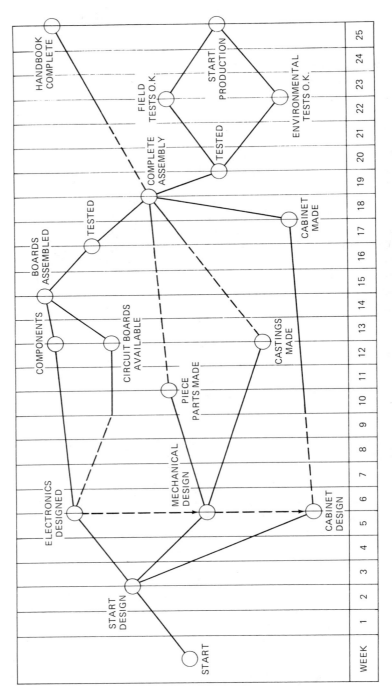

Fig. 3-3. Timed PERT chart.

week delay in the beginning can result in a delay of three weeks or longer at the end of the project. Hence, the chart helps keep the project on time.

During the course of the project, conformance to due dates and performance is vital to success. While there are important budget controls to adhere to, it is more important to complete the project on time and meet performance requirements than to miss either of those two requirements to stay within budget. If this is not immediately apparent, consider how one would defend the position of missing a key target date because the project was running very close to budget.

Reports and Projections

Tracking progress and problems is done readily if PERT–GANTT charts are used. On any given date, the project leader can find out if a given event is under way, completed, or delayed, merely by talking to the responsible individuals. Signs of trouble can be spotted and corrective actions begun. Trends in spending rates, either favorable or unfavorable, can be spotted, and projected costs for the project can be estimated.

Status reports on a regular basis have more significance if they are based on events completed as part of a planned sequence.

As mentioned earlier, the allocation of human resources is done more effectively using a project plan. It is not always a simple matter to estimate a new project's manpower requirements, but the initial estimates compared with actual hours spent in completing a given work segment provide insight that can be used to develop more accurate estimates as the project moves on.

Changing Project Dates

In a strict sense, the dates of a project are not to be changed. Dates have been set, whether arbitrarily or not, because it was judged important to complete the total work by a given date in order to have maximum impact.

However, events do not always follow a plan because secrets are not wrested from nature on schedule. On the other hand, the pressure of a deadline frequently does cause one's brain to function better and solve problems in the nick of time. This is an important reason not to change dates at all, or to change them very infrequently during the course of a project.

Should some segment be going more slowly than planned, the project leader must look ahead to determine which series tasks could be done in parallel, to allow a little slack time and prevent one missed data from impacting the project.

If dates are extended with relative ease, a tendency develops for individuals to think there is no importance or sense of urgency to the project. They may also lose some of their motivation and drive. These losses would be deadly for any major project and for the future capabilities of some of the project team members.

Thus, the project dates should be adhered to tenaciously. If problems arise, the project leader must search for ways to solve the problems and to make up the lost time. Only after all these approaches have been addressed unsuccessfully, should the alternative of a changed date even be considered. As mentioned, the pressure of a fixed deadline to meet a significant solution is often a potent motivator and mind expander that will make things happen.

If the date must be extended, the impact on all cooperating groups, both within the task force and outside in the company, must be evaluated. Various groups plan to have support people available on a certain day to begin their phase of activity. If this will be delayed for a time, they must be advised of and alerted to the changed condition.

Some Advice for the Project Leader

The project leader should operate under the dictate that it is relatively impossible to change a project date. Yet on the practical side, some problems are a lot more difficult to solve than they initially appeared to be, and much more time must be taken to solve them. If the project leader has used the time estimates and followed instructions, and has been as optimistic as possible, he will likely be in trouble because he has no available time to extricate the group from the problem without sacrificing the schedule. One way out of the dilemma is, during the project plan development, to allow time in the project for certain segments to be done over, on the premise that they will not work the first time. It is important to build in this time contingency because some things will go wrong.

The exact amount of contingency that should be built in is a matter of experience. It should be at least 20% and in some projects more than 50%. There is no way to determine how much contingency time is necessary, except that it depends to an extent on the newness of the technology used and experience with it.

While there is only a modest reward for being early on a project, there is often a serious penalty for being late. Therefore, it is prudent to plan for problems, to allow time for their solution, rather than to assume incorrectly that the project will run smoothly with no problems. People leave a job, or are transferred; crises occur in other departments that lead to reassignments, promotions, new members on a task force. Any or all of these events can happen and yet be unplannable.

SUMMARY

The basic design objective is to design the right product the right way, and complete it on time. This requires key inputs from marketing, covering performance and cost objectives. Carrying out the project successfully requires, above all else, that goal congruence be maintained—"Everyone must be singing from the same sheet music." If goals are inconsistent throughout a company or a division, the project leader is often unable to make the right things happen at the right time.

The task force approach, because it cuts across conventional functional management lines, has the potential to be effective, provided the individual managers work with the task force leader and not against him. Project planning methods using CPM, PERT, and/or GANTT charts singly or in combination have a double value. They are a thought process model of the design process and help the project leader determine where he ought to go, and in what time frame. When used in project work, they flag problem areas and aid management in following the progress of key projects. They also have a psychological value in establishing the fact that delays at any stage of a project can be detrimental to completing it on time.

4
The Design Process

INTRODUCTION

Design is the conversion of an idea into a finished product of some type. Sometimes the new or improved product is well defined, with high customer interest and tentative specifications, making the design quite straightforward. At other times, the proposed product idea is as tenuous and fragile as a soap bubble, and needs to be nurtured into what might become a technological breakthrough. This chapter deals with the total design process, from the early, conceptual phase, through early production.

Although design of a new or improved device can be considered as a number of steps, it better resembles a process, a series of gradual changes leading to a particular result. Since design usually includes some repeats and iterations, the term "process" better fits design than another word would.

One can break the design process into phases, each reasonably distinct from the others, though there may be some overlap. The phases are:

- Conceptual phase
- Experimental phase
- Prototype phase
- First production or pilot run phase
- Approval and early production phase

Some projects begin with the prototype phase, especially those of the fast follow company. Others, such as feasibility studies, may stop at the end of the experimental phase. The structure of the project may vary somewhat from one company to another, but these phases will be part of the overall design process.

Project Approvals

While project approval is not really part of the design process, it is the key that unlocks the project. Many projects will be approved before the planning phase is complete, or just after. Others will be approved after the task force leader is chosen. However, in no case can design work begin without project approval.

In some cases, project approval requires only a consensus of the engineering, marketing, and manufacturing managers plus the approval of the division or group general manager; but if the project will take a year or more for development and is expected to need extensive tooling or facilities expense, approvals of a vice-president or company president may be required.

The approval request must be accompanied by marketing and engineering estimates of time, cost, benefits, and market needs, as well as forecasts of expected returns on investment, profits, and risks, as described in Chapter 2. The extent to which these forecasts and estimates are required depends on the company's structure, the value of the project, confidence in the outcome, and the technical/financial risks involved. Once approval is received and the task force structured, the design process begins.

CONCEPTUAL PHASE

Before he can imagine new project concepts, the project engineer's first step in the conceptual phase is to review the project requirements. If any are unclear or contradictory, he should challenge them. The functional needs of the device are spelled out on the product specification sheet. It lists what is needed, not how it is to be accomplished. A good specification will have two columns—one labeled "Required," the other "Desired." These lists are particularly valuable, as discussed earlier. The "Required" column lists those performance levels and features that must be built into the device in order to make it salable and competitive. The "Desired" column lists those higher levels of performance or additional features that would make the product acceptable to a wider group of potential customers; or it may provide features that mark a significant advantage over the competition.

A thorough study and evaluation of the functional needs is essential. The adage that a problem properly defined is more than half solved is applicable here. Design is a problem-solving process, and a poorly understood functional requirement cannot be readily solved. The functional needs must be structured and their relative importance weighed. Since a good design is simple rather than complex, it must provide high performance per dollar of cost; and simple designs are usually more economical to manufacture, use, and maintain.

In this early review of requirements, the appearance and styling of the device must be looked at. Originally the domain of consumer products only, the need for improved product appearance now covers a wide range of products. Since buyers tend to form preliminary judgments based on the appearance of the product, aesthetics are a basic consideration. Form, of course, follows function. A good design should look functional, and its form should look planned. An excellent product can be much harder to sell if it looks like a conglomeration of afterthoughts that were hung on the basic device, rather than planned functions that were built in, from the beginning. Appearance does count.

The Search for Ideas

The creative phase of the project is the time when all ideas are good, none bad. It is the time when the designer or the design group become prolific generators of possible solutions to the varying problems posed by the design requirements. To put it simply, it is the time in the project when the more ideas, the better.

There are a number of ways to develop this flow of ideas. Brainstorming is still the most popular approach, whether done individually or by a group. The persons involved in a brainstorming session need to be steeped, deeply immersed, in the problem they plan to brainstorm.

The question asked to begin and maintain a brainstorming session is: "In what ways can I meet this rating, attain this size, improve this design?" Then, as ideas begin to sprout, the question changes to: "What else will do it?"

In brainstorming, an overview of each idea is needed, not the small details of the concept. Unusual ideas are to be welcomed, not scoffed at. One person's wild idea may trigger another person's novel solution.

Brainstorming sessions must be time-limited, focused on developing a list of ideas, with any form of evaluation or criticism withheld. The main emphasis is on quantity, rather than perceived quality of ideas. A quota of, say, 25 ideas in 25 minutes may be a starting point. Depending on the difficulty of coming up with possible solutions, this may be tough to do; but setting an easy quota of five or six ideas is counterproductive.

Another approach to unlocking creative ideas is to use a so-called trigger word. A list of synonyms is substituted for a given function, and the resulting phrase is examined for creative approaches. For example, if one of the functions has to do with "opening," trigger words for "open" include:

cut	explode	unravel	twist
tear	melt	fling	cut
rip	peel	swing	whirl
unscrew	bend	cleave	

Substituting a word such as "explode" would give radically new meaning to the word "open" and perhaps lead to a creative design concept.

Another approach is to select a random verb from the dictionary and try to force it into the problem, to see if a creative idea results. One opens the dictionary and blindly points to a word, such as "pencil," and spends five or ten minutes trying to see how the design problem could be solved using something that looks or functions like a pencil. Other techniques include morphological analysis, synergistics, and lateral thinking, all of which will be described and explored in Chapter 14.

Creative ideas are not an end in themselves, but are an excellent beginning. After all of the ideas have been generated and recorded, including the "morning after" ideas that often pop up when one is showering or shaving, it is necessary to evaluate and select.

The best ideas for the experimental phase will be those that are simple, answer most of the unknowns in the problem, and have the possibilities of quick phase-in. It is rare for any idea to meet all the requirements.

The ideas should be prioritized from most likely to least likely. After this list has been generated, three or four ideas should be selected from the top ten. If at all possible, this should be done by consensus of the task force because commitment is always easier to obtain when group members participate in the selection process.

To repeat, the selection criteria for evaluating ideas from the brainstorming session are effectiveness and ease of implementation.

Effectiveness means how well the proposed idea appears to answer the questions raised by the design problem. Some answers are more likely to be good solutions than others, by the extent to which they address the needs of the project. Ease of implementation, the second important criterion in selecting ideas for experimentation, is a very practical concept and sometimes a difficult one to accept. Frequently an idea meets the needs of the project, or appears to, but may be very difficult to implement. It may require expensive equipment or processes with which the plant engineers are not familiar.

With these potential problems in mind, the project leader must be cautious if the best idea will be difficult to implement. For example, a good technical answer may be impossible to implement within the cost constraints. If the project leader believes an idea will be difficult to implement but is so promising that he wants to experiment anyway, he needs to plan alternatives as well. Some team member must look for alternate ways to implement the idea; otherwise, the group may develop an elegant but unproducible product.

In searching for ideas, there are two sources that should be tapped in the early phase: competitive devices and patents. Unless one is first in the field, there are similar competitive devices already in the marketplace. It is important to have one or more of them available to see how others have provided certain functions, what low-cost ideas have or have not been used, where the strengths and weaknesses of the design lie.

A cost analysis is an excellent way to get an understanding of the processes used and how costly the device may be to manufacture. The cost analysis will not tell one how much making the device cost the competitor, but it will provide an idea of how much it would cost to make the device in-house using one's own labor and burden rates.

Patents are a second key source of competitive information on the state of the art. (Patents will be discussed later, in Chapter 15.) Their importance should not be overlooked or minimized in the assessment of competitive action. The firm's patent attorney can provide the appropriate patents for the project team's review. This should include patents issued to all companies providing equipment similar to that being designed.

Sometimes there is an aversion by engineers to spending the necessary effort in studying the patents and understanding the significance of the findings. They may prefer to rely on the patent attorney's evaluation of the patents and to do no review of their own. Some attorneys are very much aware of the patent art and can spot strengths and weaknesses in competitors' patents.

The combined team of project leader, project engineers, and patent attorney is synergistic—that is, more effective than each separately. Knowing the limitations of patents, and aided by the attorney, the task force can use its combined creativity to think up ways to avoid the claims of the patent. Not infrequently, the new ideas may themselves be patentable, and represent an advancement of the state of the art.

Engineers, especially those in new product task forces, should take the time to review the current patent art in their field. Patents are more up to date than textbooks, or technical articles, in discussing the latest advances in the field. Once the formality and unfamiliar langauge of the patent have been overcome, patent review becomes less a problem and more an interesting challenge to find out what really has been patented. Sometimes very little, sometimes very much will be found.

Patents have another useful purpose. Recent patents, including those in similar or different fields, sometimes have a trigger-word effect. Seeing a particular drawing or reading a particular set of words may unlock an idea that has been incubating in one's subconscious because of a missing element. What is commonly done in one field may be relatively unknown in another. The review of patents provides the opportunity for exposure to new ideas and approaches that may not have been thought of otherwise.

When the selection of ideas is complete, and three or four ideas have been chosen, it is time to move from the conceptual phase to the experimental. Ideas that were developed but rejected as not good enough should be retained on file. If none of the experimental ideas is successful, it may be necessary to go back to this list for a second look or to generate new ideas.

EXPERIMENTAL PHASE

The experimental phase of a project consists of designing, drafting, building samples, and testing them. It includes success and failure on the test floor. Anticipation, frustration, determination, and enthusiasm are all part of the experimental phase of a project.

Design and Drafting

The first step in the transition from idea to item is design. Here the engineer takes the penciled notes, the scribbles on backs of envelopes, the brainstorm ideas, the design sketches on cafeteria napkins, and weaves them into the first designs of the new device. Some of the sizes, shapes, and contours will remain with the parts into production. Others will change dramatically as the design evolves, so that the finished part does not resemble the original.

Design begins when the design engineer sits down in front of his drafting board, or his computer console, and makes the first lines that begin to skeletonize the device. All of his skills come into play: stress analysis, strength of materials, electrical circuitry; and if it is an electromechanical device, his newly acquired microprocessor application technology will be essential.

After the first set of sketches is finished, the calculations need to be checked. If the device must handle dynamic or static loads, the stress levels must be checked thoroughly. Electrical devices, especially those of higher voltage, have a set of electrical stress calculations to be checked out as well. Other stresses are due to electrochemical corrosion, galvanic action, ultraviolet radiation, temperature cycle resistance, and so on. The physical and chemical properties must be checked out carefully, on paper first and then in the laboratory.

Each design goes to drafting where the engineer's ideas are converted to standard sizes and shapes wherever possible. Tolerancing is applied, especially where close fits or perhaps loose fits play an important role in the device's function. Drafting the individual parts and creating some subassemblies for the model shop are the next step in the process.

Task Group Location

Since engineering plays a major part in the design task force, it is advantageous for the task group to be located in the same area as engineering. Specifically, the engineer(s), draftsmen, and technicians should have their desks and tables in the same physical space. There are many advantages to this, and few disadvantages. One disadvantage is that individuals are separated from their own groups of drafters and technicians. Also, the other engineers might inappropriately feel that drafters and technicians should remain separated, each in his own group. These feelings are wrong for a number of good reasons, and most of all they are counterproductive.

One of the major reasons for having the team together is interaction; the second is communication. When group members are working together, they share inputs, discuss problems informally, and work out solutions. When someone discusses a project matter on the phone, or receives a memo on it, it is much simpler to pass the information around to people in the same room than to route it for several days from one mailbasket to another.

When other team members can observe what is going on, the importance of their personal effort is much more apparent. They "see" how they fit into the picture rather than perceive it from another person's perspective.

The team members develop personal identification with the project and sometimes an entrepreneurial attitude toward it. This "ownership" attitude is what brings out the best in people, and yet it is difficult for management to foster. "Owning" a project yields such natural benefits as the extra effort needed to make a project succeed, the mutual commitment required to solve serious problems promptly, and the shared pride that goes with mutual success. If the project fails, there is also shared gloom because the group's best efforts were unsuccessful. However, the shared ownership in the project increases the probability of success.

In small companies, or on small projects, it may be difficult to have the team members in close proximity. They may have many varied assignments that take them in many directions. This would also occur if the project were relatively small and did not require the full-time effort of any one member. However, the manager should group team members in the same physical space whenever possible. It fosters better communications, better interactions, and better understanding of the job to be done. The advantages to the project and to development of the team members far outweigh any real or assumed disadvantages.

Model Making

Model shop effort is best done in-house, but if a model shop or tool room time is not available, test samples need to be fabricated in an outside model shop. This adds some formality to the sample-building system, but, on the other hand, it requires somewhat better attention to detail. An inside model maker may spend extra time to bring shaft diameters, for example, directly on dimension. An outside model maker, working with a time or dollar constraint, is more apt to request tolerances on key dimensions, whether they be tight or loose.

When models are built, attention must be paid not only to tolerances, but to the materials used in fabrication. If a part will be fabricated in aluminum, the model should also be made of aluminum. The same logic applies to other materials. If this is not possible, the project engineer needs to know which substitutions have been made, so that performance variations can be assessed later.

During model making, fabrication of parts brings to light errors in drafting or design. Parts cannot always be assembled if an important dimension is in error. While finding errors is embarrassing, especially to an inexperienced engineer or drafter, it teaches relatively inexpensive lessons in dimension checking and double checking. The more errors corrected here, the fewer that are apt to turn up later on, during the project.

When models are complete and assembled, they need to be checked over by the project leader. Flaws in design concept as well as flaws in, say, ceramic or plastic parts may be detected before testing begins. If the device has moving parts, the interactions need to be checked out and verified against the original specifications, not just the drawings. This pre-test sample check is a form of nondestructive testing; so the samples must be checked thoroughly, not given just a "once over lightly." After the samples have been inspected and any necessary corrections made, the initial testing begins.

Experimentation

Usually, if several ideas are being tested for the first time, the number of samples of each version is less than desired. As a bare minimum, three to five samples of each should be built, certainly never fewer than three. The reasoning is simple. If only one sample is built, and it does not meet standards because of some flaw or defect, or an unknown cause, then a second sample is tested. Suppose it performs flawlessly. If no difference can be found between the two, what assurance of success or failure is there? If a third (and fourth and fifth) can be tested, the majority of data, either good or bad, provides a greater degree of knowledge about the device—knowledge that it works, that it doesn't work, or that it is marginal.

It is important to test enough samples in the early stages of experimentation to make sure that the results can be correlated and correct conclusions drawn. A problem arises, however, if the cost of samples is high. Let us say that one is designing a large paper-making machine, a 500-ton molding press, or a steel rolling mill. It is hardly feasible to build five samples of three or four designs to decide which one works. The cost of building the "samples" might change the company's financial picture from profit to loss, even if the samples could later be rebuilt and sold.

Experimentation in these areas must be done on scaled-down models of the device, or on segments of it, where new or changed designs are planned to improve performance.

In many industries, however, the individual items being developed are small and not overly expensive. In those cases, perhaps 25 or more samples of each variation are built in order to establish the experimental data base.

Failures in Experiments

Unless the improvement in design being sought is very small, say 5%, the risk of failure is high. However, that need not and should not be a cause for concern. The failure of a design, especially in the experimental phase, is not the failure of the engineer, but of the idea being tested. However, failure is a humbling experience and sometimes a devastating one.

Most experienced design engineers accept failure as a normal part of the design process. Failure is one of the two possible outcomes of any pass/fail test, and is almost a prerequisite to eventual success of a project. Failure can occur during any phase of the project—experimental, prototype, or early production; but most failures are expected during the experimental phase because the first practical tests of a number of design concepts are being made then.

Experiments should be analyzed to obtain answers to three basic questions:

- Did the device meet the test criteria?
- If not, why not?
- Should this design concept be pursued further?

Note that the first question was not "Did it pass or fail the test?" It is necessary to make this distinction to prevent an early judgment on the merits of a concept, based only on a pass/fail crierion.

Consider this situation. A design concept passes the first, single stress test, then fails later when stresses are combined—say, thermal cycling under voltage stress. How does one evaluate the results? Pass, or fail? Or, an experimental device must withstand a critical stress level for one minute, but fails after 48 seconds. Did it pass, or fail?

In all cases, one would conclude that the device failed. But as a near-success, was the device on the minus side of a successful performance distribution curve, or on the marginally successful side of a failure distribution curve? It could be either.

The point to this exercise is that the experimental phase is too early in the design process to say with any degree of certainty that the device is good or bad. In other words, the initial pass/fail result is not an adequate criterion for evaluation. There is just not enough information for one to be certain.

Analysis of Failures

When a sample does not meet the test criteria (fails) the project engineer should find out why. The failure may not have had anything to do with the reasonableness of the design itself.

Unfortunately, most engineers view the matter differently. If one design concept works exceptionally well, and the other two or three fail, the result is predictable. The engineer will abandon those that did not work and push forward with the one that did. This could be a serious error. The mere fact that the device worked is no proof that it is a better, or even a good design.

During the experimental phase, all samples should be thoroughly checked after testing—those that did perform well and those that did not. Questions need to be asked and answered. Were any parts made differently from the drawings? Were any accidental or necessary material substitutions made? Did any electrical failures occur near a connection point? Did any mechanical failures occur near a joint, or at a change in cross section?

One must keep Murphy's Law in mind: anything that can go wrong will go wrong. Errors occur in fabrication, assembly, dimensions, test setups, and procedures. Any mistakes in these areas can cause a device to malfunction as surely as an incorrect design concept. If all parts were made to print, the mistake and resulting problem might not have occurred. If test procedures were faulty, it would be very difficult to reconstruct what might have happened differently, or why.

Analysis and investigation will prove that one or more design variations have errors. If not, the differences will point out the characteristics of a successful design compared to an unsuccessful one. While it is not likely that a serious error will be found, there will probably be several minor mistakes. These occur in experimental work because of unfamiliar parts, time pressure, and impatience. Too often samples are not thoroughly checked before testing, through lack of time. Inspections after testing come too late to correct errors. However, finding errors late is better than not finding them at all. One point to remember: the flaws or errors might be in the sample that performed well, as well as in the one that performed poorly.

Having the samples inspected is not enough. They should be inspected and checked by the responsible design engineer. Otherwise one receives a second-hand evaluation with a reduced chance of understanding just what was wrong with the samples. The engineer has the most to gain in this area; so he or she should do the evaluation and failure analysis.

After the samples have been evaluated, information of substance will be learned about why the sample failed. By asking "What else is this sample trying to tell me?" one can understand some of the other factors involved—the when, what, where and some of the hows.

The next step in failure analysis is speculation, or "what if." This encourages creative ideas that should lead to a better answer. What if a critical part were made weaker instead of stronger? What if it were made much larger instead of smaller? With some knowledge of the idiosyncracies of the design, creative speculation can generate new and better approaches to solving performance problems. These speculations may also lead to new tests for better determining the soundness of the design.

Instilling creative curiosity in the project engineer is the responsibility of the project leader, or of the engineering manager. More than lip service must be paid to keeping an open mind when evaluating failures. Design failure can be a devastating one, particularly if none of the ideas performs satisfactorily during the tests.

Angel's Advocate

One concept that helps is that of the angel's advocate. Rather than a theological concept, the angel's advocate is the opposite of a devil's advocate, discussed later, who looks for problems, errors, and shortcomings in the design.

The angel's advocate is a team member, or perhaps the project leader, who determines what is right about each design. He develops and maintains a sense of optimism and positive expectation that the problems are indeed solvable. This is not a foolish concept that maintains that "bad is good." Instead it is a combination of realism and positivism. The person evaluates all the good factors in performance, the near misses, the "almost worked" areas, and considers the positive aspects of each sample or design concept.

While the angel's advocate may actually be the person who recommends that a particular design be dropped, he or she will do so because other designs have greater merit, yield better results.

Besides keeping up the spirits of dejected design engineers and technicians, the angel's advocate brings a sense of drive and direction back to the project. This person in effect says, "Pick yourself up, dust off the dirt, and get back to making this good design work." It may be difficult at times for the project leader to bring that level of optimism to the project, but it is a necessity to its eventual success. Some may prefer to call this person the positive motivator, but because of the relatively large number of critics or devil's advocates readily available, angel's advocate seems a more appropriate designation.

Early Success

While failure is an unwelcome outcome in experimental programs, success is always enjoyed. But early success can also be a problem. Early success, when everything works right the first time, is a heady, exhilarating experience.

However, success can come from a fortunate combination of factors that are inadequately understood. The design engineer may not be able to explain why a new device worked successfully, any more than he could have explained why it failed. Perhaps it worked because the testing was not rigorous enough; or perhaps the device was overdesigned and would not be cost-effective against competition.

Surprisingly, early success in the experimental phase can be more of a detriment to long-term product success than early failure. In this stage, the "workings" of the device are usually not clearly understood, especially when they represent an advance in the state of the art. Little is known about the device's major characteristics; there are only a few thoughts about strengths and weaknesses, and nothing about its idiosyncracies.

The worst thing that can happen is to rush a "successful" experimental design into production quickly before its actual weaknesses are uncovered. By the time a problem is discovered, probably during the pilot run, a large investment has been made in tooling, substantially material inventory is on hand, and perhaps some orders have been imprudently accepted.

Then there looms a devastatingly close deadline for duplicating the success that appeared to exist months before. It has vanished, and the engineers have no idea how to achieve it again. With sheer determination, hard work, and luck combined, the deadline sometimes is met. But all too often the deadline is missed badly because the key knowledge was not learned in time.

How does one guard against the problems of early success? Being skeptical is a worthwhile beginning. The devil's advocate approach may be appropriate. To begin with, the samples need to be disassembled and checked for conformance to drawings and specs. Then another set of samples needs to be built, with parts checked before more testing. During the test program, a complete test series is required, followed by testing until failure occurs.

Not only can failure modes be determined, but strengths and weaknesses will become more apparent. If the device continues to perform well, some cost reductions can be made without losing the original performance requirements. If the new samples perform worse than the first, the samples were exactly the same as the first set, the design may be marginal. It will sometimes work, sometimes fail. That particular concept should be dropped.

In any case, more will be known about the design, and a greater understanding of the design parameters and their relationship to performance will be achieved by thorough testing.

Clearly, the project leader must not be deluded by early success; it can also lead to long-term problems. This attitude may be unpopular when schedules are tight and a new idea works; but when too little is known about a new design, the differences between success and failure will not be well understood, and unless they are, project success is in jeopardy.

Treating early success as luck rather than skill is a healthy though skeptical approach during the experimental phase.

Overcoming Obstacles in the Experimental Phase

The previous discussions centered on early experimentation because of its criticalness in building a data base, or technical foundation for the project. In the experimental phase there are several segments, each concluding with an analysis of data, review of performance, and changes needed to make the device perform better in one or several areas.

In later phases of the project it may be necessary to trade off some performance extra in one area for a deficiency in another; but in the experimental phase, it is important to achieve the performance criteria before paring down the cost or improving manufacturability.

In new product work, especially when the company is the first in the field, performance goals have been specified, but they may never have been met by any available device. If attention to meeting cost or producibility goals is the focus too early in development, and it negatively impacts performance, the end criteria may never be met. The manager must stress that his project leader must concentrate on meeting the performance criteria first. Once they have been met, effort can be concentrated on removing excess cost and making the design easy to produce. Otherwise, problems will plague the project.

This is not so serious for the fast follower, or the low-cost, high-volume companies; product capability has already been demonstrated by the leader. Knowing that a particular problem has been solved is a powerful plus in working on better or lower-cost designs.

Sometimes that concept can be used as a creative tool in moving a project forward. For example, a product task force is bogged down in the experimental phase and "struck." None of the ideas has worked well enough to meet the performance criteria. New ideas have dwindled to the vanishing point; motivation is slipping; the project may not make it. Yet the project leader or the engineering manager believes the targets are not unreasonable, and should be attained somehow.

The project leader can stimulate new thinking by asking his group to consider a hypothetical situation in which their major competitor will introduce the very device they have been working on in about six months, but there are no other details available. If this were to happen, he asks the group, "How would Competitor A be able to do it?" Taking into account their previous designs and products, what approach would this competitor use to be successful?

He is trying to turn the group around, from one that sees no way to move ahead to one that has heard that the competition is about to beat them to the market in the technology they work with so well. Rather than wonder what to do, they ask themselves what might someone else be doing to achieve the difficult goal. This new tactic, contrived and creative at the same time, can focus new creative thinking on a puzzling, frustrating problem. In some cases in the author's experience, the competition had actually announced the product; in others, the technique was used as a creative thinking stimulus, to get around the mental barrier of thinking the problem at hand could not be solved.

In most projects, however, there is enough progress that this tactic is not necessary. Most projects build from very little success in the first sample experiments to a substantially higher level of success by the second or third set of experiments. As thorough analysis leads to a better understanding of the physical (as contrasted to theoretical) factors involved, the project moves ahead. Creative speculation coupled with logical thinking improves results.

Completing the Experimental Phase

As the experimental phase proceeds, it is important to document why a particular design concept was dropped, and why another was selected. This information is important if a project runs into problems further in the development program, and the project leader wishes to review earlier ideas to see if any are worth reentering into the program. Memories are often fickle; people tend to remember, with embellishment, the good points of a design concept that did not make it but was a favorite.

The end of the experimental phase is not always clear-cut; but if the major design performance criteria have been met, and a good understanding exists about how to approach the unsolved problems, the experimental phase is completed. Design feasibility has been established; a major milestone has been achieved. With careful attention to the calendar, it should be finished on schedule, or ahead of it. If there have not been too many all-out "crash" periods, the project may have finished this phase within budget as well. As mentioned before, however, if either performance or timing was missed, whether the project was within budget is of less importance.

The end of the experimental phase may also be the end of a feasibility study, either a successful or an unsuccessful one. Since investments in effort and equipment increase exponentially with each phase of a project, this one of the lower-cost times to terminate an unsuccessful project.

PROTOTYPE PHASE

With feasibility established by the experimental stage, the project tempo increases as the most successful designs are combined or used individually in one or two prototype versions. Again, depending on the cost of samples and the cost of testing, only one design concept may be prototyped. On the other hand, one design may be so much less costly than the other that both are used for prototypes.

Prototypes are to be reasonable replicates of future production. This requires that a certain amount of redesign take place, particularly when several designs are being combined into one. As in any recombination, great care must be taken in combining designs, particularly when dynamic stresses are involved. Whether they be mechanical, electrical, or chemical, dynamic stresses can produce unsuspected effects on a composite. Whenever possible, carry a complete concept from the experimental into the prototype phase. This removes some of the uncertainties that arise when combining designs.

Before building prototypes and prior to any necessary redesign, a review of the test data and performance weaknesses is in order. The prototype testing will be more comprehensive than the experimental, so the shortcomings of design performance must be checked out and the redesign work to correct these deficiencies started, with concern for producibility as well as performance.

Manufacturing Engineering's Inputs to Design

While a manufacturing engineering (or industrial engineering) representative is usually on a task force, this valuable asset is not often utilized early enough in the design process. If this member has not played an active role during the experimental phase, it is important that he begin during the prototype phase.

It is not enough that a manufacturing engineer look at finished prototype drawings before they go to a model shop or an outside fabricator. Instead, he should sit down at the drawing board and review parts as they are being designed. Considering how they could be produced in volume, recommending what tolerances would be desirable, and determining which shapes are more easily fabricated are all part of the gradual transition from crude experimental samples into well-designed, aesthetic prototypes.

This increased activity by manufacturing engineering is so important that, on a large project, one or perhaps two engineers should be detached from their day-to-day assignments in order to work full-time on the project. The advantages of this approach are so great that it is surprising that more projects are not so structured. The more obvious benefits are:

1. Reasons for certain designs are better understood; therefore changes to reduce cost or simplify assembly are made with better awareness of the performance consequences.
2. Being in a new environment, the manufacturing engineers obtain a broader perspective about the project and increase their business horizons.

3. Designs are improved by the manufacturability built into them in the prototype stage.
4. Further into the project, the transfer from engineering to manufacturing is almost effortless. Everyone knows the project and its details so that interfacing between groups is rapid, automatic, and effective.
5. Long-term communications between design and manufacturing are improved because of personal contacts and rapport.

While these advantages are both quantifiable and nonquantifiable (objective and subjective), they are not always attainable, particularly in a smaller company, or one where many smaller projects are ongoing. It is clear, however, that the input from manufacturing engineering is so necessary to the success of the design from both performance and time standpoints that the project leader must plan for and insist upon manufacturing engineering input during the prototype development of the project and beyond it.

Designed Experiments

As prototype samples are being designed, it is equally important to plan and design the next set of experiments. Tolerances between moving or fixed parts, clearances between shafts and collars, thicknesses of major and minor parts (both metal and insulating) need to be determined with enough confidence to set later production specs. So many things generally need to be learned that a testing plan needs to be "built" to establish and verify that the key variables have been addressed and resolved through comprehensive testing. It is not enough to plan the experiments, although this must be done. Of equal importance to planning is the use of designed experiments. This is a discipline that is used to obtain unambiguous test results of high accuracy.

Experimental design utilizes four principles—replication, randomization, local control, and balance—described as follows:

- Replication means that a particular test must be applied to more than one experimental device. Any differences in performance provides a way to estimate experimental errors.
- Randomization means that any sample can be selected for any test. This ensures that all have their fair share of severe or mild testing. Sample #1 may be tested last or in the middle of the sequence, as determined by chance.
- Local control means that units fall into homogenous groups—made on a particular machine, cured in a particular oven, etc.
- Balance means that if a number of different factors, or levels, must be tested simultaneously, the test program can be laid out to assess the effect of each factor and whether any of them interact with each other.

Test designs such as Latin squares (complete and incomplete) and multiple factor experiments are used in many circumstances. The method called analysis of variance makes it possible to separate the contribution of each of the several factors affecting the variability of test results. There are many excellent texts on the design of experiments, and the subject is sufficiently complex that some course work would be advisable before one applies designed experiments to the testing of prototype units.

The method used in the ROI analyses in Chapter 2 illustrates the general approach, with a logical explanation of its power. Recall that starting from a given set of costs, sales, and investments, one could make some estimate of ROI variation by holding any two variables constant and varying the third between high and low limits. This would provide seven data points in an array resembling a three-dimensional cross, or one of the "jacks" used in children's games.

That method, while providing more data than a single point, provided no input as to what could occur if all variables changed, as happens in real life. In the proposed method, a total of nine calculations provided a cube of information, which better defined the range of changes in the ROI and whether any interactions were more significant than others. This second method resembles the layout and output of a three-factor, two-level designed experiment.

As can be appreciated, the same approach to experimentation can be very powerful. Instead of varying one factor at a time, much more can be learned if the experiment is planned to use two or even three levels of a variable to provide more information per test dollar. There is reluctance by some to consider this method because of all the unknowns, but to a large extent this hesitancy is due to lack of knowledge of the methodology and fear of the unknown. There are significant benefits to be gained by using designed experiments, which can only be realized by doing rather than by just thinking about it.

One set of benefits comes from planning the experiments. Becuaase the levels of variables must be determined before the test, the key variables themselves must be determined. Additional knowledge about performance parameters must be obtained to do this intelligently.

The second set of benefits derive after the tests have been completed. Analysis of the data will reveal which variables:

- Have the greatest effect on performance.
- Have the least effect on performance
- Interact synergistically to provide a strong (good or bad) effect on performance.

Thus, the use of designed experiments increases the accuracy of data and provides a better understanding of the effects of major variables on performance.

A designed experiment is more suitable to prototype testing than experimental testing because of the need for a large number of samples. However, an argument might be made for using a designed experiment in the experimental stage, particularly if prototype testing uncovered a major design weakness that required development of new designs and a return to the experimental phase.

As the first set of prototype tests is completed, it is once again necessary to perform a failure analysis as part of the design performance review. Prior to the review, performance in the key areas is compared to design targets in every category. Areas of missing or incomplete data are assessed for importance. If for some reason a key piece of information is missing it must be obtained before a design review is held. When the data has been analyzed, the project leader reviews it, and if satisfied, he calls for a design review.

Design Reviews

As suggested by its title, a design review is a detailed study of the major facets of the design and the design details. The participants in a design review should receive copies of the project documents, test reports and analyses, drawings, and other key information at least a week or so before the review meeting, so they have an opportunity to study it, prepare comments, and in general come to the meeting prepared to offer important ideas.

Participants include the project team members and other experts in engineering who are not members of the team, as well as representatives from marketing and manufacturing. The meeting is intended to be a thorough review of the technical progress of the design, concentrating on performance and cost. It is not a prototypes review meeting, but if the design review is successful, prototype approval will occur shortly afterward. Design reviews are covered in more detail in Chapter 9.

Devil's Advocate

As the design review time approaches, the success ratio in the testing program has improved dramatically. Much has been learned about the intricacies of the design, and a general confidence in the device and its ability to meet its requirements has been established. The feeling of optimism may have reached such heights that the project team believes its device is near perfection.

In order to provide a high level of objectivity in the design review, it is necessary in many projects to specifically assign someone to examine all the data critically. This person must look for shortcomings in the design, shortcomings in the testing, misinterpretations of data, positive inferences from marginal information, and so on.

For lack of a better name, this person is often called the devil's advocate, in reference to the Catholic church's process of canonization, or declaring that a particular person is a saint. During the process, the bishop in charge of the proceedings appoints a priest to examine the life of the now deceased candidate for sainthood, using documents and interviews. His specific assignment is to uncover negative actions and activities that could cancel the canonization.

Similar logic applies to the process of approving a design. The project team members are too close to it, and are aware mainly of its positive attributes. They may have forgotten or may downplay its shortcomings. Thus the role of devil's advocate is important to the objective analysis of the results of the project.

Specifically, the devil's advocate must study the results of the project, and must talk to the test lab technicians, engineers, and drafters. He must probe the findings and determine where weaknesses lie. His output can be handled in several ways, but the following approach is probably the most effective.

The findings are reviewed with the project leader and the engineering manager. After a critical review by them, the report is presented to the project team in a meeting. Any areas of disagreement need to be evaluated before a decision is made to accept the findings. This is particularly true if the results indicate serious shortcomings in performance that may stop the project.

The role of devil's advocate may be unappealing to some, but sought by others. To be effective in the role one must be knowledgeable about the product, its performance objectives, and the testing program that validates performance. He or she must be objective, curious, and skeptical enough not to take all statements at face value. Most important, the devil's advocate must have the intellectual honesty to evaluate all the information and to report successful results that have been missed, as well as shortcomings. A successful devil's advocate may not be universally liked, but will be respected for honesty and integrity.

This role is a necessary requirement on a major project and is an excellent training vehicle for engineers who will one day lead their own projects. It is important not to confuse a devil's advocate with a negative thinker. The former searches for truth and reports it objectively; the latter gets satisfaction from challenging others, especially in areas where information is somewhat fuzzy or unclear.

PROTOTYPE APPROVAL

When the design review has been completed successfully, when all of the incomplete technical problems have been resolved, the product is ready for prototype approval.

This is the last stage in the project design process at which an unsuccessful product can be dropped without a huge loss of money. While there has been a considerable and growing investment in engineering and testing, these sums are relatively modest compared with the expenditures for tooling and inventories that follow in the pilot production stage.

Preparation for prototype review and approval has been substantially completed as part of the design reviews. Thus, conformance to performance criteria has been well established. If there are any significant performance shortfalls, they may or may not be serious enough to close the project.

Cost review is a critical part of prototype approval. Invariably the device cost has risen as various changes have been made to reestablish the device performance. A thorough review of the original and final cost estimates is necessary to determine in which areas costs increased and how the total cost figures compare with original estimates.

Marketing estimates must be reviewed for the prototype approval meeting. Has anything changed in the marketplace that impacts the project's planned success? Has a competitor introduced a competing product? Or has one been announced? Are the selling prices estimated earlier still valid? Can they possibly be increased? Or must they be decreased, and by how much? Are new markets available for this product that were not anticipated earlier? Is the timing still certain, or has something changed?

Manufacturing inputs should have been factored into the design reviews and cost studies. As mentioned earlier, any well-run project brings manufacturing input into the project early so that the needs of design and manufacturing are considered at the same time. If this is not the case, manufacturing must have input in the prototype approval meeting. It is better to do it at this late stage of development than to delay it even more.

Failure Mode Analysis

Concerns about user safety as well as liability lawsuits make this an important part of prototype approval. As this review is discussed in Chapter 9, only an overview will be presented here.

A failure modes and effects analysis (FMEA) considers every component in the product. The function of the component is described, and the possible ways it could fail in service are listed. These could be failure from tension, impact, corrosion, electrical puncture, or tracking. Next, the expected probability of occurrence is estimated. Often a rating basis uses zero for low probability, 9 or 10 for high probability.

The severity of failure is estimated next—severity of injury to the user or the system. Detectability of a defect is considered also. From these estimates, an index is established by multiplying the three factors together. The index number determines the priority of corrective action efforts.

The net result of this analysis is an objective assessment of what failures might occur in service and how severe they might be. It is a system analysis method that can eliminate many potential service problems if used. While the final review and results are reported at the prototype approval meeting, it is clear that failure modes and effects analysis should be ongoing as the design progresses.

The Prototype Approval Meeting

After all of the checks and counterchecks have been made, it will be clear that either the prototype will be approved or it will not be, at the meeting called for that purpose. Assuming the project is approved, the project leader sees that the minutes are written, key points are covered, and the engineering manager reviews and approves the results. It may also be necessary to obtain approval of the marketing, manufacturing, financial, and legal managers if the project is a major one. Finally, in such cases it must also be okayed by the division or group general manager.

The design project has been successfully completed, and is ready for the next step, taking it into production.

PILOT PRODUCTION

A number of key decisions are made after the prototype has been approved. Most of them involve commitments of substantial resources—money and manpower—to prepare for initial production of the new device. Steps here include the preparation of final drawings and specifications, making commitments for tooling and facilities, and ordering parts and raw materials. All of these and more must be done before the first pilot production is undertaken.

The word "pilot" implies entry into production and not full-scale, high-volume levels. This is true in most product areas because it is a transition from the small-scale prototype building often done by model makers or technicians to the large-scale production done by production workers and equipment. In some cases this step in the process is omitted, again if it is a very expensive custom device or a one-of-a-kind piece of equipment; but in most projects it is necessary.

As engineering drawings are being started, or perhaps before then, it is necessary to obtain final approval for the funds necessary to complete the project. Another set of ROI calculations and sensitivity analysis (or risk analysis) forms must be prepared and approved. Because of all of the development work, cost studies, manufacturing review, and market studies, the information for the final ROI can be substantially more accurate than that for the original ROI.

Since the company is probably planning to commit at least ten times the resources to get into production that it did to complete development, this restudy is important. When one considers not only the investments in plant and facilities but also the inventory investments—raw material, finished goods, and accounts receivable—the size of the financial commitment becomes more apparent. Strictly speaking, accounts receivable are not inventory. But in the broad view, they are finished goods inventory transferred to the customer and not yet paid for.

Once the various management levels have reviewed and approved the project, manufacturing engineering has the authority to purchase tools, purchasing to order materials, and production to provide routings, standard costs, and so on.

However, none of this can be accomplished until the engineering drawings have been completed and approved. Since this can be a time-consuming process, drawings are often begun prior to approval of the ROI's. While there is the risk that the effort will be wasted if the project is turned down, the time saved in the overall project by starting drawings early is often substantial, making the risk taking worthwhile in most projects.

Drawings generally begin as engineers' sketches, which are refined into experimental drawings for a model shop to fashion ideas into parts. Finally the sketches are redrawn in final form, with ink or penciled lines, dimensions, and tolerances. However, that sequence is changing, as engineers are becoming able to turn their sketches into three-dimensional arrays on a screen, obtaining hard copy drawings that are accurate and complete with dimensions and tolerances.

As more companies are using computer-aided design and drafting (CAD), the time needed for final drawings is substantially shortened if major changes in parts are not necessary. The use of computers for design and for drafting is growing rapidly as better equipment becomes available and is progressively easier to use.

CAD is also revolutionizing the most rigorous portion of design, converting ideas on scraps of paper (dreams and doodles) to complete drawings in which fewer areas need to be fabricated into experimental parts. And in some cases (CAM), the computer can control the manufacture of the production parts themselves, so that certain machining steps can also be done automatically. The combination of CAD (computer-aided design) and CAM (computer-aided manufacturing) is bringing significant changes and improvements to the design/manufacturing process.

The process begins when the first parts have been received and inspected. Sometimes they are not exactly per print, and prompt decisions must be made. Can a deviation be allowed? Can parts be reworked? Must they be scrapped and redone? The impact of an allowed deviation in performance of the finished device must be weighed against the delays involved in reworking parts.

Some engineers prefer to err on the safe side, saying, "If it's not per print, don't use it." Others are more daring and say, "If it's at all usable, use it." Both decisions are correct, depending on circumstances. If the engineer understands the peculiarities of "his" design, he can better make a more risky approval. If not, the more conservative answer is appropriate.

To develop consistency and to avoid possible loss of control over deviation acceptance, the project leader must insist that he approve all deviations. All rejections due to nonconformance to specification also need his approval. In that way, he will be aware of the status of parts or assemblies and any impacts on project timing due to parts rejection.

Pilot Run Assembly

After all parts are in-house and approved, manufacturing will schedule and assemble the pilot run of assemblies for engineering tests. Although this is done by production workers instead of engineering technicians, members of the project team will be involved. Specifications may be unclear or incorrect. Parts that conform to print may not fit. Assemblies may not assemble, or if they assemble, may not function as planned.

With serious problems, the pilot run can be a period when extensive debugging is needed to locate and correct major errors. Or, it can be a minor-problem situation where each problem is easily corrected. Murphy's Law is often confirmed during a pilot run, but the inevitability of error can be minimized, if not avoided, by careful, thorough engineering design beforehand.

Pre-test Checkouts

Before any tests are begun, the units must be checked out mechanically and electrically to make sure everything has been assembled correctly. This is analogous to the assembly checkouts necessary earlier in the project, before prototypes were tested. The inspection or quality control group should do the actual checkout, to make sure specifications are accurate and procedures clearly spelled out.

Engineering project team members need to do their own checks, to satisfy themselves that the units are ready for test. It is not enough to rely only on the inspection group. Since engineering maintains responsibility for the product until it is actually in production, that responsibility must not be shirked, or be assumed by others.

Pilot Run Product Testing

Successful pilot run tests are the essential part of the release process. They are the focus toward which all of the project, from the very beginning, has been directed. It is vital that the testing be thorough and comprehensive.

As mentioned in previous paragraphs, testing of the pilot run must do more than establish that the product meets applicable industry standards. It must also meet or exceed the performance requirements set when the project originated. Beyond that, it must include tests that marketing or advertising believes are necessary to demonstrate where the product is better than the competition. And, the testing must include stresses beyond rating to establish how well it can take overloads and what its failure modes are.

The use of designed experiments can reduce the number of samples needed to develop the data base required and establish confidence that extreme levels have been tested and all known factors have been taken into account. The analysis of data from designed experiments, because it assesses the interactions between any test variables, is very useful in predicting performance of the device under conditions different from those under which it was tested.

Performance Analysis

After all the tests have been completed, a careful, thorough devil's advocate type of analysis is necessary. As in other test programs, the tested samples must be studied as well as the test results. A great deal of information has been learned about the devices, and a careful post-test study of assemblies will provide much information about the effects of test stresses on the device.

Any failures that occurred during the tests need to be examined carefully and evaluated. Failures are an unwanted occurrence during the pilot run testing yet will sometimes occur. There may be a tendency to gloss over failures in the pilot run, assuming they are "random" or due to assembly/production errors. This assumption can be dangerous because of possible weaknesses in the design that were not been properly evaluated originally. Failure can also be caused by an undetected error in assembly.

The actual cause of failure and the way to correct it must be determined, and agreement reached on its seriousness, before the product can be released. If it is not corrected, the product might contain a built-in defect that could cause problems in service and perhaps injure a user.

Failures must always be dealt with as shortcomings in design or fabrication, whether they occur below the device's normal ratings or above them. Too many failures in the pilot run mean an unreleasable product, no matter what the schedule or the need in the marketplace. An unreliable product is a much greater problem than a missed delivery date.

If it is necessary to redo the pilot run, the project manager must establish what needs to be corrected, what is involved in correcting it, how long it will take, and how much it will cost. There will be substantial pressure to get the problems solved, both on the project engineers and on production. Regardless of pressure, the corrections must be done thoroughly so that the retest will be flawless and the device ready for release.

Approval of the Pilot Run

After successful testing, analysis, and test reports have been completed, the project leader will call a meeting to release the product officially. Managers of marketing, engineering, and manufacturing will attend, and they or their subordinates will report on how well the pilot run went in their areas. Manufacturing will report on parts that needed modification or rework, how well the inspection, assembly, and checkout operations went, and what suggestions for improvement were generated. They will also report on actual costs to manufacture and how they compare to plan.

Marketing will review the test results from the customer viewpoint, highlight major benefits, and tell whether any compromises on ratings or features were made in order to allow release of the design. Marketing will also order their requirements for initial production units, either for sales samples or for first delivery to customers.

Engineering will review the device performance in all of the test sequences, indicating any changes required and the impact of these changes on product availability. The status of all drawings and specifications requiring change or update will be reported on.

Before asking for the managers to "sign off," indicating their approval of the pilot run, the project leader will review a "loose ends" list. In any project such a list always exists. Some drawings and specs need to be updated, test reports written, customer contacts reestablished, manufacturing routings and sequences corrected, standard costs updated, and more.

Those activities need to be completed promptly. However, they need not delay the release of the device to production. The project leader assigns these tasks to individuals in the respective areas and sets a deadline for their completion. The product can then be released to production, and the project team reassigned. The job has been successfully completed—almost, that is, but not quite.

Before the project can actually be closed, some clean-up work must be done. All test reports must be completed and routed; all engineering files must be brought up to date. One of the last duties of the project leader is to make a comparison of how well the project progressed compared to the plan. This involves conformance to time, performance, and budget requirements. That information is invaluable for estimating future projects. The project leader's report should include a performance evaluation of the key people on the project, so that the engineering manager can regard them as new project leaders, as potential leaders, or even as individuals who do not work well in groups.

EARLY PRODUCTION MONITORING

Early production must be monitored. The manufacturing engineer assigned to the new product probably will not be as familiar as the project leader is with problems that he encounters; or he may have cost-saving ideas that need to be evaluated before being phased in or set aside.

There is a benefit as well as a risk in phasing in early cost-saving ideas. The benefit is the cost reduction that will result, and in some cases will be badly needed. The risk is that the phase-in of the new parts may not be smooth because the manufacturing group is still unfamiliar with the new product. Confusion, mistakes, high scrap, and a net cost increase are the likely result of a cost reduction phased in too soon. A better approach is to wait until the first or second production run has been completed before phasing in any cost savings.

Higher unit costs and scrap rates can be expected during this early phase. Again, these are due to general unfamiliarity with the product and are not unusual. However, if costs do not drop and scrap rates stay high, some investigation and corrective action are necessary. Either something has changed, or else the initial estimates were too optimistic. The root cause needs to be determined so the appropriate corrections can be made.

As the problems are solved, and the costs and scrap rates come down, the engineering manager can breathe easier. His new product is in production and has a good chance to be a success. The project team has been reassigned to new tasks, and the manager can readdress himself to his other responsibilities.

SUMMARY

The design process is a continuing, iterative effort from the concept through early production. Some design engineering managers and their subordinates may believe that their major involvement ends after prototype approval. Nothing could be further from the truth. Following such a philosophy would lead to an expensive, difficult-to-manufacture product, instead of the economical, acceptable product originally conceived.

5
Control of Engineering Design Projects

The previous chapter dealt with a fictitious product as it went through the various stages of concept development to production release. Little mention was made of how the project was controlled and monitored from the three major considerations of performance, timing, and budget.

While a number of forms will be illustrated and ideas suggested on how to use them in this chapter, there remains a concern about excessive paperwork. Many engineers abhor paperwork with a passion, declaring to all within hearing that the paperwork jungle prevents them from doing a good job of engineering.

In other cases, engineers complain that their time schedules are so short they are unable to do their essential paperwork. Sometimes it appears that these engineers would prefer to generate reams of analysis rather than generate new ideas and keep the project moving.

The goal of this chapter is to detail the type of control information necessary for a project, how to prepare it, and why it is important. The individual manager can decide how much paperwork is essential to monitoring and controlling projects. "Essential" in the key word because too little is as bad as too much.

If there is any doubt that too little paperwork is a problem, consider this scenario. The marketing manager talks to the engineering manager about a perceived new product need, describing some of its performance features and how quickly he needs it. The engineering manager talks to one of the design engineers, gives him a few guidelines, and turns him loose. The design engineer designs the product to his perceived specifications and moves it along. When asked about its progress, he always replies that the project is coming along just fine.

Several months—or a year—later, the design engineer calls a prototype approval meeting because he is ready to take the design into production. Nine times out of ten the prototype design bears little if any resemblance to the concept envisioned by the marketing manager, who may lose his temper, question the technical competence of the engineering department, and write caustic letters about engineering not supporting the needs of the market.

It is almost certain that costs will be too high and performance too low; and the competition may have just announced a device that the marketing manager says was exactly what he wanted engineering to design.

Who is to blame? A number of people are, as is the system in which they operate. When all instructions are given orally, there is considerable room for error in the beginning. Without detailed feedback of information, the real needs of the market cannot be translated into performance ratings and costs. Without a system of engineering cost tracking, the engineering manager has a sketchy idea, at best, of how much time is being spent on the project. This is a clear case of management by abdication of responsibility. The worst offender is the engineering manager, and second, his design engineer.

There are several essential forms that the manager should have insisted on—a project initiation form, a project calendar, a GANTT or PERT chart with manpower estimates, and a project approval form set including preliminary specifications, target costs, prices and market, and ROI's. Last, a project tracking or control method is needed, to keep track of the project's progress and problems compared to all the others in the engineering portfolio of projects.

These, the essential forms required for initiating and controlling a project, are described in the following pages.

PROJECT INITIATION FORMS

The simplest project initiation form, or job opening form, is shown in Fig. 5-1. It provides a date, a project number, and a brief description of the work to be done, as well as a justification. In general, the job opening form is not a blank check to begin charging time to. Rather it is the opening of a correspondence file to collect information and a charge number for charging a limited amount of effort in developing a preliminary set of performance standards for potential product development or improvement.

In a particular company, the job opening form might allow the engineer to use up to 50 hours of time for this purpose. In another company it might be 100 hours or a maximum of $2,000 for time and materials. In still another situation, only a nominal allowance of time is provided; the file is opened only to accumulate information until a decision to proceed is reached.

All of the possibilities point to the limited action allowed by a project initiation form. It is a "first step" form without approval to proceed further. Many

PROJECT OPENING FORM

Date: _____   Proj. No.   _____

                                        Opened By: _____

PROJECT NAME: _____

_____

Description of planned work: _____

_____

_____

_____

Preliminary Timing & Budget Estimate: _____

_____

Product Specification Draft Completion Date: _____

APPROVALS: Engineering _____

                    Marketing   _____

NOTE: Preliminary work on this project to develop a product specification draft and time es-
        timate may not exceed 100 hours without authorization.

Fig. 5-1. Project opening form.

projects can be opened under this system, so that a preliminary look can be taken before those that do not meet the company's set of evaluation criteria are screened out.

Later when projects are abandoned for lack of interest, or inadequate technical or market potential, another simple form, the project closing form

(Fig. 5–2), can be generated. This form decides what was accomplished, why the work was terminated, and how much time and money were expended in the effort. It is also a signal that no more work will be done in this particular area without specific authorization.

PROJECT CLOSING FORM

Date: _____   Proj. No.   _____

                                                 Closed By: _____

PROJECT NAME: _____

_____

Reasons for closing project:

        [ ]   Released to production

        [ ]   Couldn't meet Performance Spec

        [ ]   Exceeded time limit

        [ ]   Project costs too high

Summary of Results:

Funds Expended $ _____

Authorized $ _____

APPROVALS: Engineering   _____

                    Marketing   _____

                    Manufacturing   _____

Fig. 5–2. Project closing form.

Forms Do Not Control Projects

While many of the figures used in this chapter will describe various forms used in the control of engineering projects, it is important that no one get a wrong impression about forms. One does not control engineering projects by filling out forms. That is not managing or controlling a project; rather it is a form of paper shuffling that has the appearance of control, but is a pretense.

The correct usage of a form is to distribute it to interested and involved groups after the actions have been taken. The form then serves as a communication and a written confirmation of what has been decided on or agreed to. Thus, a job opening or closing form, a product specification, or another document is a result of work done, not the sudden awareness of a change in plan or direction.

The Product Specification

In a new product project, the two most important documents are the project evaluation form, which was discussed in Chapter 2, and the product specification, shown as Fig. 5-3. While discussed before, this document is most important because in many projects the specification changes as more is learned about the market and the product as time progresses.

It would be incorrect to say that there should or must be more than one product specification. In many projects a single specification is written, covering both required and desired features, performance characteristics, and so on. As long as the product meets all of the required levels, it is acceptable; anything beyond that is superfluous.

However, some projects involve areas where the situation in the market is more dynamic than that. As more information is gathered about the market, performance levels that were considered desired are suddenly required. Or, the converse may be true; requirements may be lower than originally believed.

When these changes occur, it is necessary to update and reissue the product specification, indicating the changes. It is also important to relate the changed requirements to the engineering effort necessary to complete it. Changes often add considerably to the workload and can impact project timing and costs. For this reason, a change in specification, once a project has been approved, must be done with general consensus, after an assessment of the effects on the project time schedule and costs has been made.

Despite this negative comment, it is realistic to believe that changes in project specifications during product development must be recognized and dealt with. Products must meet market requirements in order to sell, but at the same time market research must be sufficiently accurate to discriminate between market needs and desires.

PRELIMINARY PRODUCT SPECIFICATION

PRODUCT NAME _____

PROJECT
NUMBER _____

Note: REQUIRED Levels must be reached for a saleable product

DESIRED Levels improve performance and enhance market acceptance

	REQUIRED	DESIRED
ELECTRICAL		
1.	xx	xxx
2.	x	xx
3.	x	x
MECHANICAL		
1.	x	xx
2.		
3.		
ENVIRONMENTAL		
1.	xx	xx
2.		
OPERATIONAL		
1.		
2.		
3.		
COST	xx	xx
OTHERS		
1.		
2.		
APPLICABLE INDUSTRY STANDARDS	xx	xx

Fig. 5-3. Preliminary product specification.

For engineering to change project specifications to conform to desires rather than needs is as serious a mistake as designing a luxury car for the small-car market. The product will never sell at a profit. Therefore, careful study by marketing and responsiveness by engineering will together allow product specs to change as required without the wrong product being introduced to the market.

PERT AND GANTT CHARTS

These charts, described in Chapter 3, are invaluable aids for monitoring progress of projects. They are excellent self-monitoring mechanisms for the project engineer to check on his own progress before the project leader asks how things are going.

Since planned completion dates are set up well in advance, each member of the task force knows what is expected and when. Thus, when the project leader inquires about progress or problems in a particular project task, no one can be unaware of due dates or responsibilities.

Whether the GANTT chart hangs on the project leader's office wall or is called up on his computer screen, the chart is a psychological motivator, especially for the marginal engineer or the one who needs occasional reminding or stronger direction to keep his level of effort high. When an individual is called into a supervisor's office for a review of project progress and his tasks are highlighted on the wall chart or computer screen, there is little doubt about what will be discussed. Thus psychological pressure helps keep effort focused on the end result and "makes" diligent work continue.

As each level of supervision reports to a higher level, use of the same set of charts focuses attention on areas of progress and the date by which progress needs to be accomplished. Thus one set of forms is used by all managing areas to monitor progress and maintain control of each project. This uniform format provides a greater and more consistent awareness of what is happening in the project. When the format is used for upper management, the level of detail can be reduced, and the information can be somewhat more comprehensive without any loss of value of the forms.

PROJECT TIME AND COST ESTIMATING

A precise estimate of the time or cost required to complete a project is a practical impossiblity. And, if it should happen in a particular project, it is more a random occurrence than a planned event. Estimates may be high or low, depending on how well the actual work performed matches the planned work.

To a significant degree, estimating resembles forecasting. A forecast is always wrong because the future changes; thus the forecast is revised periodically to reflect past realities and new projections of the future. Like forecasting, estimating also uses past data to form a basis for providing project costs for a future endeavor.

Project time estimates vary from a FOG (Far Out Guess) or a WAG (Wide Angle Guess) to a sophisticated step-by-step analysis of the project. All have value and limitations. The value of "educated guesses" is that the estimator uses his knowledge of results of past projects and considers how closely a new project fits others done previously.

This method of estimating is too inaccurate for a major project, but provides a good preliminary estimate of project costs in a very short time. Thus for a large number of projects that must be screened in terms of costs as well as benefits, the "quick and dirty" FOG or WAG cost estimate, based on past experience, is appropriate.

For a more detailed cost estimate, the project must be evaluated on a task-by-task basis. Similarly, time needed between events is estimated, the time required for engineering, technician, and drafting efforts. This can be done in two ways, by direct effort and by percent effort.

In the direct effort method, the project engineer doing the estimate reviews the work to be done and determines that during the four weeks provided for that segment of work the engineer will put in 40 hours of effort, the technician 160 hours, and two drafters 80 hours each. Thus his summary for the work will be:

Eng.	60 hrs.
Tech.	160 hrs.
Dfts.	160 hrs.

On the other hand, using the percent effort method his summary will look more like this:

Event 17—Design hardware support   Target time: 4 weeks

RESOURCE TYPE	TIME (weeks)	PERCENT EFFORT
Engr.	1	25%
Tech.	4	100%
Drftr. (4)	2	50%

By combining percent effort with man-weeks, a much better idea of staff needs and staff utilization can be developed. When the project detail is summarized, the project leader can determine if he can shorten time by reallocating resources (people's time) to other segments of the project. If he knows the percentages of individuals' time committed and uncommitted, he can make an accurate assessment of what he proposes to do. If man-hours are his only guidelines, he will make a number of errors in this task.

Allowing for Uncertainty

Before putting the summary together, the project leader must factor in some additional time for delays, doing things over, and solving difficult problems. One method is to use three estimates, as is done with PERT charts. The shortest time is the sum of all the optimistic estimates, the longest time is the sum of all the pessimistic estimates, and the expected time is somewhere between. If these three times are listed in the project plan, management will likely accept the shortest time and refuse to accept either of the others.

The statistically inclined project leader will use the three estimates as a pseudo distribution curve, and will factor in a mean value and a probable range for each task. The overall time has a range of "error" to which statistical analysis can be applied. This has some merit but suffers from a basic shortcoming: uncertain data cannot be made more reliable by mathematical methodology.

Another method has been used with good results, both by the author and by others who have published estimating guides. It is based on adjusting estimates with a factor based on powers of two. After the project engineer has provided the best estimates he can of the time it will take his group to perform each task, the project leader analyzes them. He divides them into three major categories, based on difficulty. Those of normal difficulty he multiplies by 1.41, those of moderate difficulty by 2.0, and those of extreme difficulty by 2.83, as the table shows.

Degree of Difficulty	Multiplying factor
Normal	$2^{0.5} = 1.41$
Moderate	$2^{1.0} = 2.0$
Extreme	$2^{2.5} = 2.83$

He computes the revised project time estimates and uses them in his project estimate.

Initially, use of a rule based on difficulty assessment multiplied by a power of two, or any other factor, may appear to be an illogical approach. However, a heuristic or rule of thumb approach such as this removes uncertainties the project leader may have about how much extra time to allow for difficult parts of the project.

Some might say that the original estimates submitted by the responsible engineer should be used, and the engineer should be required "to deliver on time." In a simple, short-term, or uncomplicated project, this stance is appropriate. It forces the engineer to develop an innovative, time-saving approach to the project. However, if the engineer is an average rather than superior performer, he or she may not be able to bring it about. Instead the schedule will slip, and the project manager will be spending an inordinate amount of time with project engineers in a last-minute effort to help them solve problems and reduce schedule delays.

These multipliers are a good starting point. Experience will show that they can be reduced for some staff members, increased for others. They are a major step in developing a realistic time estimate. With all of the revised estimates at hand, the project leader can take a close look at each task, and particularly the most difficult, to see how each should be approached. Through discussion with the project engineer, better ways of handling or even reducing the level of difficulty can be devised. The project leader and his team need to discuss the estimates, studying and revising them until they are comfortable with them.

Project estimating is the key to a good project plan. With experience, large errors in cost or timing can be minimized. The use of a heuristic method will enable the project leader to flag the areas of greatest difficulty for study and revision. Importantly, it will provide a consistent allowance for the estimates made by staff members and the time it realistically takes to complete each task in the project.

MEASURING PROJECT PERFORMANCE

Cost–Time Measurements

Measuring the costs incurred during a project is relatively simple. Each week engineers, technicians, and drafters submit a time sheet on which project account numbers are listed, and time spent on each project is totaled. The time sheets are filled in daily and then added up at the end of the week.

A computer summary of the time spent can as quickly be converted to dollars, using salary, fringe benefits, and other cost data stored in the computer. Comparing the actual data with estimates provides a quick check on how the project is doing against budget.

The project leader can observe whether all of the assigned people have been working on the project, and how much time they have spent on it. This method provides a rapid and accurate assessment of costs charged, although there is always some question as to its accuracy. In a large project where individuals charge all of a day's or week's effort to one project, there is little uncertainty. However, if people charge time to several projects, a degree of subjectivity may affect how much time is charged to individual projects.

However, these variations usually are neither large nor serious, and can be brought under control by discussing time-keeping rules with the individuals working on a project. An alternative is to have them punch in and out on a time clock. However, for technical people this is considered an onerous and demeaning task and is rarely done. In general, the accuracy obtained by using a daily time record with a weekly summation is sufficient for project costing.

Progress-Time Measurements

Measuring technical performance is as difficult as cost performance is easy. Some project engineers maintain that the project is "on schedule" or "doing fine" until it has slipped severely. Others are continually concerned that the project is going to be late until it is finished exactly on schedule. Verbal reporting of progress has its limitations, based on the persons reporting.

Expenditure reporting as a means of progress reporting has been used, in the generally inaccurate belief that technical progress is a direct function of expenditures. In other words, the project is 30% complete if 30% of the funds have been spent, and so on. This is rarely true. The spending rate often bears little relation to the project's progress. In one case, if serious problems arise, the project leader may reduce all parallel activities until the problem is solved, thus reducing the rate of expenditure. In another case, he may increase it by having all available staff work on several possible solutions at the same time.

In addition to these depatures from linearity between costs and progress, there is the general nature of design development work, which tends to follow a "saturation" or "S" curve with time. Performance against time follows an upward sloping curve until about 90 to 95% completion. Then the curve flat-

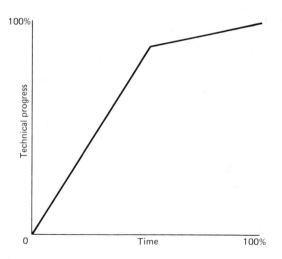

Fig. 5-4. Project progress vs. time.

tens out, and the time to complete the last 5 to 10% of the curve may be as long as from the start of the project to the 90% point. This curve is illustrated in Fig. 5–4.

The curve is a visual demonstration of all the effort necessary at the ending of a project, including tying all of the loose ends together and making sure all objectives have been reached. The project's performance–time curve, then, is quite nonlinear, particularly in some areas where simple analysis would expect it to be linear.

PROJECT MILESTONES—PERFORMANCE-ORIENTED

The best way to avoid the dilemma of how to correctly evaluate progress against cost and time, is to use a milestone chart (see Fig. 5–5). A milestone is a key event that completes a particular segment of work. It is well enough defined that there is little doubt about whether the milestone performance criterion has been met. Either it has, or it has not.

These are examples of milestone events:

1. Project initiated.
2. Preliminary specification approved.

		YEAR	1984												1985			
		MONTH	J	F	M	A	M	J	J	A	S	O	N	D	J	F	M	A
1	PROJECT INITIATED		O															
2	PRELIMINARY SPECIFICATIONS APPROVED		△O															
3	BUSINESS PLAN COMPLETED		△—O															
4	FEASILILITY STUDY COMPLETED			△—O														
5	PROTOTYPE TESTING COMPLETED					△—O												
6	PROTOTYPE APPROVED									O								
7	PRODUCTION TOOLING COMPLETED									△—O								
8	PILOT RUN PARTS READY FOR TEST											O						
9	PILOT RUN TESTS COMPLETED												△—O					
10	RELEASE TO PRODUCTION																	O

MILESTONES—
△ START
O FINISH

Fig. 5–5. Project milestone chart.

3. Business plan completed.
4. Feasibility study completed.
5. Prototype testing complete, all parameters per spec.
6. Prototype approved.
7. Production tooling completed.
8. Pilot run parts ready for test.
9. Pilot run tests completed, all results within spec.
10. Project released to begin production.

This list is not all-inclusive but it covers many of the key events that occur during the design process. There are many more milestones that could be selected—several design reviews, completion of certain subassemblies, and so on. There is no "correct" number, but there should not be so many that the list is unwieldy or so few that there is a one-year lapse between milestones. It may also be desirable to generate two milestone charts, a sufficiently detailed one for the project manager and a substantially general one for management overview.

The major criterion in establishing a milestone is ensuring that it is clear, with no ambiguity in determining whether or not it has occurred. If in milestone #5 (see list) prototype testing is complete but not all parameters are within specification, the milestone has not been achieved. The project manager cannot "look the other way" and say that the device met most of its criteria. It did not meet all of them.

In such a situation, the project manager must base his decision on the answers to these questions:

- Is it better to continue the effort and meet the specification requirement?
- Is it better to accept the result and obtain agreement to change the specification?

There are purists who insist the specification must be met, period. But the correct answer depends on timing, costs, and potential sales impact. It can be a "lose your shirt" decision for the project.

Sometimes inadequate development or low performance in an area can lead to serious customer problems and losses. Other times a product can miss market timing because of excessive development time, which reduces both early sales and future business. Such a decision is a judgment call by the project leader but should be made with the concurrence of the engineering, marketing, and manufacturing managers.

In addition to providing a key reference for project progress, the milestone chart can also include the spending plan. While it is often difficult to provide this information on the chart itself, a subsidiary chart (Fig. 5-6) can be generated that ties actual spending to planned spending at a given milestone. Major emphasis must be placed on meeting performance targets on time, but project costs and budgets must be respected and lived within.

Project Control by the Project Leader

Control of a project involves direction and accountability for the many tasks and subtasks comprising each project milestone, and it demands a willingness to cooperate on the part of all involved.

The project leader has responsibility for the project, but the engineering members of his task force also report to the engineering manager, marketing to the marketing manager, and so forth. When the project leader concentrates

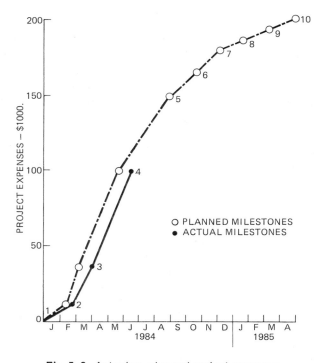

Fig. 5-6. Actual vs. planned project expenses.

only on moving his project forward, conflicts are likely. Other project and non-project areas important to the company are not being attended to.

The resultant conflicts will not be resolved by accusing others of being uncooperative. Rather they are solved by realistically striking a balance between the groups' needs. For example, the project leader needs to define, in advance, not only the tasks and subtasks but also the schedule by which performance is expected. There must be a common understanding of what the task is, who is expected to do it, and by when. Thus, the project leader is not performing adequately if he "farms out" the work and waits for results. He must take an active role in seeing that the work is done on schedule.

Control means more than reporting on events that have occurred and managing by exception. It is necessary that, on a weekly or semimonthly basis, the project leader compare program progress with his control chart. In addition, he must make forecasts of what is likely to happen and detect trends in ongoing critical activities, especially those that may have an adverse affect on project timing.

The project leader must confer with his project engineers, alert them of his findings, and obtain their feedback. Whether he reports any suspicions of schedule delay to higher management is a matter of judgment. If additional resources are to be required, beyond the project leader's scope, it is necessary to obtain them. On the other hand, a project will not be delayed by every test floor failure or every time an engineer is temporarily stumped by a problem.

The role of the engineering design manager regarding control of a specific project is one requiring tact and good judgment. He needs to have trust in his project leader and must foster mutual confidence.

His natural curiosity as an engineer as well as informal chats with engineering staff members will often uncover unsuspected problem areas on a project. It may turn out that the project leader had decided it was not vital to pass this information upstream because he was still evaluating it. He may have thought the problem was readily solvable without higher management concern, or perhaps he was optimistic that it would work itself out.

Whatever the case, the manager must tread a narrow path between intervening in the project and reducing the project leader's control, on the one hand, and waiting for the project leader to take action and risk having the project go out of control, on the other.

Regular, objective discussions between the manager and his project leader are a good way to avoid these situations, particularly if confidence and mutual trust have been established. In addition, agreement must be reached ahead of time to alert the engineering manager to potentially serious problems early on, so that corrective actions can be determined and implemented promptly and effectively.

Multiple Project Control

Much of the information that follows has been excerpted from a *Harvard Business Review* article.* It remains of the best discussions on the subject. The method it proposes is simple and concise, its application straightforward.

Individual project control methods were discussed earlier in the chapter. By breaking the project into major milestones and milestones into related tasks, progress can be monitored well, except in those areas where the results are quite uncertain because the problems are new and unique.

However, when an engineering department takes on a substantial number of projects, it is more difficult to keep track of them because of differences in reporting styles, shifts in priorities or staffing, and the desire to keep all projects on target and "full speed ahead."

Requirements for Project Control.   The initial requirement for project control is that each project have a written program plan prepared by the responsible project leader. As described earlier, it defines the project objectives, the design approach to be taken, the resources needed, and the time schedule.

In addition, the plan considers the effects of several contingencies, including funding limitations, vendor default, and unexpected technical problems. The plan must be complete and thorough, although not elaborately and minutely detailed. Once it has been approved, the project leader has the authority to begin, execute, and control his project.

It must be recognized that projects are dynamic and that change is normal during a project's lifetime. Customer needs change; project scope changes; unexpected problems often necessitate updating of the program plan. However, in order to maintain top management control, performance against the plan must be reviewed periodically so that appropriate direction and guidance can be given to the project leader.

Control of Multiple Projects.   One effective project control system consists of using a status board for a number of projects, with a monthly calendar across the top, as shown in Fig. 5-7.

Following the title of each project (or its number) are six rows (see figure). They are headed by letters that have the following meanings:

P—Performance, technical status
T—Timing, schedule status
B—Budget, cost status

* R. Howell, "Multiproject Control," *Harvard Business Review* (March/April 1968).

PROJECT		1983 J	F	M	A	M	J	J	A	S	O	N	D	1984 J	F	M	A	M	J	J	A	S	O	N	D	1985 J	F	M	A	M	J
"A"	P	●	●	●	●	●	●	○	●	●	●	●	●	●	●	○	●	●	●												
	T	○	○	●	○	○	○	○	○	●	●	●	●	●	●	●	●	○	●	●											
	B	●	●	●	●	○	○	○	○	○	●	●	●	●	●	●	●	●	●												
	F	●	●	●	●	●	●	●	●	●	●	●	●	●	●	○	○	●	●												
	R				■				■	■	■	■	■	■		■		■		■											
	A				□																										
"B"	P				●	●	●	●	●	○	●	●	●	●	●	○	●	●	●	●											
	T				●	●	●	●	●	●	○	○	●	●	●	●	●	●	●	●											
	B				●	●	●	●	●	●	●	●	●	●	●	●	●	●	●	●											
	F				●	●	●	●	●	●	●	●	○	●	●	●	●	●	●	●											
	R						■	■	■	■	■			■	■		■		■												
	A						□																								
"C"	P													●	●	●	●	●	●	●	●	○	●	●	●	●	●	●	●	●	●
	T													●	●	○	○	●	●	●	●	●	○	○	●	●	●	●	●	●	●
	B													●	●	●	●	●	●	●	●	●	●	●	●	●	●	●	●	●	●
	F													●	●	●	●	●	●	●	●	●	●	●	●	●	●	●	●	●	●
	R														■		■		■		■		■	■		■		■		■	
	A													□																	
"D"	P				●	●	●	●	●	●	●	●	○	○	○	○	○	○	○	○											
	T				●	●	○	●	○	○	○	○	●	●	○	○	●	○	○	○											
	B				●	●	●	●	●	●	●	●	○	○	○	○	○	○	○	○											
	F				●	●	●	●	●	●	●	●	○	○	○	○	○	○	○	○											
	R					■	■		■		■		■	■	■	■	■	■	■	■											
	A					□																									

P = PERFORMANCE STATUS   R—■ = REVIEW MEETINGS   ● = GREEN STATUS
T = TIMING STATUS   A−□ = PROGRAM PLAN   O = RED STATUS
B = BUDGET STATUS   APPROVED
F = FUNDING STATUS | = PROJECT START
                        AND COMPLETE

Fig. 5-7. Multi-project status report form. Reprinted by permission of the *Harvard Business Review.* An exhibit from "Multiproject Control" by R. Howell (March/April 1968). Copyright © 1968 by the President and Fellows of Harvard College; all rights reserved.

F—Funding status
R—Review meetings
A—Program plan approved

Each month, when the project leader reviews his projects with his manager, he simply gives a yes, a no, or a qualified yes or no, depending on the current problem. While answering yes or no forces a simplistic answer to many a complex situation, it is a necessity to assess project status.

If the performance level is per plan schedule (yes) a green circle is placed in the "P" space for that month. If the schedule itself is per plan, a green circle is placed in the "T" space. However if the budget status shows an overrun, a red asterisk (*) is placed in the "B" space. If in any category the project is in danger of serious trouble and attention is needed, a yellow square is used.

Thus, green means the project area is moving forward, red means the area has stopped or is in trouble, and yellow means potential problems. Although books such as this one and computer printouts are not normally printed in color, the master control chart often is a wall chart and can use color for contrast. However, the distinctive shape of each mark is unambiguous and easy to interpret, so that color is not a necessity.

In addition, project initiation is indicated by a heavy vertical line marking the month when the project began. Formal review meetings are marked with a plus sign as they occur. The planned completion date is similarly marked with a vertical line on the chart.

When a project is in trouble, or potentially so, an explanation of the situation and proposed corrective action are necessary. This is normally in the form of an oral report; the written summary is in the monthly report.

Implementation.   The first reactions to implementation of such a multiple project control method are usually disbelief, cynicism, and some hostility. However, if the projects that are in trouble receive management support and assistance for corrective action, cynicism changes to confidence.

In one company's experience, the first month of reporting showed 37% of projects were green, 30% yellow, and 33% red. About one-third were in each category. With emphasis on "getting the red out," the 33% was reduced to 11%, with no substantial increase in green. The reduction in red meant an increase in yellow. While the project situation was not so critical, more than half the projects were still on shaky ground.

The major reason was a lack of initial project planning, showing up in unrealistic schedules, budgets, performance, and so on. To bring the situation under control, a system of program planning was initiated, and all new projects required careful planning. This served as a beginning of change, since many of the ongoing projects had not been carefully planned or provided adequate resources.

After the planning was implemented, the percentage of red projects moved lower, the percentage of yellow projects declined dramatically, and the percentage of green projects increased. The totals were 77% green, 23% yellow, zero red. Two years later, when the method was introduced in a second division, and when monthly status report and corrective actions were emphasized, within a year 67% were green, 28% yellow, 5% red.

Evaluation of the Method.   While the initial objective (above example) was to reach a point where all projects were in the green, that was determined to be an undesirable goal. To achieve that level, standards of technical performance would not be stringent enough. The company believed in very tight control of budgets.

It was agreed that with such an overriding constraint it was realistic to expect that all projects could be completed to the last technical specification, on schedule, and within cost. While there was no way to determine the optimum level, the company's management felt reasonably comfortable if 75% of projects were green, 20% yellow, and no more than 5% red at any one time.

Supporting this position was the realistic awareness that different managers press the panic button at different times in their search for help. Also, project managers are optimistic or pessimistic about the progress of their projects. Some managers prefer to keep classifying a project as yellow or red throughout its life, presenting it as green only at its conclusion. Other managers keep reporting their project green and do not let it become yellow or red unless the bottom falls out.

This system does not eliminate these differences in reporting projects, but is intended to eliminate both overoptimism and overpessimism. A manager can look foolish if he repeatedly classifies his project in trouble and does not require any corrective actions; and the consequences are worse for an overly optimistic manager who will not own up to his project's problems.

Though a replanning program is often needed, a formal planning system need not take precedence over the reporting and control aspects of the system. Reporting and control force a manager to evaluate progress specifically against his plan, whether it be in his head or on the back of an envelope. Obviously, the reporting and control system inevitably forces improved project planning, but it is doable to start from "right now" rather than wait till all projects are formally planned.

This system is relatively simple and inexpensive. Instead of intimidating project managers, it encourages them to accept responsibility for the outcome of their projects. Project managers gain a much better idea of what higher management wants to know about their projects; and management is now able to review quickly a large number of projects and apply its energy to those projects in need of attention.

SUMMARY

Control of engineering design projects is an important yet sometimes difficult area for the manager. Financial measurements are relatively easy but quite inaccurate. Milestone measurements provide a strong basis for establishing checkpoints for performance, as well as schedule and budget. Any inaccuracies due to subjective evaluation are limited to those within the boundaries of a particular set of milestone tasks.

Multiple project control can be done by a relatively simple system that brings top management focus to those projects needing attention and allows each project manager to assess his project's performance and problems on a regular basis. It is a system that starts at a given point in time and calls for increased planning inputs afterward, rather than as a condition for implementing the system.

6
Organization and Functions
of an Engineering Department

INTRODUCTION

An engineering department will be organized according to its functions if it is to be effective. Organizations and function must be considered together, another application of the adage that "Form follows function." This chapter deals with several organizational structures, and the functions that they perform well, including their strengths and weaknesses.

TYPES OF ORGANIZATIONS

Engineering departments are organized in several ways, the original and traditional mode being the triangular type. There are two other variations: the task force and the matrix organization. All three types will be discussed in this chapter.

Traditional Organization

The engineering manager is at the top, with several chief engineers reporting to him and several layers of senior engineers and engineers reporting to them, much as is shown in Fig. 6-1. The traditional organization resembles a triangle, or perhaps a pyramid.

This general structure can function well if there are several technologies involved, or mechanical and electrical groups. In some engineering departments technicians and drafters report directly to engineers who supervise them. In others, as shown in Fig. 6-1, they report to a chief engineer responsible for engineering services to all the technological groups. Both reporting methods can be effective, depending on the workload of the department, both for projects and for product engineering.

Given a choice, most engineers would prefer having a technician and a

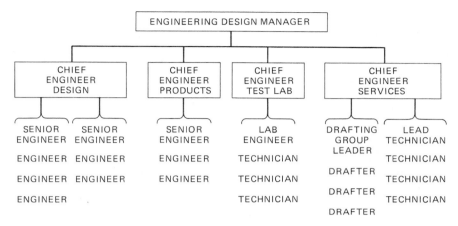

Fig. 6-1. Functional organization of an engineering design department.

draftsman reporting to them. In that way, sketches can be prepared promptly, and parts can be assembled quickly or obtained from the plant or a vendor. The availability of support staff to an engineer can add significantly to his or her effectiveness, or, more accurately, can multiply it. This effectiveness is very important in a project with a tight time schedule and a heavy workload. This mini-team pitches in and can get much more accomplished than the individuals themselves can.

On the other hand, if an engineering department is operating under tight budget constraints, these support personnel are part of a pool of technical people in engineering services. Under that mode of operation, the engineer must request technician or drafting effort, either on a project basis or a test request (or drafting request) basis, or by some other formalized method of obtaining the needed services. This has the advantage of better utilization of fewer individuals to do all of the necessary work. It has the disadvantage of requiring a written or verbal request each time help is needed, and in many cases trying to quantify a general type of need.

Some engineers adapt to the system and use it to their advantage. Others find it difficult and prefer to work alone, doing technician or drafting work themselves rather than requesting the help needed. This approach of course is wrong, but some engineers balk at paperwork.

Strengths of the Traditional Structure.   The orderly, hierarchical arrangement is the major strength of the triangular structure. In it, a young or inexperienced engineer can be trained in one or more product areas, be exposed to one or more engineering technologies, and become more experienced in time.

He or she can advance in technical proficiency and be promoted from junior engineer to engineer after only a few years.

In the long run, the positions of senior engineer and chief engineer are attainable; and even the top job, that of engineering manager, is a viable goal for the right person.

On the other hand, some engineers do not aspire to engineering management but prefer to become specialists, engineering experts in one or more technical disciplines. Many companies provide for this parallel path by positions called staff engineer or consulting engineer. These have a salary level equal to or higher than that of the chief engineer. Frequently a staff engineer will report directly to the engineering manager, and provide technical assistance to him. He also may be a technical consultant on specific projects, or be assigned special projects that utilize his particular skills and talents.

Another major strength of the triangular structure is the unity of command from top to bottom. There is no question as to authority, even though it may be delegated downward to specific subordinates. Throughout history, this structure has been used to rule kingdoms, fight wars, operate churches, and direct governments, as well as manage businesses. In entrepreneurial, owner-operated businesses it is a very natural way to run an engineering department, or the business, for that matter.

Many businesses use the same structure, with a president, executive vice-president, or general manager in the top spot. Reporting to the manager are the heads of marketing, finance, purchasing, and engineering, with their respective subordinates reporting to them.

Despite the strength of the traditional engineering department in carrying out its responsibility, particularly in large organizations with multiple product lines, it has difficulty in providing adequate support for projects that require significant inputs from many parts of the organization. To overcome the difficulty and provide greater organizational flexibility and response, two offshoots have developed: the task force and matrix management. They are somewhat similar to each other in that they utilize the talents of one person to coordinate or lead the project, and they both cut across the functional structure of departments and use the talents of people in many departments.

Task Force Organization

The task force organization is an offshoot of the traditional structure. It is shown in Fig. 6–2, where the vertical lines represent the functional departments, and the horizontal line shows the project coordinator's project authority. Often, the project coordinator will be a member of the engineering department, designated by the manager to coordinate a particular project from design through production. He will usually be selected from engineering

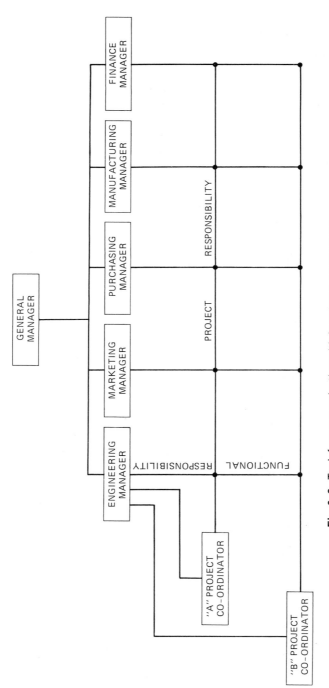

Fig. 6-2. Task force organization with functional departments.

because the project design effort begins there. The other functional managers will provide task force members from their own groups.

Since the engineering effort requires the major commitment of time during a project, except for tooling and the pilot run, the other members generally spend only occasional amounts of effort maintaining their portion of the project. Thus the task force members are not wholly committed to the project. However, the task force leader or project coordinator functions effectively only if he or she can bring together the diverse functional elements and resources.

A problem that often occurs is that the functional manager assigns task force membership to a subordinate as an ''add-on'' assignment in addition to his regular duties. When the requirement to meet an assignment from the functional manager conflicts with a task force assignment, the latter usually suffers. ''Lack of goal congruency'' is the usual complaint of a project coordinator whose task force members miss meetings, miss assignments, or in other ways have ''more important'' things to do.

If persuasion fails, the coordinator can go to the functional manager in an effort to restore a sense of project urgency, and solicit his cooperation. Another alternative is to review the problem situation with the engineering manager and ask him to work with the functional manager.

This is obviously a tenuous situation that relies on the persuasive and assertive skills of the project coordinator as well as the commitment of the members to get the job completed.

The fact that many, if not most, engineering task forces complete their assignments successfully is due to the actions and commitment of the people involved, rather than to any organizational strength or weakness. The task force structure has no inherent strength; there is only a dotted-line relationship between the project coordinator and the members. If they can commit to the project goals and see individual growth and development in meeting these goals, then a task force can work. If functional groups are understaffed and overworked, task force membership becomes a difficult task, and task forces flounder. Congruent goals, set by functional managers and adhered to, are a vital factor in the success of a task force.

The task force was and is a natural outgrowth of the functional structure, designed to place emphasis on completing particular projects more rapidly than by normal procedures. The major weakness of a task force headed by a member of engineering, marketing, or manufacturing is lack of authority on the part of the task force coordinator plus a lack of goal congruence. Sometimes the problem is due to the ''today'' orientation of production support groups that must solve current problems to keep the plant running. At other times, there is a lack of awareness on the part of functional managers, including failure to understand their own role in product development.

Matrix Organization

The matrix overcomes some of the major weaknesses of the task force by removing the project coordinator from the functional department, calling him a project manager, and having him report to the division manager. (See Fig. 6-3.) This deliberate focus on the project manager provides him with greater implied and somewhat greater real authority than in the other organizational structures, because he reports to the top.

This type of organization was pioneered in the aerospace and defense areas, and has a history of spectacular successes, such as the Polaris missile program, as well as some failures.

Since the program manager reports directly to the general manager, his requests for assistance and support from the various functional departments carry great weight. More important, the project manager's reporting to the division manager means continuing interest in the project's progress at top levels and awareness of it by functional managers. The task force problem of goal congruence is greatly reduced, although not eliminated. The project or program manager still does not have full authority over the functional department specialists on whose performance the project depends.

As before, the success of the project depends on the abilities of the project manager to work well with people toward a common goal. Because the matrix structure does not significantly disrupt the existing triangular organization structure, its strengths are still maintained.

One of the subtle strengths of the matrix is that future managers can be tested and developed by being managers on projects or programs of both small and large scale. The traditional structure does not provide for this at all, and the task force approach allows it only to a limited extent.

Variations of matrix management are employed in organizations to fit specific situations. A wealth of articles and books, both on matrix management and more specifically on project management, are available. This overview points out only a few facets of alternative structures, both for engineering and for the other functional departments. In times of change, the matrix enables business to deal with the changes needed in a structured yet flexible way. When correctly used, the matrix utilizes this needed flexibility to effectively manage systematic change in a business, whether it be a single project or a major thrust in a new direction.

STAFFING AND BUDGETING

The engineering staff size and budget level are variables that depend entirely on what the company is currently doing, how much effort in engineering is needed to support the ongoing efforts, and where the company plans to go.

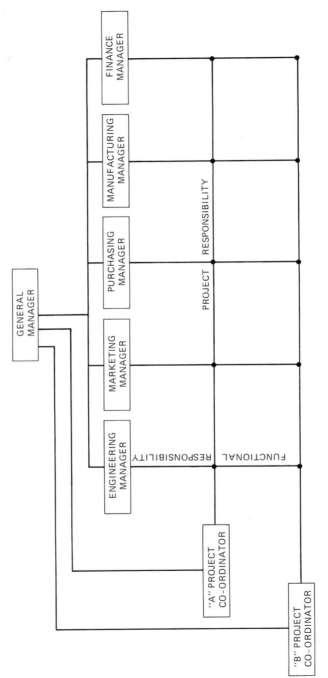

Fig. 6-3. Matrix organization (project management).

Does it plan growth in any areas? Decline in any areas? Will any areas remain status quo, or will some need to respond promptly to meet a competitive threat?

Staffing

If new products are to be designed and developed, there is more involved than just the cost of the persons assigned to the project. They must be sufficiently trained and familiar enough with the technology to do an effective job on the project. Hence, the costs of keeping a staff relatively intact and of bringing new people in for training and development are part of the costs of operating an engineering department.

Whereas engineering efforts are looked on as an expense that as such can be readily decreased in bad times and increased in good ones, this analogy is far from the truth; engineering is more like a capital asset whose time frame is unique and to a degree unexpected.

For example, if a large number of people in an engineering department are terminated because of company or industry economic problems, the loss will not be felt for some time. With good vendors delivering quality parts and machine shop areas fabricating assemblies, the short-term effect of an engineering staff reduction is almost zero. In fact, the profit and loss statement will be improved because of reduced engineering expenses. However, if vendor problems occur and one or more new vendors must be selected, problems will begin because subjective (experience-related) criteria as well as objective criteria become involved. Also, the company will find it difficult to respond promptly to competitive challenges in lower-cost products or in new designs.

If the company then decides to increase staff rapidly to meet these challenges, it has the other problem of staff training and development to handle the technologies involved. In some cases, this problem can be reduced by hiring experienced engineers; but if their experience is not in the same product lines, only their general background will be of immediate value.

To illustrate, an engineering staff could be said to resemble an electrical R-C circuit consisting of a resistance and a capacitance in series across a DC battery, as shown in Fig. 6-4. The voltage across the capacitor, $E,$ is analogous to the engineering effect; and the resistance, $R,$ to both the complexity and the variety of technologies within a company.

In Fig. 6-4(A), if the original level of engineering effort is reduced, simulated by opening the switch, the effect is not noticed immediately. For a time, things continue as they were; but then, the loss of engineering is realized, and the damage increases exponentially.

In Fig. 6-4(B), assume that the company decides to restore its engineering

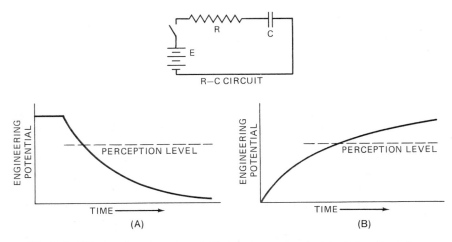

Fig. 6-4. Effects of engineering staff size changes—electrical circuit analogy.

effort, or to increase it. This change is symbolized by closing the switch. While skills are building and training is going on, improvement in capability does not appear immediately. However, after some period of time an improved level of performance capability is achieved.

The time lag between engineering staff size change and the perception of improvement or reduction frequently masks the harm done to an organization's technical capability by staff reduction.

Budget Size

Recognizing the inherent risk in rapid changes in staff size and the need to be able to apply effective engineering effort to the project and problem areas, one must ask again, how much should a company spend on its engineering effort? There are several possible answers: as much as the competition, whatever is needed to support planned growth, and whatever is needed to provide all necessary functions. These will be considered separately.

As Much as the Competition. It is not easy to know how much a particular competitor spends on engineering effort. However, various industries publish data on various types of expenditures, and it is possible to get industry averages from trade associations.

Similarly, one can get Dunn and Bradstreet reports or call a stockbroker to obtain recent annual reports or other P&L sheet data on a specific competitor. In most of these reports, some estimate of engineering can be obtained, although it is frequently masked in an array of other data. However, knowing the structure of one's company's annual report and the actual level of engineering expense, it is possible to make an educated guess about the competition's level.

While it is true that companies' engineering costs vary from as low as 1.0% to as high as 10%, that 9.0% range means an order-of-magnitude difference from highest to lowest! As a general rule, however, a company should—in order to retain its present market position—spend its industry average level, assuming it has an average range of products.

Whatever Is Needed to Support Growth.   Increasing market share is an expensive way to grow. It is more easily done in a new and growing market than in a mature stable market. Making product improvements, by adding features and higher ratings in order to further penetrate the market and increase market share, can mean doubling the engineering effort. This is in addition to a significantly increased marketing effort covering promotion, pricing, and sales.

Use of a general rule, such as doubling the engineering effort to improve market share by half, is, of course, inaccurate. Each situation is unique. The size and growth of the market, the maturity of the products already in the market, the current market share and its trends, all are external factors that will have a substantial effect on the amount of engineering effort to be expended.

No matter how negative the foregoing statements, if growth in existing products is needed, adequate engineering effort is necessary to support it. This may include new designs of existing products, design improvements, or totally new designs of brand new products. Each requires a substantial amount of engineering effort.

Whatever Is Needed to Provide All Necessary Functions.   The best way to establish what the engineering budget ought to be is to determine everything the engineering department is currently doing, what has been directed to begin, and what needs to be done that has been set aside for lack of anyone to do it. Then, the costs of providing those functions need to be determined. Generally, the engineering manager knows what the various functions are, and he with some of his subordinates can use engineering charge numbers and budget data to arrive at reasonably accurate information about current spending, as well as what is needed for future efforts.

Depending on how this information is summarized, it will probably include these categories:

Customer assistance and application
Product cost reduction projects
Product improvement projects
New product projects

Exploratory or feasibility studies
Miscellaneous efforts, meetings, etc.

Since each company has a different emphasis on new product versus existing product development, one cannot say what is the "right" blend of these efforts. However, in several companies the author observed the split between new product effort and all other efforts to be 20/80. Approximately 20% of the engineering effort was spent on new product and feasibility studies, and the remainder was directed at everyday activities to keep existing products going, and to improve them. Some companies refer to this ongoing effort as RTS (Readiness to Serve). Others call it GPE (General Product Engineering).

Usually the realization that 80% or more of engineering effort goes into "keeping the wheels turning" is a surprise to upper management. It often seems that most of the engineering effort goes into new product work, and that time schedules should be much shorter. In one company, the engineering manager argued unsuccessfully that the great majority of his engineers' efforts went to maintenance or improvement of existing products. Only when he installed a formal engineering timekeeping system and summarized his findings six months later could he establish that 70% of his staff effort was, in fact, in general product engineering.

In a new company with few if any existing products, the ratio of new product work to product engineering is reversed: 80% or more for new products, 20% or less for product engineering. However, as products reach the marketplace and are manufactured in greater quantity, more engineering effort is involved—in approving additional vendors, cost reduction efforts, and producibility improvements. In some product areas, additional application engineering is needed as new uses for a product are possible and the limitations of these new uses must be established.

In a mature company, in a mature industry that is stable or growing at a slow rate, the expenditure for new product work may be close to zero. The company may be in a "harvest" mode and be satisfied to make an acceptable product at the lowest possible cost, with no plans to obsolete the product or redesign it.

Since each situation is different, the engineering manager has no norms for comparison. Industry statistics are difficult to analyze, and company annual reports often lump engineering costs with R&D, or, worse, with marketing, sales, and administrative costs.

Zero Base Budgeting

The best way for an engineering manager to assess what his department budget should be is literally to start from zero and develop his budget from a base of zero dollars. This methodology was introduced in 1970 and is a tool for expense control, decrease, or increase. It enables the engineering manager to deliniate exactly what his department does, and how much is spent doing it.

With that information, he can determine the need to continue certain activities or discontinue others, and the relative importance of each major activity. Priority ranking is a natural outgrowth of this zero base budget approach. Engineering activities usually are quantified, have time charge numbers, and, in a project area, have approved funding. Thus, they are relatively easy to break down into the "decision packages" required for zero base budgeting.

A decision package is a document identifying a discrete activity in such a way that management can evaluate it and compare it to others. In addition to individual projects, a decision package could comprise product engineering effort for product family A, another for family B, and so forth. These packages must not include fractional personnel, as they are difficult to evaluate. The package includes the relative significance of the activity, to the goals of the company and to the division. A statement on the effects of discontinuing the specific activity is necessary as well.

When the engineering manager has his total set of decision packages, and the sum of all of the costs anticipated for the year agrees with his budgeted amount, he can prioritize the packages, based on expense level.

The results often indicate that considerable engineering effort is being spent, as mentioned earlier, in supporting products currently in production. Probing further, he may find products whose sales are declining but that have proliferated into many product variations needing frequent drawing updates, vendor updates, and shop floor modifications, all of which consume engineering time. These particular products are candidates for culling from the product line, since they may not be generating profits consistent with costs.

He may also find that some projects have not been receiving enough effort to bring them on stream as planned. This often occurs in medium-priority projects. They are not sufficiently large to warrant a project team to carry them to completion, yet are important enough to be done.

The evaluation of all the engineering projects by the manager, on a zero-base-budget basis, provides him with a quantifiable summary of all of the major engineering activities as well as projects. By analyzing them, he can find indications of which products or activities should be deleted because of their marginal value to the department and the company. By prioritizing them, he can decide in advance which trade-offs would be least painful to the company should an economic downturn require a reduction in engineering effort and staff.

The concept of zero base budgeting for engineering has merit for properly assessing where engineering effort has been spent. It is also a powerful tool for establishing an engineering budget consistent with the work that needs to be performed, on both projects and ongoing essential activities.

If budget cuts are necessary to trim the engineering effort to meet a target level, zero base engineering budgeting allows the manager to make those cuts that hurt the key engineering effort the least. At the same time he can assess the effect of these cuts on the engineering program and communicate this to upper management.

Development of the Engineering Staff

Within a relatively stable engineering department, different levels of expertise develop over time. Some persons become specialists, preferring to concentrate on a single facet of technology; others become generalists and have technical competence in several technological areas. Some engineers stay in one company for their entire career; others leave within five years and take with them the accumulated knowledge and experience of product, process, and procedures.

There is no way to keep all members of an engineering staff from becoming dissatisfied, from having ambitions beyond the manager's capability to satisfy them, or just from wanting to change jobs for more money, different experiences, or to relocate. However, a manager can mitigate the serious effects of losing well-trained competent people through an ongoing staff development program.

In a large company, this can be done automatically because the company has training staff at the corporate level, runs internal technical and nontechnical seminars, and has numerous opportunities for the younger or less experienced engineer to better understand his company and his job.

In a small or medium-size company these training services are either limited or nonexistent. Yet the need for development is no less critical. Staff members may leave at any time, and new people must be brought up to the necessary competence level as quickly as possible. In addition, it is often desirable to have staff members sufficiently adaptable that they can switch from work in one technological area of a company to another, and thus be competent in more than one area.

Regardless of company size, training courses are a great help; but by no means are they the best or only approach to staff development. Development is best achieved by assigning the engineer to a project slightly beyond his or her capabilities. Successful achievement of the project goals will be a significant learning experience as well as a technical achievement. If a senior, more experienced engineer is available as a project consultant, some pitfalls can be avoided, and progress speeded.

The Value of Experience.   Experience still is the best teacher, and challenging job assignments are the best way to obtain that experience. A certain amount of failure and false starts must be allowed and even expected. The developing engineer will have different and perhaps new ways to approach a design. If the more experienced engineer acts as a judge and predetermines which ideas have the best chance of working, the developing engineer learns little. Some of his or her own ideas need to be tried, even if there is reason to suspect they will not work. Sometimes a wrong path leads to a better, more elegant solution than reworking the same design concepts again.

If the engineer is allowed to devise some of the new concepts and test, evaluate, and prove them, creativity is developed and innovation fostered. This topic is covered in greater depth in Chapter 12, but it is important here to recognize the value of individual ideas as a part of the development process.

Failure as a Step to Success

Since some new ideas will fail because the engineer will make wrong choices, it is also important that the supervisor or engineering manager be somewhat tolerant of failure. Just as failure in experimental phases of projects (Chapter 4) is usually a stepping-stone to success, so is failure important in the development of the engineer, whatever the type of design project involved.

It is important, of course, that the developing engineer not be allowed to make a million-dollar mistake on a one-of-a-kind device as a learning experience. Such latitude on the part of the supervisor would be inexcusable.

However, in perspective, allowance for mistakes and failure must be made. In addition, the manager must gradually become intolerant of failure. If failures lead to understanding and new knowledge, then future activities should lead to a gradual reduction in the number of mistakes and failures. Otherwise, learning from experience and from correcting failures will not occur.

How Long Does It Take to Develop Competency?

One might ask how long it takes for a new engineer to become a competent performer in a particular area. There are, of course, no conclusive norms. Each person and each industry is different. So the following comments, based on the author's observations and those of other managers he has consulted with, are somewhat subjective.

A new engineer will begin to be a contributing member to a group about nine to twelve months after joining it. After two to three years, he or she will be competent in the technological area; and in five years many engineers are very competent and a few are expert in the particular field. In more complex, rapidly changing fields it may take ten years; but about six to eight years is the average length of time for a person to become an expert, or an authority, in his or her field.

After that length of time, it is important to provide different assignments to the engineer, to avoid stagnation. The stimulation of a new technology, the challenge of needing to learn something new, are part of the renewal process. The engineer develops additional skills, sharpens the ones he already has, and becomes a more valuable employee. There is reasonable concern that such an employee might be inclined to leave for another job, but the risk is probably less than expected. If the engineer is well paid and has challenging, varied assignments, he is likely to feel needed and appreciated. Another job may not seem as appealing as the present one.

Development of engineers can be embellished by seminars, short courses, or college classes at the graduate level. Depending on the company's product and technology base, outside education may be of great benefit in development, of modest benefit, or a waste of time. No general rule applies.

However, development on the job through varied and challenging assignments is still the best way to bring out the best capabilities of staff members, whatever their age or experience level. It is not easy to provide this because the manager must emphasize certain projects, retard others, and, above all, keep the operation going. If he remembers that the need to develop people is as important as his other assignments, he will figure out ways to provide these challenges and assignments.

If the manager does this, everyone benefits. If he does not, in the long run he will probably be the loser—of his job, perhaps. There is great value in developing staff members who have ten years of experience or more, compared to staff members who have one year of experience repeated ten times.

How Many Projects to Attempt

Although this topic was covered in Chapter 2, the emphasis there was on the project load of the individual engineer. The following discussion deals with the manager loading his design engineering group.

There rarely is a scarcity of projects to undertake in any company. Marketing finds many unfilled customer needs; products within the company need to be rejuvenated, or excess costs eliminated. Further technological applications need to be explored. In most situations, the design engineering manager has more design-related projects than he can readily assign and hope to have completed.

Prioritizing, already discussed, is the key to ensuring that the most important projects are done first because of their value to the company. However, one can have a list of five top-priority projects but still find it impossible in most cases to handle five at one time.

Some specifics need to be kept in mind. In almost every project, limited resources are the constraint. There are both people resources and facilities resources to be applied to a project. If a manager attempts simultaneously to address all of his top projects, it is almost certain that all of them will come in later than planned. A much better approach is concentration of resources, and focus.

Consider the following, oversimplified example. The engineering department has ten qualified staff members available for projects and ten key projects, each of which will take two man-years to complete. There are several choices open to the manager. He can assign each member a project; all of them will be finished two years in the future. Or, he can assign five members to the top two projects, finish both of them in 0.4 year (about five months), and get them into production. Then he can assign those individuals to the next two and get them finished in another five months. In less than a year, four key projects will be in production and developing revenue for the company.

Admittedly, the example is oversimplified. It may not be possible to complete a project in five months with five people. Restrictions on lab facilities, fabrication, tooling, and parts procurement may make it impossible. And five people may get in each other's way.

On the other hand, several people working together on the same project may work synergistically. In one afternoon of discussion, they might solve problems that would take one person a month. Thus, there is great potential benefit from assigning several people to a project. As a general rule, every project undertaken should have the maximum number of people available assigned to it.

A number of activities can be done in parallel rather than in series—a great time saver any time it can be done. Sometimes there is risk in doing things in parallel, but with a well-planned project the risk is minimal.

Several key benefits may be obtained by concentrating efforts on a few projects rather than doing many at the same time:

1. Development time is reduced.
2. The new product appears on the market sooner.
3. As priorities change with time, the need to do all of the projects is negated.

Reduction in Development Time. It is a truism that ten hens setting on one egg do not hatch the egg faster than one hen; and some phases of development work do require significant periods of time for generating and incubating ideas. However, most stages of a development or new product project can be handled faster and better with several knowledgeable people working on it together. The 1 + 1 = 2.5 (synergistic) effect often occurs. Ideas of one person building on ideas of another usually develop a better result than one person alone does. The major result is a better product in a shorter time.

Earlier Market Introduction.   This benefit cannot be overemphasized. It derives from reduced development time. There are very few products that do not benefit from an earlier market entrance. New product introduction six months ahead of a competitor developing a similar product can establish a better market position, improve sales, and even change customer buying habits. In many businesses, competitors have similar ideas simply because they are trying to solve the same problems, and the first one to the marketplace has a distinct advantage in intangibles—prestige, leadership, and so on—as well as the tangible one of increased sales.

In established product lines that are undergoing cost reduction or value analysis studies, earlier market appearance means increased profit dollars. With an established market, the return on a high-quality, cost-reduced improvement is immediate. Conversely, delayed market appearance of a cost reduction reduces profit.

For example, an electrical equipment manufacturer successfully completed a value analysis study of one of its high-volume fuse lines. The market was well established, and the device was recognized as a quality product. The value analysis team developed an electrical contact that looked better, performed better, and was substantially lower in cost. The immediate annual saving was in excess of $100,000. Because of a heavy project load, and attention given to many projects rather than a few, the improved design was provided relatively few resources, and it did not become available until two years later than originally planned. The product line continued to be well accepted, but in the interim the company missed out on $200,000 of bottom line profit that would have accrued if resources had been applied to achieve an early market entry.

Change in Project Priorities.   The priority of a project is a measure of its perceived value to the company. However, priorities change with time and with the market. Sometimes a high priority changes because customer needs change, or because an existing product can be adapted to meet a need. Priorities also change because of errors in the original assessment.

Other reasons aside, the principal advantage of working on a limited number of priority projects at a time is the avoidance of waste. If engineering is attempting to work on the top ten and two of them become unnecessary after a year, a significant amount of effort has been wasted—as would be the case if this happened when engineering was concentrating on the top two projects. However, engineering concentration on the top projects requires equal concentration by marketing. Focusing on the real needs of the market reduces the number of these errors.

Summary

Focusing on a few vital projects, by putting as many resources as possible on each, is the way to establish how many projects an engineering department should be doing at one time. Every project undertaken should have the maximum (or optimum) number of people assigned to it. Experience with design projects teaches every engineering manager that time estimates tend to be optimistic. With fewer projects, greater emphasis can be placed on applying full resources to get them done on time, regardless of unexpected problems.

SUPPORT FUNCTIONS

Depending on the structure of his department, the engineering manager may have laboratories under his direction. Drafters and technicians may also be grouped in subunits, headed by a chief drafter and a technician supervisor, respectively; or they may be directly assigned to specific engineers. This section discusses the important services these support functions provide, how they can be optimized, and some of the problems that can be encountered.

Engineering Laboratory

An engineering laboratory is more a combined workshop/laboratory/development area than it is a test lab. Whether its primary functions are mechanical, electrical, electronic, chemical, or combinations, the laboratory is a place for testing ideas. Usually this is done in a preliminary way, to evaluate them—or screen them—before an expensive, involved testing program is undertaken in a larger testing laboratory, in the company or on the outside.

An engineering laboratory can provide these services to the design engineering group:

1. Perform preliminary tests on new materials that are being considered.
2. Evaluate field return units.
3. Develop hybrid materials (compounds) for in-house processes.
4. Provide a workshop area for assembly/teardown of test devices.

Types of Laboratories. An engineering lab can be elaborate or simple, depending on the needs of the department. It must be adequate to meet the testing and other requirements of the group; yet it must not be so large as to replace major testing labs within the company.

The lab's chief engineer will influence the development of the lab; so choosing this person is especially important. If he or she is too laboratory-oriented, the lab will concentrate on testing, seeking more and more test equipment. In some cases the requests for testing equipment will exceed the real needs of the department. If this individual is too workshop-oriented, the lab will have plenty of tools for assembly and teardown but too little equipment to perform anything more than rudimentary tests. A balance of both orientations is needed.

Effective utilization of the lab is the primary goal. If this is left to chance, the immediate needs of the engineers who are having work done in the lab will predominate; and some engineers will "take over" by having lab technicians become more and more involved in their project work.

One way to avoid lab tie-ups is to use test requests for lab-related work of any type. A simple form (as shown in Fig. 6–5) is very useful. The requesting engineer fills out the form, indicating what needs to be done and the objective of the test. The project number is recorded, and an estimate is made of the work to be done.

Using a test request system allows the lab chief engineer to schedule the lab's technicians, provide optimum service to all engineers, and keep track of what is planned and what is ongoing. The use of estimates and actual hours is a continuing check of (1) the effectiveness of the lab staff and (2) the understanding of the engineer requesting the test of the amount of work that needs to be done. Estimates should be reviewed by the chief engineer before he schedules tests. Any major difference between an original estimate and his revision should be checked out with the project engineer. This avoids unpleasant surprises later, when the project engineer learns that he is being charged 40 hours time for the "short" test he thought would take less than a day.

A side benefit of having an engineering lab is that it provides a training ground for new staff members, both engineers and technicians. The variety of products and components requiring testing and the variety of tests performed provide a first-hand opportunity for the new person to quickly learn the products and the testing methods. Whenever possible, new employees should spend a portion of their training program in the lab for its real benefits to them and the company.

In summary, an engineering lab is a near necessity for most engineering departments. It can vary from a tiny workshop or assembly area to a sophisticated testing facility. The lab provides many necessary services for engineering and can be a training ground for new staff members. While there may be a tendency for some project engineers to monopolize lab facilities, this can be prevented by using a system of test requests to define work or test segments to be done by lab staff.

TEST REQUEST

Date: _____   Test Req. # _____

TITLE: Proj. No. _____

Requester _____   Appvd. _____ Priority _____

                                          Estimated Time _____ Actual Time _____

                                          Completion Date Desired   _____

                                                               Actual   _____

TEST LAB: Elec. [ ]   Mech. [ ]   Envir. [ ]   Mat'ls. [ ]   Thermal [ ]   Other [ ]

Speciman Description: _____   No. Furnished _____

Purpose of Test: _____
_____

Testing Required: _____
_____
_____
_____
_____
_____

Special Procedures: _____
_____
_____

Applicable Standards: _____
_____

Fig. 6–5. Test request form.

Engineering Technicians

Technicians can be effective supplements to the engineers in a department, or when assigned to a project. A competent technician not only relieves the engineer of some of the less technical tasks, but brings practical experience to the assignment. While formal technical training is becoming more common, often interested, motivated persons can become effective technicians through on-the-job assignments, outside reading, observation, and asking questions.

A technician's duties vary with each company and with the technical group to which he or she is assigned. These are some of the duties a technician may have:

- Assemble and check parts before tests.
- Perform specific tests on parts or assemblies.
- Witness testing and take notes on performance.
- Collect data and write brief test reports.
- Assist in after-test disassembly and postmortems.
- Prepare simple engineering sketches for parts or for test arrangements.
- Assist in preparing samples for display or marketing presentations.
- Fabricate simple parts for test samples.
- Act as expediter for engineering purchase orders.

These duties may vary considerably, particularly if a company specializes in mechanically oriented products, or chemicals, or electronics, and so on. Simply put, a technician provides the important function of an engineering aide.

Selecting Technicians. In many cases, technicians are hired from the outside, after completion of technical training courses beyond high school. In other cases, they are transferred from the shop, after having applied for a technician opening or requested consideration for a future opening. These latter cases should be taken very seriously, for some can make an excellent transition from work force to technician.

Some characteristics to look for in a prospective technician are these:

- Above-average performance in the present job.
- Reasonable grasp of technical matters.
- Ability to express ideas clearly, orally and in writing.
- High degree of initiative.
- Ability to work effectively with others.
- Good level of mechanical dexterity and skills.

Opinions about each candidate should be sought from the immediate supervisor; and each should be interviewed to assess his or her personality, thinking ability, and extent of technical knowledge. A brief written test is appropriate to evaluate mathematical and language skills.

The successful candidates usually work out well. Over time, they develop a comfortable relationship with the engineers and become valuable project contributors. With increased experience and good performance, greater freedom should be given them in planning their work assignments; and the accomplishments and contributions of technicians should be publicized frequently, to give them recognition for jobs done well.

Some problems can be expected. Certain engineers treat technicians as expediters and "go-for's." Others give technicians difficult and complex assignments far beyond their capabilities. Both extremes lead to unhappy, dissatisfied employees. While challenging assignments are always a desirable goal, they do not realistically occur all of the time. A balance of duties and good communications between engineer and technician can resolve most assignment problems.

Should a technician be assigned to an engineer, or be part of a talent pool, available to work on a variety of projects? There are good reasons for either arrangement. Since the subject relates to the utilization of drafters as well as technicians, it will be discussed in a following section.

Drafters

Drafting is another key function in an engineering department because drawings and sketches are the key communications tool to transmit concise, accurate information on parts and assemblies to the "outside world." This outside world includes purchasing, vendors, manufacturing, and manufacturing engineering. Without engineering drawings, the department's design effort would go nowhere.

Drafters vary in skill level, depending on both background and experience. The beginning drafter, recently out of high school or technical training, will become a detailer, who makes and dimensions simpler drawings. Moving upward in skills, the detailer becomes a drafter, then a layout drafter, capable of making large-scale layouts of complex equipment or machinery. Some companies also develop the best layout drafters into design drafters, who have responsibility for developing designs on paper from engineering specifications and other inputs.

On a design project, drafters of more than one skill level can be utilized, depending on the phase of the project. In an early phase a design draftsman may start with some "back of an envelope" or "cafeteria napkin" sketches, to develop a first version of the device. He or she may design several versions in response to various features to be added or emphasized. In a later phase, drafters and detailers put the final touches on the many drawings that are completed and checked before the engineering release.

In many companies, drafters are in short supply, even with good salary scales and modern drafting equipment. It is not clear whether this is a trend, due to many years of low pay scales and limited opportunity for advancement. The tendency toward computer-aided drafting and design will exacerbate the shortage. Most drafters will adapt themselves to the use of machines, whereas others will gravitate toward smaller companies that are unable to afford the computers.

Effective Utilization of Technicians and Drafters

While their skills and project inputs are different, these support personnel play an important role, both in project work and in day-to-day activities. Two questions relating to their utilization need to be addressed:

1. Should technicians and/or drafters be assigned to individual project engineers or be part of a talent pool, to be shared by all the engineers?
2. Is there an optimum ratio of support people to engineers?

The two questions are interrelated. With an adequate level of technicians and drafters in the department, some can be assigned permanently or on a long-term basis to project engineers. Others can be assigned to a pool where their services are allotted to short-term testing, sample building, updating production drawings, or whatever assignment is appropriate.

From the project engineer's standpoint, the assignment of one or more competent technicians and drafters to his or her project means faster progress, better response to problem situations, and additional people to discuss project snags with. From the technician's or drafter's standpoint, the project assignment provides greater continuity and is preferable, particularly if there is a good rapport between that individual and the project engineer.

However, if new technicians or drafters have been hired, and their experience does not mesh with the company's needs, a period of training is necessary. This has a dual purpose, familiarizing the new employee with engineering activities and determining whether he or she is an employee worth keeping. This is almost impossible to do in a project environment, where the emphasis is on performance and meeting goals. The need for training and employee evaluation is an important reason to have a pool of technicians and drafters; but if a pool consists only of inexperienced employees, because the experienced ones are assigned to projects, training is more difficult and slower.

Also, consider the situation where the number of technicians/drafters is limited, and all are assigned to projects. An engineer does not have anyone assigned to him, and, lacking a pool from which to request assistance, finds his efforts diminished. He must not only act as engineer—he must draft and be his own sample assembler and tester.

This happens in some companies for budgetary reasons, where staffing is inadequate to handle both project and non-project design effort. It is an unfortunate waste of engineering talent, and is counterproductive in the long run. The engineering manager must be aware of situations where engineers function as part-time technicians and react promptly to them when they are brought to his attention.

How Many Support Persons?

The question of how many support personnel are needed for each professional is an important one, with a variety of answers, depending on the situation. In a laboratory, a professional employee may have four, five, or more technicians reporting to him for assignment and guidance. An application engineer, by contrast, may only need a drafter or a technician occasionally, to develop graphs or perform limited tests. A project engineer may need one of each for most of the duration of the project, and may need more of them at certain times.

As a starting point, the support staff of technicians and drafters should at least equal the size of the professional staff. If there are 20 engineers, there should be approximately 20 support personnel, probably 10 technicians and 10 drafters. This level of assistance provides a technical base of support for multiplying the capabilities of the engineers. There may not be an equal number of technicians and drafters. Depending on the assignments, 12 technicians and 8 drafters may be preferred; or the numbers may be reversed.

If there is less than one support person per engineer, efficiency and effectiveness will suffer. The engineer will be doing too little engineering and too much support work. His assignments will be completed late; he will be frustrated and demotivated.

An ideal number, in many circumstances, is two support persons per engineer. This frees the engineer from doing the support tasks, which can usually be done better by the technician or drafter. At the same time it provides them with opportunities to do more varied tasks to improve their skills.

Technical Pool Versus Assigned Individuals

Since many engineers function in a product-support or readiness-to-serve capacity, having a drafter and technician assigned to them is a disadvantage. Sometimes they need all the support possible, but usually they need only their personal skills to think through a problem or analyze a production system. For their needs, the pool concept of drafters or technicians has merit. Being able to request needed services on an individual-assignment basis serves their needs adequately. A particular person may be requested, if he or she has done the work satisfactorily before, or a particular skill may be requested.

A talent pool has the advantage of providing a means for training new employees or giving more experienced employees the opportunity to work in new areas if they enjoy having variety in their work. Use of a "generalist" technician or drafter is also advantageous to a project where several technical skills may be needed.

On the negative side, the pool does not appeal to those engineers who want to have support persons handy to do the varied jobs that arise, and do them in the manner the engineer prefers. With a different technician on each assignment, the engineer must spend considerably more time providing instruction on how to do the work.

Also, the pool is not the most secure environment for technicians and drafters if workload is down, or business slows. The support person who has absolutely nothing to do will enjoy the freedom from stress for half a day, or perhaps all day; but the next morning he or she will be asking for assignments to stay busy with. Without the security of a project, the support person will feel vulnerable to layoff when business is slow, even if he or she is a satisfactory performer or an excellent one.

One unfortunate consequence of this feeling is the tendency for the competent "worrier" to begin looking for another job—and if one is found, to leave the present job for one that appears more secure. Then, the personnel who would have been more likely to be laid off or terminated will be retained, and the better ones will have left.

In spite of its shortcomings, the pool concept offers some advantages in training and providing varied assignments. Both scheduling options need to be considered in the light of what is best for the engineering department's short- and long-range functions. With heavy project loads, the assignment of project technicians and drafters has great merit in building a strong team that works well together and can tackle tough technical projects. When a number of staff members have individual assignments, as in customer service or application engineering, part-time use of a technician makes sense. If one cannot be "borrowed" from a project engineer temporarily, it is necessary that a drafter or a technician be available from a pool.

The pool also is needed when a project has ended and a new one has not yet begun. At those times, the project engineer is busy writing final reports and wrapping up loose ends. There is relatively little for a support person to do. Assignment to a pool would make his or her services available to others, although the project engineer might be unwilling to give up the chance to use that technician or drafter on the next project (if he or she were a competent team worker).

Overall, a combination of individual assignments and a small pool of technicians makes the best sense. It maintains most of the above-cited advantages, and reduces the disadvantages. It is somewhat more economical, in that fewer people are needed overall. Instead of two support persons per professional, the ratio could drop to about 1.5 to 1, and effective technical support would still be provided to projects and to engineers on general assignments. If the ratio drops much below 1.5, it is difficult to provide technical assistance to professionals. If the ratio drops toward 1.0, project work suffers as much as individual engineers who need assistance occasionally.

There is a level of support above the 2.0 ratio where there are too many people to be managed effectively; people figuratively are stumbling over one another. This inefficiency level depends on the individual circumstances, the type of work assignments, the capabilities and experience of the support people, and group size.

One can hypothesize a very effective small group, with eight to ten support persons and two engineers carrying out assignments in a low- or medium-technology area—a ratio of 4 or 5 to 1. In a large group, with 20 engineers and 80 to 100 technicians and drafters, many would be unproductive with that ratio. So while the 2:1 rule is a good starting point for initial planning, the actual ratio that provides effective support depends on the work situation.

INTER-UNIT ENGINEERING RELATIONS

Multi-product companies frequently have several technologies in-house. The engineering department will have experts in many diverse disciplines, including electronics, mechanisms, chemistry, and thermodynamics, just to name a few. When these specialists are working on the same projects, communications and information exchanges are easy to accomplish. However, when several product lines are involved, each with different technical disciplines, the engineers in large departments may barely need to have a working relationship with one another. Individuals may have only the vaguest notion of what their counterparts in other engineering sections are doing.

Each engineering group tends to be an island unto itself, demanding its own share of resources, concentrating on its own projects. A person may reach a dead end in a department and leave for a job in a different company, unaware of opportunities for developing his skills or interests in another group in his own company.

The manager of engineering is a major factor in fostering good relations between groups, and the individual group leaders play an important part as well. There are a number of ways for engineering groups to become more aware of each other, to be less competitive and more supportive of each other. Inter-group meetings, technical report exchange, and personnel transfers are all potential avenues of improved inter-group relations.

Inter-group Engineering Meetings

On a regular, perhaps quarterly basis, representatives of each engineering or technology group meet to exchange information on the progress of key projects. The projects may or may not have impact on other groups, but they will reflect the directions in which the company's technologies are moving. Depending on the technical level of participants, the meetings can be useful vehicles of information exchange, or boring reviews of technical data that are poorly understood or ineffectively presented. The respective chief engineers have a major interest in seeing that positive results are achieved.

Other topics of mutual benefit should be discussed, such as market trends or economic conditions as they impact the business. A guest speaker from marketing or finance could be invited. However, using the meeting as a basis for fostering exchange of technical ideas and new concepts is a better way to start meaningful dialogue between groups.

The meetings can be held during or after business hours. If they are held afterward, the atmosphere is more relaxed, and there is some socializing between individuals. This encourages new contacts between them and improves working relationships.

Technical Report Exchange

Subsequent to technical meetings, an exchange and group routing of the projects discussed at a previous meeting makes good sense. The representatives of each group can serve as a focal point, distributing the reports within their own group. They can also act as the channel for passing on questions or comments that may be generated by a thoughtful review of the reports.

Routing should only be to interested parties in each group. Otherwise the exchange of reports will be a mere exercise in paper shuffling, and the reports may be so slow in routing that the exchange becomes a waste of time. Selective routing preserves the positive aspects of technical report exchange as a move to further inter-group relations.

Personnel Transfers

In an engineering group with several technologies, some technical groups will be expanding, and others will be relatively static. The expanding groups need more staff, on either a project or a permanent basis. Getting them from one of the static groups may not please that group's leader, but it often is a very good idea. Not only does the acquiring group receive a trained professional familiar with the company, but the individual selected has an opportunity to broaden his or her technical base, learn a new technology, and develop new skills. From the engineering manager's position, the transfer provides inter-group training, broadening the skill base of staff members who might otherwise feel they were mired in a dead-end job.

FACILITIES FOR DESIGN ENGINEERING

The facilities where engineers work have an important effect on the quality and quantity of their output. However, experts do not always agree as to whether the effect is major or minor. The facilities are part of the total environment, the climate in which they function. Topics of interest here are open versus enclosed office areas, individual computers, and CAD/CAM, which was discussed earlier.

Enclosed Versus Open Office Areas

Having a private office, with a large desk and one's own telephone, is the dream of most design engineers. The privacy afforded by four walls (and sometimes a door) provides an environment conducive to uninterrupted concentration and superior work, or so it is said. However, the need for frequent communication with the engineer will have people coming into the office often. Frequent business chats often digress into long nonbusiness discussions, which are counterproductive.

Using an open office with sound-deadening movable barrier walls has an advantage in terms of low-cost office rearrangement when it becomes necessary to shift people's locations. However, the higher noise level associated with open areas, particularly with hard-surfaced floors, negates much of the advantage of flexible office walls.

Observing some basic considerations of office or facility design may be helpful in planning an office. For example, extremes of temperature, light level, and crowding are detrimental to employee performance. As noted earlier, privacy can be an important factor for both job performance and job satisfaction.

While privacy is a comprehensive concept and difficult to quantify, in terms of business privacy it could be broken down into speech confidentiality, visual boundaries, ability to control interruptions, and reduced levels of distraction and noise. These terms suggest the use of private offices. However, small groups of work stations or semiprivate offices shared by two to four people, especially those on the same project, offer greater value, in terms of both function and privacy, than the private office.

Various studies have debated the value of offices against the open bullpen of several desks. There is little disagreement; the bullpen is probably the worst environment because of noise, distraction, and lack of privacy. However, the enclosed, private office tends to develop isolation and loss of easy intercommunication although it fosters privacy. A combination of semiprivate offices with a few private offices may be a blend of both that has more pluses than minuses.

For professional workers, an adequate amount of space is needed for files and samples. A reduction of two or three feet in an office may not appear to be much, but if the engineer feels so crowded that he has no room for his "necessary samples," his attitudes and productivity may be negatively affected.

Above all, the area must be functional. It must have adequate space, good-quality furniture, a table and chairs, and adequate lighting.

In addition to the functional areas needed for work, each engineering department needs one or two conference rooms for meetings. These rooms should seat at least ten people, with comfortable chairs and an oval or nonrectangular table so that participants in a meeting can see one another.

Each room needs a board, preferably a "white board" with erasable colored markers rather than a chalkboard. It needs lighting control to dim the end(s) of the room so slides or films can be projected on a screen or the white board. Besides being invaluable for meetings or group presentations, conference rooms provide private locations for one-on-one evaluation meetings or small group meetings.

Computers and Calculators

Few remember the slide rule which nearly every engineer carried to mark his status as a technical professional. With a few slides of a "slip-stick" the engineer could quickly reduce an array of calculations to a three- or four-digit number, and then spend several seconds locating the decimal point and the size of the exponent.

The slide rule has been replaced by the calculator. These ubiquitous devices range from the tiny business card type that does basic arithmetic, to the shirt pocket version with trig functions, logs, and exponential capability, and beyond, to the slightly larger calculators that fit neatly into a coat pocket or attaché case, are programmable, and pack unbelievable computational power in a hand-held device.

Few companies provide calculators to employees, but their low cost and the variety of devices available have facilitated their widespread use. In only a few years the calculator has become an indispensable engineering tool.

One criticism of calculators must be aimed at the user rather than the device; and it would not apply to the older calculator, the slide rule. It relates to accuracy and precision. In multiplying and dividing numbers that represent test data of, say, four-significant-digit accuracy, even a pocket calculator will generate answers with eight or more digits, depending on the numbers involved and the size of the display field. There is a great tendency to use all of these digits in a published answer, regardless of the faulty logic of doing so.

For example, assume one multiplies 20 times 300 and divides by 70. The three entry number are integers, with nothing after the decimal point. But, a pocket calculator will develop the answer 85.714285, and many an engineer will record some version of that number, depending on how many post-decimal-point digits he enjoys using. The most correct answer is 86, rounding up from 85.7 and reflecting the integer accuracy of the inputs.

However, despite this small shortcoming, the pocket calculator has rapidly become a powerful engineering aid. It packs in a shirt pocket device the computational capabilities of the very expensive, huge computing machines of the mid-1960s.

Desk Computers

In many companies, one computer terminal per engineer is rapidly becoming the norm. In industries where extensive computation is needed to develop a workable design, a computer terminal in the office, capable of interfacing the mainframe computer on a time-sharing basis, is a necessity. The time saved by communicating with the computer directly—compared to preparing data inputs sent to the computer center for batch processing—is a key item in accomplishing complicated computational assignments in a short time. Of course, substantial monthly rental costs, or purchase costs, must be weighed against the value of the individual terminal.

As a corollary, and occasionally as a competitor to the terminal of a time-sharing computer, is the small, stand-alone microcomputer with "friendly" access. It would be difficult to predict the future of this powerful computer accurately, but decreasing costs and increasing flexibility may make it as commonplace as the pocket calculator.

One of the tools of the microcomputer that adds substantially to its value is the spread sheet program, known under a variety of names, many of which end in "-calc." Using a spread sheet or similar program, the engineer can do budget planning for projects and for the department. Relationships between variable inputs and outputs can be expressed as functions, constants, or calculations; and use of this versatile approach allows a broad range of computational possibilities. Moreover, knowledge of computer languages such as BASIC and FORTRAN is not necessary to utilize the capability of these computers.

The shortcoming of microcomputer proliferation in an engineering department is that programs may be incompatible between machines. This can be a serious pitfall, leading to many obvious communications problems. If a given program can only be run in a particular engineer's computer, the versatility of having many "micros" is suddenly lost. If a program is accepted by another machine but processed somewhat differently, the results can be in error. This may happen in subtle ways, not always easily detectable.

The best solution to this problem is to install identical microcomputers in the department, or to make certain the computers will accept and accurately process programs and diskettes from another machine. This type of computer compatibility began in the early 1980s, and it is likely that the trend will continue.

CAD/CAM

Computer-aided design (and drafting) is perhaps the most revolutionary new concept to supplement and improve engineering design creativity and productivity. Because of the difficulty of extrapolating the present into the future, any predictions about the impact this concept will have in the next few years are likely to be inaccurate. However, most rapidly developing technologies expand faster than predicted. Thus, the estimates are likely to be surpassed, rather than remain "blue sky."

Computer-aided design (CAD) and computer-aided manufacturing (CAM) are necessarily linked to obtain maximum benefit. However, each is a powerful tool by itself. Simply stated, CAD/CAM is the integration of the computer into the design and manufacturing process. In 1981, the National Science Foundation said that "CAD/CAM has more potential to radically increase productivity than any development since electricity." While this is a strong statement, it appears to be correct. Users typically find productivity increase of at least 2 to 1, and in the majority of cases 4 to 1.

Using the computer in the actual visualization and design of a part is a natural extension of the use of a computer for its number-handling capabilities. However, it requires a substantial investment, both in equipment and in training. While costs of equipment have dropped dramatically, one does not or should not approach CAD/CAM casually. It is important to ascertain the engineering department's needs, and how well they are being met, before opting for CAD/CAM.

Its greatest benefit lies in areas where there is a regular amount of repetitiveness in product design. This occurs in two situations: when a high-volume product requires extensive tooling costs (and lowest part cost is a requirement), and when similar designs are customized by building up standard parts in various configurations. In both of these cases, the computer improves accuracy, shortens time to design, and reduces the chances for layout errors where parts may unintentionally overlap.

The area of design creativity cannot be overlooked. By developing three-dimensional models on the computer screen, the creative design engineer can better spend his or her time devising shapes, forms, and contours to fit the models and not be concerned about detail dimensions. This capability has been greatly enhanced by solid modeling, which stores the solid nature of a part within the computer. It is a substantial improvement over the previously used finite-element and surface-model approaches.

Computer-aided design leads naturally to computer-aided manufacturing. The data in the computer can be fed to machines that punch holes, weld fittings, and cut large plates to shape. More important, the data can help the tooling engineer to design punch press tools, and the mold design engineer to design molds and locate feed spurs and gates for optimum material flow, minimum use of material, and maximum productivity.

Additional uses of CAD/CAM are found in simulation models, quality assurance, inspection,.and kinetic design analysis. It provides better use of standardized parts, and better utilizes the expensive resources of design engineers and drafters in the engineering department and their counterparts in other areas. It can save significant time in design and drafting and improve communications between departments.

Successful integration of the computer into the design engineering department will be done by careful assessment of needs, objective evaluation of costs and benefits, and effective communication of the changes to occur in the department.

This will require a far-reaching commitment because it will change the department's fundamental approach from "design, build, and test" to "design, model, and simulate." With this fundamental change in departmental strategy, it is important for training in the department to include the engineering manager and his key subordinates, as well as the drafters, engineers, and other operators of the equipment.

SUMMARY

The ideal organizational structure and composition for a design engineering department probably exist only in theory. However, the engineering manager needs to be aware of the capabilities and limitations of his present organizational structure. Knowing how other structures function well, he may decide to experiment on a small scale, particularly on a key project, in order to expedite completion in the shortest possible time. With advance preparation and planning, good results can be expected.

PART II
THE DESIGN MANAGER—
SUPERVISOR AND
TECHNICAL LEADER

7
The Design Engineering Manager, Supervisor and Technical Leader

The manager of engineering design has responsibilities different from those of most, if not all, nontechnical managers. Not only must he meet the overall company goals and objectives; he must also develop a climate where creativity and innovation can be achieved. It is not enough that he be a competent manager; he must also manage a mix of creative and noncreative staff members and bring out the best in each of them. This chapter will cover many of the aspects of being a good manager of engineering, including supervision, goals and objectives, behavior modification, decision making, effectiveness (versus efficiency), and management of time. Creativity and innovation are sufficiently important that Chapter 14 will focus on those subjects.

SUPERVISION OF TECHNICAL STAFF MEMBERS

Successful supervision of technical people depends to some extent on the technical skills of the manager; but it depends also on the three factors of visibility, measurement, and accountability, in his interactions with his subordinates.

Visibility

The physical presence of the manager, spending time away from his office tasks and with the technical people, is a strong productivity factor. His visibility in the department improves the overall productivity level because most subordinates will not loaf if the manager is around. However, if the presence of the manager is the only reason that people are productive, the department is in trouble, and those subordinates should be in deeper trouble.

On the positive side, the manager's presence encourages a good-quality job. Each staff member needs to know that the boss takes an intelligent interest in his work. By stopping by the employee's desk and not only inquiring about which project is being worked on at the time, but asking "How's it going?," the manager invites communication on the project. If it becomes clear that he wants more than a one-word response such as "Fine" or "OK," a dialogue will usually begin. Sometimes a subordinate will say "Terrible," which means that he feels terrible, or he has a problem he cannot solve, or he wants to find out if the manager was really listening. If the manager responds "Sounds good" to all three of the subordinate's comments, it is obvious he was not listening at all; and his lack of real presence will be duly noted and passed on to the other subordinates. While many subordinates are self-starters and work well on projects, some have a tendency to "visit" with co-workers. While these visits can be lengthy and time-consuming, the physical proximity of the manager, on a regular basis, reduces their frequency and duration. He does not actually look over people's shoulders, but his presence discourages long, often unproductive visits.

A large measure of the manager's credibility is lost if he pays only lip service to this aspect of the job. His staff would be better served if he stayed in the office all of the time, rather than take a token walk through the department.

In any design department, it is difficult to keep track of the detailed work of many people. Handling ten persons is about the practical limit. This limit goes back to biblical times when Moses gathered his people into groups of ten, leaders of ten people, leaders of ten leaders, and so on. Limiting the span of control to about ten people is necessary for proper management. Some staff members need criticism; others need a lot of encouragement. Some need advice, and others need the stimulation of argument or spirited technical discussion.

To be able to handle such diverse needs, and do it consistently, requires that the number managed be manageable. Some managers can handle a few more than ten subordinates, some a few less. Whatever the number, it is important for the manager to keep in mind that each subordinate needs to know that the boss takes an intelligent, sincere interest in what he or she is doing.

The manager's visibility must extend beyond contacts with the group of subordinates who report to him. On a nonroutine basis, he needs to visit the lab, the drafting group, the model shop, and any other support groups that report to him through subordinates. He must be sure not to issue direct orders nullifying those of the immediate supervisor (except in an emergency). He shows, by his presence, that he is interested in what is going on, what equipment has been added or removed, and who the people working in these groups are. In the future he may need to make judgments on who should be promoted, or whether one of these groups should be expanded or disbanded. It is important the he make first-hand, frequent observations of what is going on, who is doing what, and how well.

Measurement

Work measurement is generally difficult in an engineering design department. Certain areas, such as running a routine test or detailing drawings of simple or semicomplex assemblies, can be estimated and measured with reasonable accuracy; but completing a design phase, writing a test report, developing a project plan, and many other tasks are much more difficult to measure.

Though direct measurements are difficult, indirect measurements are not. The use of time estimates and deadlines can become an effective measurement tool for the subordinate to monitor his progress and productivity, with or without the manager's close observation. This can be done in the following way.

Instead of giving a specific assignment to an engineer and telling him to work on it, a better approach is asking him to estimate how long it will take to complete the work. This estimate would be necessary in a dual form—elapsed time and effort; for example, three weeks of elapsed time during which he will expend a total of 50 hours completing the assignment.

After he estimates the completion time, a realistic date must be set. If he says the task will be completed in three weeks, that means by August 27 if the start date is August 6. He will either then agree to the date or hedge a bit and say it may take a little longer. After some iteration, he will agree to complete it by a certain date. That calendar date becomes his deadline for completing the task, whether it be August 27 or September 1.

The manager jots the date down and lets the subordinate know he is expected to meet the deadline. If he has confidence that he can do it, he probably will. With most subordinates, the stimulus of a deadline provides an incentive to meet the time constraint or beat it.

By using a deadline set by the subordinate, performance measurement becomes a better evaluation tool than work measurement. The latter tends to measure the effort expended, not the results obtained. As the subordinate develops more estimates and keeps records of tasks accomplished, the accuracy of his estimates increases and the number of missed deadlines decreases.

On occasion, the manager should impose a tighter deadline, particularly if he believes a particular task is not as difficult as estimated or the subordinate should be stretched to do more. In most cases, a blend of subordinate-imposed deadlines with a few manager-imposed deadlines is more effective than either alone.

Deadlines must be treated seriously, or they will be missed most of the time. The subordinate may find he cannot meet the deadline—he has underestimated the work required, or has run into an unexpected problem, and so on. The manager considers that a deadline is inflexible and the subordinate must do his best to meet it, regardless of circumstances. It is far better for a subordinate to miss a deadline and explain why he missed it than to get a deadline extension merely by asking for it.

Accountability

Just as the manager is accountable for the performance of his staff, each staff member is accountable for his areas of responsibility. In the relationship between manager and subordinate, accountability includes an interchange between the two when things are not going right as well as when they are. Confronting the subordinate with one or more missed deadlines, or inadequately completed assignments, is that third essential of supervision. Accountability combines the functions of visibility and measurement into a meaningful whole. When the accountability of a subordinate is evaluated, the manager learns how well the subordinate accepts responsibility for his actions as well as how he delivers on his promises. In the same way, the subordinate learns how much (or how little) the manager demands in the way of effective results.

If the manager does not require full accountability from his subordinates, he will not get it. Instead, their output will drop, their morale will sag, and complaints will increase. Worse, the manager's credibility will sag and eventually disappear. His presence in the department will be a cause for jokes, good fellowship when he is around, and sarcastic comments when he leaves. His measurement methods—estimates and deadlines—will be gratuitously provided but pointedly ignored. The department will slip out of control.

To avoid this situation, which could lead to his transfer or termination, the manager must maintain control. He does this by holding his employees accountable for their actions. This requires him to confront each situation, to be understanding where necessary, but to be firm and unyielding almost all the time.

He must follow through. If a piece of work is not on his desk by the end of the agreed-upon day, the subordinate should be called in to the manager's office to explain why not; and, more important, why he did not contact the manager several days before the deadline to explain his problems and indicate that he might miss the deadline. This rational approach to firmness is handled well by some, overdone by some, and avoided by others.

Being firm and following through on assignment completion is important for another reason. If the subordinate realizes a deadline is important, then he considers that the work itself is important and needs to be done well, in addition to being done on time.

A design manager is a technical leader when he shows his subordinates that they need to be concerned about doing their assignments well. By his visible presence, his measurement of their output, and his follow-through on their accountability for performance, he demonstrates the importance of what they are doing.

Further Comments on Responsibility and Accountability

In order to develop a subordinate through the method of his assuming responsibility and accountability, one additional factor must be considered—the value of the chance to fail and learn from it.

Many jobs are structured with checks and balances that supposedly avoid errors and failures. Authority and responsibilities are limited to a subordinate's presumed capabilities so that failures do not occur. However, if a position and its checks and balances are so well structured that failure "cannot happen," the system may be so tightly controlled that progress will not happen either.

In development work, described in Chapter 4, the importance of failure to successful product design was discussed and analyzed. The same is true about failures in nondesign tasks, whether they be administrative or technical. Nothing else teaches so well as the mistake from which one is encouraged to learn. Some mistakes are so memorable (and embarrassing) that they and the lessons learned are passed on to subordinates in the form of things to do, or not to do.

Most mistakes do not have the catastrophic consequences that many imagine—unless management officially frowns on them. In that case, mistakes are often kept under wraps until they burgeon into a dangerous situation.

When a manager purposely decides to give his subordinates some elbow room to make right/wrong decisions and act on them, his method of control must change. He must give the subordinate the freedom to make mistakes and correct them, but, at the same time, he must not put the organization in jeopardy. This may appear to be a fine line, but it is usually less risky than not allowing a subordinate to develop.

The manager must remain sufficiently in control to stop an activity if it becomes too dangerous. He must know enough of what the subordinate is doing to prevent a catastrophic mistake. A manager should be more concerned if a subordinate makes the same mistakes again and again than with various mistakes made only once or twice each. He should also be concerned if the subordinate makes no mistakes at all. Such a subordinate may be unwilling to try new ideas to master the job and develop himself.

GOALS AND OBJECTIVES

The previous section touched on objectives in a general way, treating them rather like specific assignments. However, good supervision has the built-in requirement of managing people by a system of goals and objectives. This in turn allows the subordinates to manage their own efforts as well as those of their own subordinates.

Goals

Goals are pluses a company wants to achieve, such as developing an annual return on investment of 20% on new products, reducing costs on an established product line 9% by year's end, and so forth. Objectives are the steps by which progress toward a goal can be measured and compared. A simple analogy is taking a trip by car to a distant destination, the goal. Progress is measured each day by the distance driven. Each day's objective may be to reach a certain city before dinner. Confidence in one's ability to reach the objective could be measured by the number of advance, guaranteed reservations made at hotels or motels enroute.

If a company or a design department had only one goal, its objectives would be easy to quantify and steadfastly followed by all. Unfortunately this situation does not exist. Many goals exist at the same time, some unique to one department, some initially unique to one department then broadened to include others. This is particularly true of design projects. They begin in engineering but at about the time of prototype approval they branch out to include marketing, manufacturing, finance, and other groups in the company. Problems often occur because these groups have their daily workloads and emergency workloads in addition to project workloads.

Objectives

Recognition that other activities in other departments may inhibit achievement of engineering objectives is an important part of setting objectives. If an engineer receives an assignment from his manager that requires inputs and assistance from marketing and/or manufacturing, he may struggle alone and be frustrated because the others refuse or are unable to help out because of their own objectives.

The lack of common, or congruent, objectives between departments is one of the major problems in management by and with objectives. While most approaches to this powerful tool for accomplishing worthwhile goals assume common objectives, that situation does not always exist. In some companies—those floundering and in trouble—it seldom happens.

Before agreeing with a subordinate upon an objective that involves more than one department, the design manager must make sure that the other department managers are concerned with meeting that objective and will support it with manpower. If not, the objective is worthless unless the manager wants to use it as a test of the subordinate's ability to get things done regardless of roadblocks.

Some problems with setting conflicting objectives may be the design manager's own. If he gives his lab manager the objective of operating without overtime during the first quarter of the year, and assigns a project leader to complete a test program (in the lab) by the end of the first quarter no matter what it requires in time and effort, he generates a possible conflict of interests and objectives in his own department.

Unless each noncongruent objective is addressed and resolved in each major project or assignment, the subordinate responsible for achieving the particular objective may have an impossible task—a challenge with high built-in failure probability. A prudent manager will ensure that such incongruities are eliminated before the assignment is made.

Individual Objectives.   Each subordinate must have objectives toward which to work. Like a project, each should be quantified as to performance, dollars, and time. An objective that cannot be stated in those terms cannot be worked on with any degree of certainty as to outcome; it is desired outcome rather than an objective.

The number of objectives varies with the individual, the effort involved in completing the objective, and the staff size. Some objectives can be completed in a few weeks, some in several months, and some not until a particular event of breakthrough occurs. Objectives should be few rather than many. A person with two or three objectives can devote maximum effort to see that they are completed before others are tackled. It is far better for a subordinate to ask for an additional objective for the quarter because he has completed his original objectives ahead of schedule, than to show only moderate progress on a dozen or more objectives that have to be kept moving.

Reviewing Objectives.   Objectives should be reviewed on a quarterly basis, for performance evaluation during that period. There is no magic about a quarterly review, but it is consistent with the way in which a company views its performance during the year. Some objectives will have been completed during the quarter, some are to be completed by the end of the quarter, and others will still be in process.

This review is another way in which the subordinate's accountability is reviewed by the manager. The objective is either completed or not. For one that has not been completed, an estimate must be made of present status, why the objective was not completed, and what steps are necessary to complete it promptly. With many subordinates a monthly review of progress toward objectives is better than a quarterly review, to catch problems early so that key objectives can be met on time.

In some companies, the quarterly evaluation of performance against objectives determines a bonus, so objectivity and quantification are especially important. Good performance against objectives can be rewarded, while the reward is denied for poor performance. If a company does not have a financial bonus system, there are discretionary rewards a manager can utilize, such as lunch with the boss, dinner for the subordinate and spouse, an extra day or two of time off, attendance at a seminar or trade show, and so forth. If budgets are tight, these "perks" may be difficult to fund. But if the objectives completed were of significant value to the company and the department, they should be rewarded by more than a pat on the back or a word of praise. These verbal rewards are not to be demeaned because they do have value, but exceptional performance should be rewarded by more than words.

More could be written here about management by and with objectives, but the interested reader may consult many excellent books on the subject—some in favor of it, some not. Management by objectives can work well if the objectives are quantifiable, if they are congruent within the company, and if there are rewards for their accomplishment.

BEHAVIOR MODIFICATION

Behavior modification is a form of applied psychology, and in one sense does not belong in a chapter dealing with supervision. Yet a manager must deal with different types of behavior from his staff and encounters a substantial amount of behavior he would like to change if possible. Since he does deal with behavior in one way or another, a manager needs to know how to encourage desirable behavior and discourage the undesirable.

A fundamental axiom of human behavior is that people are likely to repeat an action if the consequences are pleasant; and its corollary is that people are unlikely to repeat an action if the consequences are unpleasant. Behavior can be modified by four means: positive reinforcement, negative reinforcement, punishment, and ignoring it (interaction).

Positive Reinforcement

When an employee's performance on an assignment is very good, it is recognized and rewarded. The reward can be as simple as a "nice work" comment, a note of recognition from the manager, or a type of financial reward, depending on the significance of the performance. The reinforcement must specifically relate to the particular assignment that is done well.

Negative Reinforcement

This is the removal of something negative after desirable (or good) performance is demonstrated. If a lab technician cleans up his area at the end of each work day to stop his supervisor's angry complaints, the negative reinforcement is the lack of complaints. Or, an engineering manager's budget is frequently checked by senior management, and during that year he stays within budget; then the next year he is told his budget will be reviewed only if any item exceeds budget by 10% or more. Management thus encourages better budget management by removing an unfavorable constraint.

Punishment

The employee's undesirable performance (or behavior) is followed by a distinctly unpleasant event. An employee late on a key project date is not allowed to present a paper at a technical meeting; or a manager is called to task for poor performance by his manager at a top-level meeting.

Although punishment is used often, its real value is doubtful because it may make an employee defensive and hostile. Punishment should be used selectively because of the negative side-effects of such resentment and hostility.

Ignoring Behavior

The undesirable performance is neither punished nor recognized in this form of behavior modification. After a time, the undesirable performance ceases because it fails to yield a response. An employee's memos to a supervisor are ignored. He never comments about their contents and never refers to receiving one. Eventually the employee stops sending them for lack of response.

If it is done properly, ignoring unwanted behavior will be effective, and the undesirable behavior will cease. But if it is done accidentally through carelessness or forgetfulness, ignoring behavior patterns will also cause the manager to forget to respond to desirable behavior, and it too will be discouraged. The manager will lose more by failing to respond here than he will gain by ignoring unwanted behavior.

Reinforcement Planning

If the manager wants to use behavior modification techniques, it requires some pre-planning, thorough execution, and follow-through. These pointers may be helpful in using behavior modification as a form of personnel development:

- In a new project, apply more reinforcement in the beginning, when the subordinate is furthest from the goal.
- Reward even small pluses and reinforce behavior frequently. This applies to both positive and negative behavior.
- The performance and reward must be closely coupled so they can be clearly associated with one another.
- The reward must come after the behavior; otherwise its value is lost. And it must come immediately after (shortly after) the desired response. Annual raises do not have the desired effect, for example, because they are too far removed from day-to-day performance.

When the desired behavior has been established, the response–reward pattern should be shifted from regular reward to a random reward pattern after desirable performance. Random reward is effective because a fixed, or periodic, pattern of reward will eventually be preceded by a fixed pattern of improved behavior.

The importance of reinforcement concepts cannot be ignored. They are real. Most managers use them effectively as a result of intuition and planning, or use them poorly for the same reasons. Knowing the techniques and how to apply them effectively could reduce performance problems by as much as 50%.

DECISION MAKING AND DECISION IMPLEMENTATION

Successful engineering design is the result of a succession of well-made decisions, effectively implemented. Goals and objectives must be set—some accepted, some rejected. Projects must be selected and initiated; directions must be changed; jobs must be completed, and projects terminated. Problems are discovered an solved. All of these areas require decisions, some of them more important than others. The effective design engineering manager will not only make effective decisions himself, but will train and coach his subordinates to make them.

Making a decision involves selecting the best alternative available to meet a set of objectives and initiate some type of change. Implementing a decision is putting the decision into practice. While related, these are separate functions. The best decision, poorly implemented will not yield effective results. A poor decision, effectively implemented will usually not yield effective results either. This subject merits further discussion later.

The major difficulty in decision making is having to choose a course of action to meet a specific objective without being certain that it is right. With all the facts at hand, reducing uncertainty to zero, the correct action would be clear, and anyone could make the best decision.

Realistically, there is no way for a manager to be certain that the action he has decided on is the best alternative to select. He knows he does not have all the facts and never will have them all. In spite of this, an effective decision maker will use all the information inputs he can gather in the time frame he has to work with, will choose the best alternative he can, and will implement it vigorously.

When confronted with incomplete facts, some managers will postpone making a decision. They hope that either more facts will become available, or the problem situation will somehow solve itself. This will rarely happen. Most of the time, neglect will allow a problem to become worse, not better. Too long a delay can create an out-of-control situation. By not making a decision, the non-decision-maker is making an unintentional decision for no change in the status quo. Whether right or wrong, a decision is far better than a non-decision.

However, there are sevaral situations in which a decision should not be made. Consider these possibilities:

- Should someone else make the decision? This does not mean "passing the buck," but, for example, the design manager should not make a marketing decision, nor should the marketing manager make a manufacturing decision, or a design decision.
- Has the correct problem been identified? An early phase of decision making is identifying and analyzing a problem situation. If the manager determines that the wrong problem is being worked on, there is no choice but to go back to square one and start over.
- Is a decision needed? Sometimes the situation is being corrected by someone else, and no decision is needed. Since each decision is an intervention in the course of events, an unnecessary decision could create problems.
- Has time run out? Often a decision must be made by a specific time. Be sure time-dependent decisions are made before the deadline date. Post-deadline decisions will usually be ineffective and indicate that the manager did not know what time it was.

Decision-Making Methods

Some managers ponder the facets of a decision by themselves, sifting facts, considering alternatives, and reaching a decision. Then they share it with subordinates, and direct them to carry it out. Others will discuss facts and alternatives with subordinates or other managers affected by the decision.

Searching for hidden facts and new ideas to improve the decision, they also are searching for a consensus.

Each method has advantages and disadvantages. Many managers will restrict their decision making either to the individual approach or to the consensus approach. The astute manager will utilize both methods, depending on the circumstances, objectives, and time. The following are some of the advantages and disadvantages of each method.

Individual Decision Making.   An individual decision maker can be quick and decisive. An able analyst, he can slice through incomplete data and inconclusive facts to the heart of a situation. In a few days, he will reach a decision that would otherwise require several meetings and discussions, and take several weeks. If his decision is successful, he will receive most of the credit; similarly he will take the blame if it is not.

There are two major disadvantages to being a lone decision maker. One is that all too often that individual ignores or does not solicit the input/advice of peers or subordinates. Second, lone decision makers have trouble implementing their decisions. Since those who must bring about the required changes were not brought in prior to the decision, they may not like it, and may see its loopholes and risks. The decision maker has an uphill struggle in motivating his subordinates. Often they resent not being consulted, and show their displeasure by providing the least possible effort during implementation.

Yet, there are several situations where a manager needs to make key decisions alone. These are some examples:

- Some decisions are delicate, involving either a subordinate's promotion or, in the other extreme, termination. In such cases the manager must decide on the basis of facts and opinions he has accumulated and perhaps his overall grasp of the situation.
- If a decision has two alternative courses of action that appear to have equal merit, the decision maker must decide between the two, sometimes on the basis of his personal preference.
- There are situations where hunches, intuitions, and past experience are the decisive ingredients, especially when facts are skimpy or inconclusive. This is appropriate because hunches are usually the result of one's subconscious mind evaluating intangible factors. When a manager feels strongly that a particular course of action is correct, he should go ahead based on his hunch and not dilute his judgment with peer or subordinate input.

Consensus Decisions.   In many respects the consensus decision is the best. Prior to reaching a decision, the decision maker solicits the views and opinions of his peers, and particularly the subordinates who will be affected by the decision. He considers their ideas, favorable and unfavorable, before reaching a decision. The aim of a consensus decision is to arrive at a course of action representative of the best thinking of all the persons involved.

A decision reached by consensus will be warmly received by those whose views were reflected in the decision, with a cooler reaction by those whose views were not. The manager can do much to reduce or eliminate any resentment by explaining, both before and after the decision was made, that he needed inputs to illuminate all facets of the situation before reaching a decision. And he should reassert his expectation of the others' full support. Consensus decisions can be much more successfully implemented than individual decisions because the logic and rationale are known to those who implement the decision and are affected by it.

There are also some shortcomings to a consensus decision that must be kept in mind:

- Some subordinates may have a vested interest in the decision and may try to force their views on others to establish an internally manipulated consensus.
- Others may try to perceive the majority consensus and support it, rather than develop and express their own views.
- A subordinate may believe he is doing part of his manager's job and should be better compensated for decision making.
- The manager may be viewed by some as weak and indecisive because he asks for counsel on a decision.
- Meetings and discussions needed to develop a consensus take time and slow the decision-making process. Deadlines may be affected if too much time is taken.

Manipulation instead of Consensus.   The manager may himself be a cause of problems if, instead of seeking views for a consensus, he intends to have his own decision be perceived as a consensus. As he talks to his subordinates, he actually is searching for a viewpoint that matches his own. When he locates it, he will comment on the excellent approach. More than likely the group of subordinates will swing into line with the idea the manager likes best.

If used infrequently, this approach can be quick and effective and develop a high degree of support from the subordinate group. And the "originator" will gain satisfaction from having his idea accepted. However, if it is done often, the group will recognize that they are being manipulated. Their flow of ideas and their analytical perspective will shrink. In future situations they will try to manipulate the manager to find out his view so that they can support it and get back to work.

It is best for the manager not to use the pseudo-consensus, manipulative approach. It usually does not work, and in the long term it will destroy the rapport and mutual respect that the manager and subordinates should have for each other. If he believes it is necessary, the manager should ask himself why. In answering this question, he may find he is acting from personal motivation to be in charge and control people, rather than to be the leader of a group of competent subordinates from whom he is trying to develop a consensus.

The consensus decision can sometimes cause problems, as described. However, it is a powerful method that develops superior communication and working relationships between manager and subordinates. In addition it fosters a solid commitment to implementing the decision.

Implementing a Decision

A good decision is easily distinguishable from a poor one. The good decision succeeds; the poor one fails. It is not only the quality of the decision that makes it so successful, but the commitment to implement the decision, and make it work.

After a decision has been made, the focus shifts from analyzing options against objectives to developing methods for reaching the end result. Commitment to implementing a decision requires a plan, with an "implementer" and a good idea of the resources needed to execute the plan. The plan must incorporate several check points and be amenable to change if conditions require it. Implementing a decision is like managing a project. As in a project, progress is measured by performance, costs, and time.

Frequent follow-up is important. At the first sign of delay, reasons for it must be determined as well as how to correct the variance. At the second missed milestone, corrective action must be taken to bring implementation back on schedule. This requires prompt problem identification and correction. Assignments may need changing, or additional people may have to be brought into the effort. At all stages of implementation, appropriate managers must be informed as to what is happening and what problems are being addressed.

The critical factors in making an implementation plan work are direction, priority, and communication. Communication is often the most difficult, particularly if the subordinate implementing the decision has misunderstood it. If he thinks he is following the manager's instructions but has misinterpreted them, his implementation could be ineffective. That is why follow-up and feedback are important.

In most cases, a second-best decision implemented with first-rate commitment is much more successful than a first-rate decision implemented with second-best commitment. The reason is not often obvious. Since most decisions are made with incomplete information, they tend to be suboptimal. Also, many decision situations have many possible answers, more than one of which could be correct. The Romans had a saying ''All roads lead to Rome.'' A similar but less inclusive statement can be made about decisions.

Since more than one possible decision can yield the desired result, quality of implementation is important. As the decision is implemented, sometimes more information is learned about the original situation. Or, the process of implementation provides additional discoveries that aid in understanding and solving originally unknown problems as well as those that were known. Thus the decision maker does not worry as much about making the best decision possible as about having it implemented the best way possible.

Additional Comments about Decision Making

While facts and information are vital to making a good decision, the decision maker needs to decide in advance how he is going to analyze and use the information he has and receives. It is useless and time-wasting to gather more and more information in the hope that the sheer weight of facts will illuminate the best decision. That will not happen.

In a way, this is similar to a designed experiment. Before any experiment is performed, its designer determines what data he will need from how many samples, how he will test them, how he will analyze the data, and what statistical significance the data will have. In a similar but less structured fashion, the decision maker must decide what information he needs to make a decision and how he will use it, and then must go out and get the information needed.

Alternatives in Decision Making.   While evaluating alternatives is critical, they must not be evaluated only against each other. More important, they must be evaluated against the objectives of the decision. Before generating alternatives and enjoying the creative fun of developing blue-sky and down-to-earth alternatives, the manager must perform the more difficult task of setting objectives—defining the problem, as it were. Only after the objectives are set and prioritized can the alternatives be evaluated. Usually, one's mind prefers to jump quickly from the more difficult task of setting objectives to the easier one of generating alternatives. If that happens, the manager should jot down the alternatives, in order not to lose them, but should stick to the primary task of setting objectives.

Risks.   Every decision, every alternative, has an element of risk. The decision may be wrong; it may go awry in the implementation stage. There may be external threats, or internal adverse consequences posed by each course of action.

Risk needs to be assessed from two aspects: the probability that the threatened consequence will occur, and its seriousness if it does. Although it is subjective and simplistic rather than sophisticated, risk analysis is very important. The adverse consequences could be a serious financial loss to the company, risk of lawsuit, or, on the personal side, a reprimand, or perhaps even termination for a wrong decision in a critical area.

At the same time, the benefits must be analyzed because the risk at times may be far outweighed by the benefit gained. At other times it may not. The willingness of a decision maker to accept a certain amount of risk has considerable bearing on which alternative will be selected. Since willingness to accept risk and its consequences varies from person to person, there is no right or wrong approach. Some managers will decide on a low-risk approach even if its benefits are small, whereas some will accept high risk if there are high benefits. Others ignore risk and decide to bull the decision through and somehow make it work. Regardless of which type of decision maker an individual manager is, he is wise to evaluate the risks involved before deciding upon an alternative.

Decision Making—in a Few Words.   The process of decision making has seven steps:

1. Determine the objectives.
2. Generate a number of alternatives.
3. Evaluate each.
4. Consider adverse consequences and risk.
5. Make the decision.
6. Delay the implementation overnight.
7. Implement the decision.

All the steps but number 6 have already been discussed. In any important decision, many people want to "sleep on the decision" before going ahead. This is a wise practice. After making the decision, one could mow the lawn, paint a room, work in the garden, play racquetball, shoot a round of golf, go fishing, or engage in some other diversion that allows him to consciously forget the decision. In the meantime, the subconscious continues to analyze the information, reviewing the decision and its consequences. Quite often a new insight will develop, or a forgotten fact will be remembered, which will have significant bearing on the decision.

Sleeping on the decision allows the subconscious mind a second review of it. The subconscious, overnight, may come up with some new insights that require a change in the decision. If not, the decision should be implemented in the morning. One's subconscious thinking process is a valuable asset that should be utilized in any decision process.

EFFECTIVENESS IN DESIGN MANAGEMENT VERSUS EFFICIENCY

Being an effective design manager is not an easy task because it encompasses making all of one's subordinates more effective, not just one's self. Effectiveness can only be measured by results, both on-going and future. Much of this section will cover effectiveness rather than efficiency because, of the two, effectiveness is of greater importance.

Efficiency in management is a good end, but it rarely produces excellence, breakthroughs, or leadership development. A qualitative measure of efficiency is the standard engineering equation:

$$\text{Efficiency} = \frac{\text{Output}}{\text{Input}} = \text{Doing things right}$$

The more output per man-hour, the greater the efficiency. The efficient manager has work done on all of his assigned projects, and knows how many hours it takes to make a drawing, design a gear, or assemble a project's paperwork. However, it is pertinent to ask whether he is spending enough time on the key projects to get the results his company needs to move ahead in key areas. The answer may be no.

An effective manager's performance is different from an efficient manager's. Although it is somewhat difficult to quantify, effectiveness can be measured on a qualitative basis by this equation:

$$\text{Effectiveness} = \frac{\text{Desired output}}{\text{Actual output}} = \text{Doing the right things}$$

This relationship measures how much of the total output is in the most important areas. It is difficult to assign numbers to these qualitative ideas because the desired output may be a key cost reduction, an innovative design improvement, a new product introduction. In other words, the desired output may be only a few items that overshadow the bulk of everything else done by the design department.

Effectiveness is managing by purpose, managing for results, managing by concentration. It focuses on specific opportunities to bring about significant improvement in the status quo. The basic mission of a design group is to move technology into products, followed by moving products into markets, and this can only be done by focus and concentration.

Pareto's Law

At about the turn of the century, in 1897, an Italian economist/engineer named Vilfred Pareto studied the distribution of wealth in Italy. He found that a large percent of the wealth was concentrated in the hands of about 10% of the population. His thinking and speculation led to the equation for a hyperbola, in the general form of:

$$Y = \frac{A}{x}$$

where

Y = income level
x = number of persons at each level
A = a constant relating the levels of each

Despite being published in 1907 as Pareto's Law, this insight went unnoticed for years. Shortly after World War II, inventory control analysts, when plotting inventory items on a cumulative basis, found some interesting results, which they had intuitively expected. They observed that about 20% of the items in inventory accounted for 80% of the inventory dollars (see Fig. 7-1).

Some years later, in 1954, Joseph Juran, one of the fathers of modern quality control, described Pareto's Law as a universal that was applicable on a broad scale to management planning and control. In other words, both empirically and practically, the 80/20 rule applies to just about everything: 80% of sales are generated by 20% of the items sold; 80% of production problems are caused by 20% of the products manufactured; 80% of the output of a department is generated by 20% of the staff; 80% of personnel problems are generated by 20% of the staff (hopefully, not by the same 20% that is generating 80% of the output).

Effectiveness, then, is concentrating on the "vital few and not on the trivial many," a phrase coined by Juran. This focus on opportunities will improve the operation of the design group. Instead of attempting to do all things, to take on all potential assignments, the manager will start saying no, will insist on good marketing forecasts, before committing to a project. When he does commit resources (people) to the project, he will commit an adequate staff to do the project in the shortest sensible time, as discussed in Chapter 5.

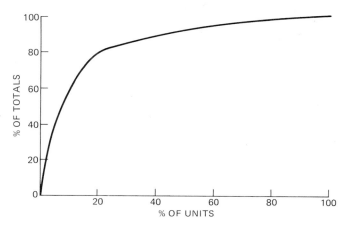

Fig. 7-1. Pareto's Law distribution curve.

The essence of focus is, of course, determining which are the vital few because they are easily lost among the trivial many. The selection criteria of near-term profits, sales potential, long-term market penetration, and suitability for production all enter into the screening process. Most important is the manager's orientation toward the vital few, his concentration and focus on these areas. This, in turn, motivates his staff to action, following their manager.

Effectiveness is a learned trait, and can only be learned on the job. Effectiveness includes more than focusing on the vital few, but that is the keystone. Other important facets of effectiveness are development of subordinates and delegation, speed versus promptness, staying on top of the job, and time management. Together, these qualities make an effective manager and an effective design group. The two are joined and inseparable. A manager cannot be effective without effective subordinates, and subordinates will not stay effective very long with an ineffective manager.

Development of Subordinates and Delegation

A subordinate is developed on the job by the assignments he or she receives, the amount of responsibility/accountability given (both to make mistakes and to do the right things), and the amount of authority the manager has delegated.

Training courses have a place in subordinate development, but too often the courses do not relate to the work at hand, or else they are too general to be of specific value. Training courses come at specific times, according to a schedule convenient to the trainees, and are available to subordinates on the basis of budgets and the convenience of other personnel. On the other hand, the opportunity and need to apply the training tend to occur at random.

Stated another way, subordinate development occurs as a response to a challenge: a problem exists, a crisis develops, or some situation needs correction. Through successful responses to such challenges, people develop. Since management means getting people developed through their work, it is a natural requirement that they have the responsibility, authority, and accountability for this work. Then the subordinates have an opportunity to be successful.

Since problems occur at random (or at the worst possible times, according to Murphy's Laws), problem-solving approaches learned in a classroom may be too remote in time to be of much value. Therefore, subordinates must be given the freedom to respond to problems. Developing a subordinate is something like sailing. One can teach the fundamentals of how to sail, how to anticipate storms, sail before the wind, or tack against it. However, the skill of a sailer is tested not in smooth weather, but in storms, when the lessons learned are applied poorly or well. And, praraphrasing the words of an old captain, "After a storm, no one asks how rough the seas were, only if the ship made port."

Delegation Process

Subordinates are developed when the job environment provides for delegation. A manager must delegate a certain amount of work to his subordinates. He does this partly to develop them into competent engineers and future managers, and partly to multiply his efforts through others. This leaves some time for him to think about the future and plan to meet it.

Delegation does not come automatically and is not a blank check for the subordinate to act completely on his own; and some managers refuse to delegate at all—for reasons that will be discussed later. However, the effective manager will delegate important responsibilities in a gradual, planned method that enhances the subordinate's chances of success and reduces (but does not eliminate) the chances of failure.

There are several possible levels of delegation, from determining and reporting the facts of a situation to making a decision and implementing it. These levels are progressive in the degree of freedom of subordinate action, as the capabilities of the subordinate develop.

Investigate and Report.   This is the lowest level of delegation and is used for training. The subordinate reports the facts, and the manager analyzes them, makes a decision, and takes action. The subordinate is advised of the action taken.

Investigate and Recommend Action.   The subordinate investigates, analyzes, and recommends a particular action. The manager evaluates the recommendation, but makes the decision and takes action. If the decision is different from that recommended, the manager explains why he did not follow the subordinate's recommendation. This level encourages the subordinate to react to the information he has gathered, and to decide which course of action would be in the company's best interest.

Investigate, Decide, and Advise Action To Be Taken.   Here, in a more advanced level of delegation, the subordinate goes through the entire decision-making process and develops an implementation plan to carry it through. The manager reviews the subordinate's decision and the implementation plan, and approves or disapproves his acting on it. As in the previous level, the manager must indicate the reasons for not approving the actions. When he approves them, he should point out any reservations or areas of concern he has, so they can be monitored closely.

Investigate, Decide, Take Action, Advise Action Taken.   The subordinate moves through all of the delegated steps without having the manager check him on the way. As he begins to implement the decision, he advises the manager so that he is aware of and can monitor and/or alter what is occurring. The subordinate is in control of his decided course of action, subject to review by his manager.

Take the Appropriate Action.   In this full-delegation level, the subordinate has full control of the action from investigation to implementation. He periodically keeps the manager informed of the control points and key milestones he has developed, but, for practical purposes, the subordinate is in charge.

This last step, of full delegation, is the goal of both manager and subordinate. It is the result of a several-level growth and development process whereby the subordinate takes a progressively greater action role in each level, as he demonstrates his ability to accept more responsibility. The guided experiences lead to more independent, responsible actions.

It is possible that in the last two steps the subordinate may be acting as manager and training a subordinate in the first-level actions. It is axiomatic that the teacher usually learns more than the pupil; so, by teaching his own subordinate, the upper-level subordinate understands his own job much better.

In summary, delegation can be successful if the manager sincerely wants it to be successful, and if he has developed a high level of confidence in his subordinates. While full delegation is the ultimate goal, a manager may have each of his subordinates operating at a different level of freedom. Some may be operating at lower, some at higher levels. Some may never progress beyond the second level, to investigate and recommend. Others may push to move beyond it quickly.

This process depends on how well the subordinate develops and what risk the manager is willing to assume during the development of judgment as well as skills by his subordinates. Through the whole process, the manager needs to maintain control, to be on top of things and stay in charge. While he can delegate authority and responsibility, he can never delegate accountability. While he can and must hold his subordinates accountable for their actions, the ultimate accountability is that of the manager to his superior. It remains unchanged.

Delegation Guidelines

1. When delegating, require the end result, not the method of reaching the results. The subordinate needs to develop his own methods.
2. Set performance standards to measure accomplishments.
3. Give the subordinate all relevant information. Do not restrict him to that information the manager believes to be important.
4. Delegate to each subordinate at the level where he can perform. Subordinates are different from each other and may not all be able to handle the same degree of freedom.
5. Establish early warning controls so that a decision or implementation plan can be adjusted if it begins to go wrong.
6. Establish an environment within the department that encourages open communications. This can be of vital importance if a decision or its implementation is failing.

These guidelines make delegation a success-oriented approach. The manager develops confidence in his subordinates, and they, in turn, develop it in themselves. Each subordinate develops a vested interest in his own decision-making success, and there is enough risk of failure to keep him on his toes. More than a learning experience for each, this is a development method more powerful than a series of training courses. By effective delegation, the manager has more time to do his job of managing.

Why Some Managers Do Not Delegate

Despite the clear benefits of delegation, some managers will not delegate. They will never allow a subordinate to go beyond providing a recommendation, and often not that far. Why is this? There are a number of reasons, including these five:

1. Lack of confidence in subordinates. A manager may believe his subordinates' judgment is faulty, that he can do the job better and faster than if he delegates to his subordinates.
2. Aversion to risk taking. Delegating an assignment to a subordinate involves some risk of failure; something could go wrong. Some managers are too upset by this possibility to give their subordinates a chance to succeed as well as to fail.
3. Fear of subordinates as competitors. This fear, which may be real or imagined, causes a manager to be overly critical of his subordinates. He may play down their achievements, or force them to compete in a win–lose situation, where one of them looks bad. He may misuse their talents in low-skill jobs. He will never delegate any authority.
4. An inflated self-image. Some managers make all the decisions in their department; their way is the only way. Such a manager checks all details personally. He makes sure that only he has an overall picture of what's happening. He fosters dependence and gets it because his stronger subordinates chafe under his dominance and decide to leave.
5. Equating action with productivity. A manager may need to hold onto work because delegation may leave him with little or nothing to do. Such a hyperactive manager will do, rather than think and plan. He cannot or does not wish to appear to have any time because he likes to be overworked. His subordinates will have difficulty getting meaningful assignments from him.

Delegation is the means by which a manager's abilities are multiplied and leveraged through his subordinates. If he is an effective manager, he has brought them up through several levels of progressively greater delegation. He has allowed them to make small errors, from which they have learned. Delegation has enabled them to develop, and has increased the effectiveness of his department. There were some risks involved, but the benefits far outweighed them.

Managerial Efficiency

While effectiveness is of greater benefit to the company than efficiency, there is no reason to believe efficiency is wrong. Efficiency is performing actions the right way, in a reasonably short time. Many routine tasks must be done efficiently because they need to be done in the first place.

Drafters may take an hour or two to do a lettering job that could be typed in minutes. Engineers may make copies of technical articles they need to study when a clerk could copy them more cheaply. Other engineers write long and detailed memos, which sometimes only reflect their thinking process, when a phone call or two could serve as well in one tenth the time.

Constant vigilance and pressure are the only means of preventing routine jobs from becoming inefficient. Parkinson's Law, that work expands to fill the time available to do it, applies particularly to routine jobs in the design engineering department. If a shade too little rather than too much time is allowed for a job, people tend to become more efficient, although cutting down time too much leads to overwork and ineffectiveness. Time pressure, the force of deadlines, and having time goals to meet can be powerful motivators to develop a more efficient organization in the routine areas, much as they do in the effectiveness-related areas of project work.

TIME MANAGEMENT

Another measure of a manager's effectiveness is his skill in time management. Time is a unique resource, and the limiting resource in many projects and programs. Time cannot be accumulated, reused, or recycled. Everyone spends it at a rate of 60 seconds per minute, 60 minutes per hour. Everyone has exactly the same amount of time available each day.

Since the supply of time is inelastic, its management becomes important in everything the manager does. As stated earlier, effectiveness is doing the right things, and efficiency is doing things the right way, in the shortest possible time. If time were used efficiently, each item on the manager's "do" list would be done quickly and according to plan.

Unfortunately, a manager's day is rarely structurable. Interruptions, conferences, and phone calls are the rule, not the exception. Also, some tasks require fairly large blocks of time. For instance, a first draft of a report requires a great deal of review of notes, and thinking about structure and form, before the dictation equipment is picked up, or the pen set down on paper. Attempting to write such a draft in 15-minute periods once a day is usually an exercise in futility. Most of each quarter hour of allotted time would be spent in reviewing the previous work and deciding what to write next. Therefore, time must be used effectively.

Time management is a way of helping a manager use time more efficiently and more effectively, but the term "time management" is not literally correct. Time cannot be managed; instead, the manager must manage himself with respect to the clock. However, the term "time management" is in common use, and will be used here with the understanding that it applies to self-management with respect to time.

This process consists of learning where one's time actually goes, and doing something about it. As a three-step process, time management consists of:

1. Recording time spent (over a period of days or weeks).
2. Managing the use of time.
3. Consolidating time.

Recording and Analyzing Time Spent

In order to get a "handle" on where time goes, one needs to keep records by means of a log sheet. One example is shown in Fig. 7-2. Tedious as it may seem, one cannot determine where his time goes unless he logs his daily ac-

GOALS:		DATE _____	
1. _____		4. _____	
2. _____		5. _____	
3. _____		6. _____	
TIME	ACTION	PRIORITY 1. IMPORTANT AND URGENT 2. IMPORTANT, NOT URGENT 3. URGENT, NOT IMPORTANT 4. ROUTINE	COMMENT
8:00			
9:00			
10:00			
11:00			
12:00			
1:00			
2:00			
3:00			
4:00			
5:00			

Fig. 7-2. Daily time analysis.

tivities as a starting point. One week is the minimum time to keep such a log. There probably is no such thing as a manager's typical week or day; so unless one is away at a seminar or on a business trip, any week is a good week to start.

Analysis of a time log will provide some interesting surprises. One of them may be the relatively short time that can be spent on one project or item without interruption. For many managers, 20 minutes is the longest such time. Another is the number of activities that comprise a typical day. In a study by Lakim, managers performed over 34 activities in a day, each lasting 3 to 20 minutes. None had significant, uninterrupted stretches of time to study major segments of the business, or to do long-range thinking.

Managers find that they spend much less time than they anticipated on high-priority projects, because of interruptions of all types. Many of their major goals receive only perfunctory attention.

Analysis of time spent, after the data have been gathered and consolidated, is facilitated by posing some diagnostic questions, such as these:

- What would happen if this activity were not done at all?
- Which activity could be done as well or better by a subordinate?
- Which meetings, phone calls, visits, and so on, waste the time of others without contributing to their effectiveness?

When these questions are directed at the activities, the motivation to become more effective in the use of time has more direction. A manager learns which activities he could drop, or reduce, which he should have delegated and now can. And he learns that a great deal of his day has time-wasting activities that are relatively unimportant.

Time Wasters.   Time is wasted in innumerable ways. Some time wasters are generated by others, and some are self-generated. This is a brief listing of both types:

Generated by Others	*Self-Generated*
1. Problems of employees	1. Inability to say no
2. Telephone interruptions	2. Lack of delegation
3. Lack of priorities	3. Attempting too much
4. Too many meetings	4. Procrastination
5. Poor communication	5. Involving everyone
6. Mistakes	6. Snap decisions
7. Routine tasks	7. Unrealistic time estimates
8. Conflicting priorities	8. Failure to listen
9. Visitors	9. Bypassing the chain of command
10. Training new staff	10. Personal and outside activities
11. Crisis situations	

This listing is by no means exhaustive, and a few of the items from the list of outside-generated time wasters could well appear on the self list and vice versa. A surprising result of time analysis is that before making a time analysis, one believes that most time wasters are externally generated; but afterward, managers concede that most time wasters are self-generated. That means they are relatively controllable, by directed action.

Consolidating a Manager's Time Use

There are many effective ways to consolidate time use, and to reserve chunks of time for all important activities. Focusing on major goals, setting priorities, concentrating on the important activities, and eliminating time wasters are the essence of consolidation.

A manager may report to work at a later time once or twice a week because he spent an hour or two of interrupted time at home writing an important first draft.

A manager who has an open door policy may also have a closed door policy for a specific hour or two each day, when his secretary takes all calls and defers all visitors so that he can concentrate on his important priorities.

"Things to do today" lists, which are forward-planning tools, are useful. They may include a listing jotted on a pad of paper on the desk, or a pocket or desk time-planning book. None of these devices is effective, however, unless it is used.

One easy way to list the key items for the day is to use a time-planning book. Another is to buy a 3″ × 5″ spiral-ring notebook, which is small enough to be carried in a shirt pocket. The list can be started on a Monday, or any other day, with the date heading the top of the page. A list of 10 to 15 numbered items will fill the page. They can be asterisked or otherwise marked to indicate their priority. As each is completed, a line is drawn through it. The left-hand page can be used for any notes or detail reminders that are important. New items can be added until the page is full.

Each time a new page is begun, it is dated, and all the uncompleted items are copied on it before new ones are added. In this way, the book becomes a running record of what was done and when. Recurring items, which must be rewritten time after time, need to be examined. They must be done, delegated, or dropped.

Setting priorities, with A's as top, B's as less pressing, and C's as desirable, is always a good approach. Based indirectly on Pareto's Law, the priority system allows a manager to determine and concentrate on the vital few. A priority system must be flexible enough to allow for shifting. For example, a report due in six weeks may be a B or C priority today, but in two or three weeks must be elevated to one's A priority level to ensure its completion on time. Priorities change; some items become very important, gaining A priority today, that were not even known about yesterday. One must be flexible and realistic about priorities, whether self-imposed or system-imposed. Priorities do change, and one must work within the dynamic framework of change in business.

Why Some Managers Cannot Avoid Wasting Time

It seems incongruous to speculate on the point that some managers waste time because they cannot "help it." Yet, a *Harvard Business Review* study of managers showed that the anxieties of many managers' jobs have forced them to use "busy work" as a way to relieve the anxiety.* Completing activities, regardless of whether they could be done by subordinates, becomes a measure of success to them. Only about half their time is spent on managerial duties.

Some of this anxiety is triggered by top-level dictates that they must produce better results without additional resources. This causes a tendency to initiate programs without a clear strategy. The call for "action now" may be the unintended vehicle for increased busy-ness rather than for results.

The best way out of the "busy-ness trap" of time wasting is to attack the major problems head on, rather than be intimidated by their size and vagueness. This strategy consists of several elements:

1. Breaking long-term, complex management assignments into several sequential, short-term well-defined projects.
2. Bringing work planning discipline into these projects—timetables, clear-cut accountability, sharp definition of goals, measurement methods.
3. After achieving subgoals, expanding the process and capturing a larger share of the overall job with measured work plans.

Substituting results-oriented effort for busy-ness, one step at a time, is of major importance in focusing on managing a manager's time. The incremental approach does more than improve the use of time; it makes a manager more effective by viewing edicts for "action now" as the basis for setting up plans for taking effective action in incremental steps. These measured steps can determine whether the action plan is yielding the desired results. Focused efforts eliminate the need for busy work for its own sake as an anxiety reducer. A much greater feeling of accomplishment is produced when each measured step toward a goal is taken and achieved.

---

* Reprinted by permission of the *Harvard Business Review,* from "Managers Can Avoid Wasting Time" by Ashkenas and Schaffer (May/June 1982).

Time management is one of the essentials of effective management because the manager focuses on his own use and measure of time. He manages himself and uses time more effectively. Peter Drucker states simply that time is the scarcest resource, and unless it is managed, nothing else can be.

SUMMARY

Effectiveness in all aspects, from supervision of subordinates to decision making, to behavioral aspects, to time management, is a continuous learning experience. One never becomes effective in all things, but must continue to work to be. The old saying that "Practice makes perfect" is not true. Any talented musician or athlete knows that practice does not make perfect; practice makes one better. In the same way, continued practice makes a design engineering manager more effective.

The effective manager concentrates on being effective—doing the right things. The ideal manager, who exists perhaps only in theory, is both effective and efficient. He does the right things, and does them the right way.

8
Design Experimentation

INTRODUCTION

Experimentation is the sometimes bumpy road along which a design progresses on its way to completion. While it is part of the design engineer's responsibility, experimentation also falls under the jurisdiction of the engineering manager, who must oversee the experimental programs to make sure they meet technical requirements and stay within budget.

The rate of technical progress must also be measured against the project timetable. Sometimes the tough decision to terminate a project is necessary simply because progress is too slow and the project dates slip too much. This chapter considers these circumstances and provides the design engineering manager with some insights into more effective experimentation by his group.

Experiments provide information. The results of experiments are called data, and data collection is another name for information gathering. Experiments lead to conclusions if they are done with forethought and planning.

Before an experiment of any kind is begun, the test should be performed in the engineer's imagination. This brings creative insights into play, and enables him to thoroughly plan and execute the experiment. Of equal importance, it provides a sound base for analyzing and reviewing actual results. An imaginary experiment can also establish whether an experiment needs to be run.

Failure of a test device to function properly occurs often, especially during the early phases of a development program. Since failure has a strong personal impact, as well as a functional one, it must be handled with as little emotion as possible. However, in the course of product development, failure must give way to success as sufficient knowledge and insights are obtained to correct the problems encountered. If not, the future of the project is in jeopardy.

Serious review of the project and its potential product must be made to determine whether the product is vital enough to the company to be continued. Sometimes a project must be terminated; at other times it must be continued in a strong effort to force success. Each of these possible directions involves risk and uncertainty. This chapter will discuss experimentation and how the manager must juggle the opposing requirements of gathering all the information needed and keeping test budgets under control.

EXPERIMENTS—IMAGINARY AND CREATIVE

Experiments are conducted to find out how and why things work, or do not work. There would be no need to experiment if outcomes were certain. A device would work properly, once designed, and there would be no need to test it.

Since this does not happen except in the imagination, it is a good design approach to perform imaginary experiments before any tests are run. How will the device perform? How well will it sustain a particular loading condition? If it fails, which failure modes are likely? What kind of tests will be realistic, and what kind too severe?

These and more questions should be raised and answered in a speculative review of the types of experiments that can be run. These imaginary experiments are very inexpensive to "perform"; they cause no damage to the devices; and they open the designer's mind to the options available for testing, and prevent him or her from just plowing ahead and trying anything to get the tests started.

Another benefit to using the imagination before experimenting is that the possible outcome can be reviewed under various test conditions, and the number of "I don't know" areas can be listed as possible areas for learning.

It is sometimes difficult to work with imaginary experiments because one may have imaginary test equipment of a type and cost that in reality would be unaffordable. If a design can creatively explore how to use existing equipment and facilities to perform the real tests that imaginary experiments have recommended, then the best of both worlds exists.

PLANNING AND PERFORMING EXPERIMENTS

Planning is the most important part of an experiment because it dictates what is going to be done on how many samples, which versions of samples will be tested, the facility to be used, and the type of data to be logged on each. It is a model of the actual experiment.

In order to plan an experiment, one must determine in advance what information is needed and what results are expected. This is why the imaginary-experiment concept is valuable. It provides clues as to what may actually occur.

There are at least two ways of planning experiments, depending on the number of sample variations to be tested and where one is on the project. In the experimental stages, the design engineer is learning about the physical phenomena he wants to use in the end device. He takes the information available from, say, an R&D summary and wants to reconfirm part of it and explore some new areas of interest.

These early experiments are performed to gain new knowledge and test the premises initially made. While he is looking for specific results, he is not always sure what they may be. In other words, experiments do not always follow predictions; they obey the laws of nature. The engineer experiments, in this early stage, to establish cause-and-effect relationships. Often the results will show multiple possible causes for each effect, and several variables that appear to be causes. For these reasons, not only must the early experiments be planned carefully, but the results must be studied thoroughly to extract every possible bit of information from them.

In later stages of the project, when most of the variables have been established and the key cause-and-effect relationships determined, it is necessary to perform statistically planned experiments, called designed experiments. This is frequently done during prototyping.

Designed Experiments

To some extent, all experiments are planned, and all experiments are designed. The extent to which they are well planned and designed determines how efficiently the end results are obtained. Poorly designed experiments are a waste of time, effort, and money.

The comfortable way of designing experiments is to change one variable at a time; the design engineer believes that all the other variables will remain constant. If one variable is changed through extreme points, the information gathered provides some insights as to the capability of the device. Then additional experiments are run on a second variable, a third, and so on, until a substantial amount of data has been gathered. Conclusions are drawn about the state of nature as observed in the device.

Although the designer may feel comfortable collecting information in this manner, the one-variable-at-a-time method is not cost-effective or efficient, nor does it provide enough information. In Chapter 2, ROI calculations were discussed, and the method of sensitivity analysis was illustrated. It was shown that if the extreme values of cost, price, and investment were simultaneously varied, the results when analyzed provided greater insights than individual variance.

A review of this approach with an experiment in mind will be useful. For example, consider an assembly or component that needs to be tested for performance under different levels of tensile stress, ambient temperature, and voltage stress. Figure 8–1 shows the one-at-a-time method of testing. The

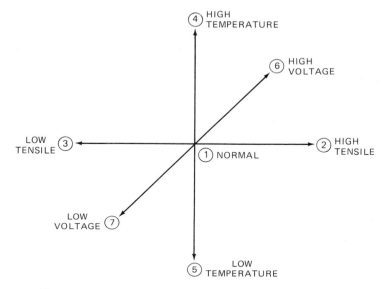

Fig. 8-1. One-variable-at-a-time, three-dimensional test plot.

design begins from a base-line set of tests at room ambient, normal applied voltage, and normal tensile stress. The tests tell the project engineer that the device meets some output criteria that are measured with various instruments, and they indicate satisfactory results.

However, with the other variables held constant, no estimate can be made of performance that would be obtained if two or more major variables were at a high or low level at the same time. A better approach is to measure the performance effects under controlled conditions where variables are purposely changed to extreme points.

Figure 8-2 shows the cube of test results that can be thus evaluated. The top and bottom surface of the cube represent high and low temperatures, the front and rear surfaces low and high voltage, and the right and left surfaces high and low tensile forces. With numerical outputs representing data taken at each of the eight corner points, the response of the device can be determined. By considering one surface of the cube as a constant, the experimenter can analyze the effects of the other two changing variables.

The amount of information to be obtained from this type of test methodology is substantially greater than the additional effort necessary to obtain it. For example, let us assume that, in both the one-at-a-time and the multiple-factor experiments, it is necessary to perform each test twice to reduce the experimental error. In the first case, with a total of seven points, 14

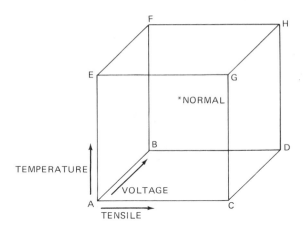

Fig. 8-2. Multi-variable analysis "cube" test data plot.

tests would be necessary to establish the three-dimensional "star" of information. In the second case, with eight corners of the cube and a ninth center point, it is necessary to run 18 tests. This is an increase of 29% in the amount of testing needed.

However, the information gained is at least doubled. In the first case, one develops some knowledge of the effect of each variable, nothing more. In the second case, one also develops some knowledge of the effects of each variable, and in addition can determine if an interaction between any two variables, or all three, has a significant effect on performance. "Significant" is used in the statistical sense, which means it appears to be a real difference and not part of normal experimental variation.

Thus, for a modest amount of additional testing, a substantially greater amount of additional information is generated; and the analysis of data is done mathematically rather than intuitively. This statement is not intended to denigrate the value of intuition in experimentation, but indicates rather that mathematics is an objective tool in analyzing complicated data.

Designed experiments are structured to generate maximum data when analyzed. They require that certain rules be observed, in both the setup and the sequence of testing. The rules include random selection of each sample for the test series and an adequate sample size to ensure that valid conclusions are drawn from the results. There are numerous texts on the design of experiments, and it is important that those who plan these experiments be familiar with the methodology.

Undergraduate courses both in statistical analysis and in design of experiments provide important background knowledge. While it is possible to follow a checklist approach in designed experiments, it is preferable for effective design and analysis for the engineer to have some background in this area. Then he can determine when designed experiments are useful, and when they are not, and be able to benefit from this excellent approach to experiments.

Performing Experiments

If an experiment has been well thought out and carefully designed or planned, performing it is just a matter of carrying out the plan. This is particularly important in using designed experiments because the analysis depends on it.

In early feasibility of pre-prototype studies, the plan should be followed, but the engineer should be prepared to alter it if some new or unexpected results are observed. Then it may be appropriate to stop the experiments in order to decide how to proceed further, or the engineer may decide on the spot how he wants to proceed. Experiments should be simple and straightforward, and every effort should be made to extract the maximum amount of information from each one. Deviations or unusual results usually provide clues to a better understanding of the physical phenomena at work.

REVIEWING AND ANALYZING EXPERIMENTAL RESULTS

Experiments must be planned to reduce to a minimum the probability that a wrong conclusion will be reached. If this planning is not done well, there is a high risk that analysis of the data collected will lead to incorrect conclusions. However, on the assumption the experiment was carefully planned, the results are reviewed to learn what has happened.

A design engineer who has participated in, or witnessed and observed, the tests has a good general idea of what has occurred; but the results need to be reviewed prior to analysis to make sure that all of the required tests were performed according to the test plan, that all the data have been collected in proper form, and that all test sample identification has been maintained for analysis, despite any damage that may have occurred to the samples.

It is important also that any comments by test observers be reviewed with the persons who made them. Sometimes a cryptic comment is of major significance but passes without recognition because the analyst interprets it as nonsignificant.

Analysis of test information will probably be started at the same time that it is reviewed, but it is a good idea to treat analysis and review as separate entities. Test analysis will be faulty if the data have not been carefully reviewed, and clarified where necessary, prior to any analysis.

If tests have been run as part of a designed experiment, data analysis is straightforward. A series of mathematical analysis called ANOVA (an acronym for analysis of variance) is performed on the data. From this analysis, the effects of each variable can be determined, as well as any interactive effects between two or more variables. These results usually are calculated with a 95% confidence level favoring their statistical significance (95% confidence that the observed differences did not come from chance variations between sample performance).

These analyses are limited to test circumstances where either the sample performance can be measured by some numerical output, which could vary between zero and any other value, or where the test result can be expressed as either pass or fail. In both of those cases, the outcome of the test is clear-cut and unambiguous. Typically, this analysis occurs as part of a prototype-approval sequence of tests, or approval of a pilot run of a new product.

Subjective Analysis

During early stages of a program, even when it is desirable to study the major effects and interactions by means of designed experiments, the results may not be completely good or completely bad. Some ambiguity may exist, which a quantitative evaluation by itself cannot adequately address.

Consider this situation: A sample test includes withstanding a static force or stress for a period of time after the dynamic test has been completed. Let us speculate that the test requires a one-hour withstand after the dynamic shock load has been delivered to the samples.

Depending on the approach and temperament of the project engineer, a withstand of less than 60 minutes could be reported to the manager as a failure or a near success. Some engineers have such a conservative nature that they call all shortcomings failures; others are so optimistic they consider all near misses as "almost" successes, or "successful failures."

In such cases, the manager must probe more deeply into the actual results, particularly if it is an important area and a wrong decision could be expensive. It is relatively easy to manipulate facts or shade the analysis enough to make either a pass or a fail impression come out of the report. The manager needs to ask what, why, and how much questions in order to get the facts straight and home in on the truth.

Optimists, Pessimists, Pragmatists

Did the device pass, or did it fail? In the above example, the device did both. It passed the dynamic shock test but was so weakened that it was unable to withstand for the entire test period. The real question to speculate on is: how should the manager evaluate the data, or better still, how should he train his key subordinates to evaluate such performance? Figure 8–3 illustrates two approaches to analysis, the optimistic and pessimistic. The horizontal, time axis is logarithmic, plotted from .01 second to 10,000 seconds, or for seven orders of magnitude. The vertical axis is percent performance, from 0 to 100.

One hour (3600 seconds) is the benchmark. The pessimistic analyst gives any performance less than one hour a zero rating and awards a 100 for one hour or more. The optimist feels better each second the device withstands, and his confidence builds faster than the withstand time of the device. If it withstands 6 minutes of the 60, he considers that to be a successful performance by the device. He may be right, but chances are that he is incorrect. The pessimist understates the design's capabilities, and the optimist overstates them.

There is no correct answer, but the pragmatic approach lies somewhere in between, and is a better first approximation than either of the other two. Figure 8–4 illustrates the pragmatic approach.

In this approach, the ability of the device to last for at least one second after test is increasingly important. But, whether it lasts one second or 1000 seconds is a measure of only slight improvement, perhaps 25 percent. After it has withstood more than 1800 seconds (50% of the time), the equivalent performance rating goes up at a more rapid rate.

Comparing all three on the same graph (Fig. 8–5), the pragmatic approach is between the optimist and pessimist, and for good reason. Some withstand is

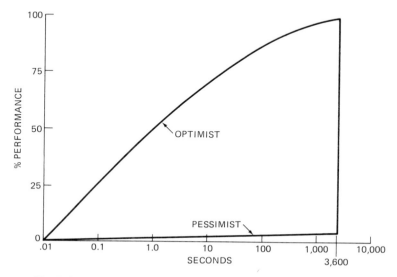

Fig. 8–3. Relative performance estimation, optimist vs. pessimist.

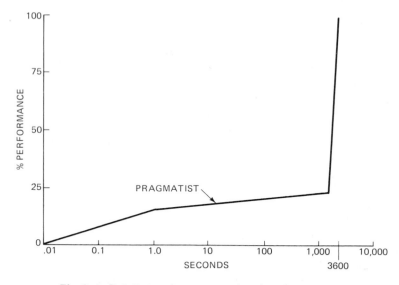

Fig. 8-4. Relative performance estimation, Pragmatist.

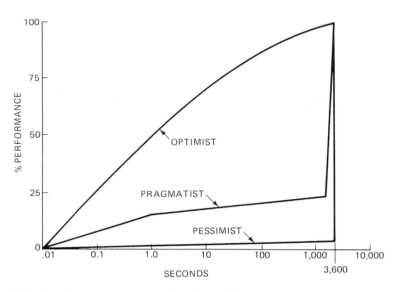

Fig. 8-5. Relative performance estimation, optimist, pessimist, and pragmatist.

definitely better than none, although the pessimistic approach does not recognize this. However, the optimist overvalues the withstand time, not realizing that a physical failure process is possibly going on within the device.

The pragmatist withholds any substantial enthusiasm about performance until the device has withstood at least half the time specified. By then, he reasons, the physical damage process has become relatively stabilized, and the failure mechanisms may have been stabilized within the material. Beyond that time, the device has an increasingly good chance of passing the withstand requirement. Should it fail, the task of redesign would not be so difficult as the original design.

This approach and analysis could be argued to be incorrect or an oversimplification, but are presented as an alternative for a manager to consider when a device's performance is not quite up to specification. Both the optimistic and pessimistic approaches are likely to be wrong. Neither of them addresses the reality that degradation or failure mechanisms are at work within the device, and that unless somehow they are recognized, analyzed, and brought under control, the device will never work. Sometimes, the withstand test should be stopped before completion and the samples dissected for signs of incipient failure, to assess this.

The manager, despite his desire to have the device work well, must avoid the traps of the optimist and the pessimist. He must train himself and his subordinates to be pragmatists, and to analyze and respond to each situation as unique to a degree. He can be neither overly pleased nor overly discouraged when a device nearly works. He needs someone to get more answers and find out what is happening in the device.

WHEN TO TERMINATE A PROJECT

No one wants to terminate a testing program or terminate a project. The project engineers believe they are close to an answer to a knotty problem; if only they can have some additional time and money, the remaining problems will be solved. The project leader may feel very strongly that he personally can get the project back on course if some leeway can be obtained on the specification requirements. Usually, since no one wants to make the decision to stop a project, the engineering manager must evaluate it and decide.

A project is generally terminated because of a clear shortcoming that cannot be corrected in one of the three major areas—performance, cost, or time.

Performance

The objective of a project is to design and produce a product with certain minimum performance characteristics to be acceptable to the market. Some of these are easily met, and some are met with difficulty. Sometimes they cannot be met at all. The lack of a key idea, the unavailability of a suitable material at any cost (or at a reasonable cost), and the lack of full understanding of a technical problem are all possible reasons why a key performance target cannot be met. When that circumstance occurs, the engineering manager must make the hard decision to terminate the project and advise his peer managers as well as his superior.

There is a natural lack of enthusiasm about making such a decision because it reflects negatively on the department's ability to solve technical problems, but a decision is often the only way to avoid pouring continuing amounts of resources down the drain.

Costs

Each project has a budgeted cost. People's salaries and fringe benefits are a major factor. In addition, there are facilities usage costs, outside-purchased materials, rental of equipment, and numerous other small items. If a project runs out of money before the performance goals are reached, there may not be any additional funding to allow it to continue. The company may have internally or externally imposed financial constraints. Or, perhaps in the perception of the project's sponsors, it is only worth a certain amount of money and no more.

A budget shortfall often can be anticipated early, and measures taken to control the spending rate and still maintain a good project pace; but sometimes, it will be clear before half the money is spent that the project was underfunded and will not make it. The project leader may believe he can solve problems by throwing money at them—in other words, by applying more and more resources to them. Occasionally, such overspending is successful; more often, it only wastes money. However, whether the project is underfunded or overspent, it must be terminated if it lacks sufficient funding to continue.

There is one more reason not to continue a project. That occurs when the project has run out of funds but its proponents are seeking additional funding because of the substantial investment already made in the project. Frequently the comment is made that funds are needed to "protect the investment already made from being lost or wasted." The argument has a ring of sincerity and truth, but if that is the only reason to continue the project, it is better closed than continued.

Any time additional funds are requested for a project, it is necessary to decide if that is the best use of available funds. A different project may be a much better recipient of the funds. The term "sunk costs" is sometimes used to describe this situation. It means the funds that have been spent have already been sunk. There is no way to recover them, as they have been spent. They cannot be protected by any additional investment of funds. The project must have a valid technical or performance reason to be continued. The previous expenditure of a high level of funds is not such a reason. The simple thought that "sunk funds are sunk and not recoverable" is an easy way for the manager to remind himself not to be sympathetic to the plea for additional funds to protect the investment already made.

Time

Time, as is often said, is an asset that cannot be apportioned, stopped, or repeated. Time-out occurs only in the sports arena, not in the real world of projects and product development. Time can be wasted, squandered, or used wisely; but it can never be reused, recycled, or recharged, and certainly never reversed.

Sometimes a project must be stopped because it runs out of time. The several meanings of that statement will be discussed below. They include running out of time, missing a time window, and having competitors beat one's time.

Running Out of Time.   If a project must be completed by a certain date and it is not complete, it may have to be terminated. Follow-on projects using the same engineering resources may be waiting; or lab facilities may be committed for other projects and may no longer be available.

Lack of timeliness may be the fault of the project leader, the design engineers, or key problems that resisted solution; but regardless of why a project is late, it may be terminated for that reason alone.

Missing a Time Window.   The time window puts limits on a project's running time. In some cases, a project must be completed by a particular time in order to meet a particular opportunity. It may be a large-order annual bid time, or the project may need to be displayed at a national or international exhibition.

There are a first date and a last date within which the project must be finished. If it were finished much too soon, it would have to wait before it could be tooled up. If it were finished too late, there would not be time to complete needed tooling and prepare the project for marketing.

If a project misses its time window, it is probably doomed to be dropped, the effort substantially wasted. In an unusual case, the project may be put on hold for later introduction; but in most cases this is ineffective, and it must be terminated.

Competition Closing the Time Window.   One of the more frustrating time constraints occurs when the actions of a competitor cause the project to run out of time. Consider the situation of a high-volume, low-cost company striving to be the third producer of a product sold to a limited market. This occurs in oligopolies, where there are a limited number of users and a limited number of suppliers.

A company introducing an improvement or a lower-price version on an existing product must enter the market before too many of its competitors are there. Otherwise the market share for the new entrant will be quite small, and the profit opportunity even smaller.

If this company proceeds to develop a significantly improved product and finds that a fourth competitor has just announced its version of the product, a decision must be made. If the competitor is a worthy adversary—a good-quality, low-cost producer—the only available strategy may be to terminate the project. There is no time left to complete the project and enter the market; the marketplace will be saturated with sellers. Pricing will deteriorate, and profits will be marginal.

Some will opt to continue the project and stay in the competitive arena to maintain a full product line; but the better choice is the difficult one, to close the project because time has run out. Competitive actions have closed the time window.

Other Reasons to Close Projects

Three major categories of reasons to close projects relate to the previously discussed internal factors of performance, cost, and time. These are external factors, including the market, company strategy, and personnel.

The Market.   The market may have changed since the project was initiated, or an incorrect assessment of market potential may have been made. The customers who appeared to be interested in a new device may have found it too unwieldy or expensive to use. The fast follow company may not have any better success than the leader, or the market size or growth rate may have been too optimistically assessed. After a few months of development work, marketing advises that a more accurate assessment of market potential is only a fraction of that originally estimated. Realistically, the project should be closed.

Company Strategy.   Survival in a competitive environment requires adaptability to change. If a company strategy changes, an entire group of products may no longer fit the plan of maintaining market position in a certain area. While there still is room for a few small companies making buggy whips, the businesses supplying accessories for horse-propelled human conveyances long ago decided their business would decline rapidly. So too did the manufacturers of radio and TV vacuum tubes, in more recent times. A change of company strategy to adapt to major business trends or conscious decisions to enter different business areas will force a particular project to be closed.

Personnel.   If one or more key people decide to leave the company, their project usually must be terminated. In a small or medium-size company, there are seldom backup people for every technical specialty. So if one or more persons decide to leave for an opportunity elsewhere, a project might not be able to continue. This is another tough decision for the engineering manager, who may want that project continued for its value to the company. He may elect to continue the project for a time, hoping that the competence of a new project leader will be sufficient to maintain its direction and momentum.

If not, the manager must close the project and reassign the people— promptly. Otherwise there is serious risk that the project will falter, use funds at an excessive rate, and miss its major objectives.

Problems in Stopping a Project

Two significant problems occur when a project is terminated. These relate to staff morale and final documentation. While quite different from each other, both must be dealt with.

Staff morale will drop because the project staff were immersed in the project and developed a strong personal interest in making it successful. Its termination is an emotional letdown for each member. Thus, when it is necessary to close a project, the staff members must be reassigned quickly. The challenge of a new assignment does much toward reestablishing a high level of interest and a feeling of job security and personal worth.

The last members to be reassigned are often the project leader and one or more of the key project members. Their assignment is to document the project—to make sure there are adequate and accurate records of what was complete and incomplete. Project documentation is a thankless task even when the project has been completed successfully. If it is terminated early, the motivation to document it is much less.

The manager must insist that this documentation be done completely and in a reasonably short time. Should the project ever be reinstated, these records will provide a firm basis for a new start. If not, they will provide a complete record of what can go wrong in a project, and it is hoped that they would suggest how not to do such a project later on.

WHEN TO CONTINUE A PROJECT

Should a project that might be terminated ever be continued? Yes, sometimes. There are two circumstances where continuation is worth considering, and may be the best action to take. In one case, there are clear signs that the technical problems are substantially solved; in the other, a broader time window exists than was initially perceived. The two possibilities may occur together or separately.

Solving the Technical Problems

Most project engineers remain confident about their ability to solve the problems that frustrate them and cause the project to lag; but the manager must be perceptive, and determine when the problems are really under control and when they are not. If they are not under control, new facets of the problems will continue to pop out like a handful of worms.

One way to assess this situation is to develop a pseudo-mathematical relationship between problems and solutions, and use it as a guide. In the early stages and in the difficult project phases, more problems develop than solutions. Or, one could say, the ratio of new-problem generation to new-solution generation is large.

This can be expressed as:

$$\frac{\text{delta } P}{\text{delta } S} \rightarrow N$$

where P stands for problems and delta P is the rate at which new problems are developing; and similarly S stand for solutions, and delta S is the rate of new solution development. In the early phases of a project, N is a large number.

Later in the project, when new problems become fewer and new solutions proliferate, delta P becomes smaller, and delta S becomes larger, so that:

$$\frac{\text{delta } P}{\text{delta } S} \rightarrow -N$$

The minus sign before N indicates that because the number of new problems is decreasing, the rate of change is negative. When that happens, significant progress occurs because solutions that prevent new problems have been devised.

While much of the mathematics is intuitive rather than deterministic, the pattern can be discerned if the manager closely follows the project's progress. If there is a need for the project's product and the time window has not closed, the concept of negative rate of new problems to new solutions can be useful. It is also clear that the turning point in the project occurs when the rate of change is zero.

Figure 8-6 uses a sine wave graph to illustrate problems and solutions. Figure 8-7 is the derivative, a cosine wave. This analogy adds the insight that

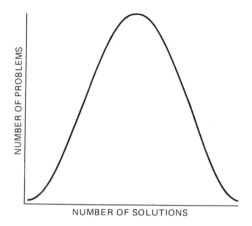

Fig. 8-6. Problems vs. solutions in a project.

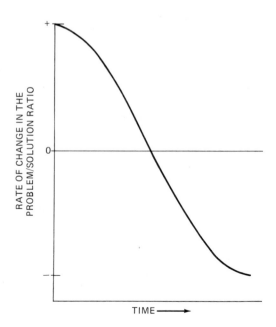

Fig. 8-7. Changing rate of problems vs. solutions in a successful project.

the turning point in the project may well be when the greatest number of problems has occurred, since that is when the rate of problem development shifts from positive to negative with respect to solution development. Of course, few, if any, projects have a sine wave distribution of problems versus solutions. On the other hand, this relationship reflects the saying that "It's always darkest just before dawn." Thus, the approach has some merit, at least conceptually.

Broadened Time Window

The marketplace is a dynamic entity and variable because of its dynamics. What appears to be constant may be variable, and vice versa. Time windows are constants that occasionally become variables. If this occurs, a project does not require termination, particularly if the constraints of a limited time window were the major reason for it.

Time windows can be extended when customers' requirements for a particular availability data have changed. Or, competitors may not have introduced the product that market intelligence predicted they would. Also, additional customers may have been found who want the product at a later date. All of these factors point to a variable, rather than fixed, time window.

The variable time window is not under the control of the engineering manager, or of his project leader. Nor is it under the control of the market manager, per se. However, the engineering manager needs to periodically determine if the original time window still exists, or if it has changed because of some external factor in the marketplace.

If the time window has been shortened, that is an important negative factor to be made aware of. Conversely, if it has been lengthened, a good project that might otherwise be closed could be given a much needed time extension.

The decision to continue a project is, in some respects, a tougher one than the decision to close it. The needs for extra funds, extra people, extra time, or perhaps an easing of a performance requirement all carry an element of risk that the project will not be successful, that the sunk funds were sunk and so was the project. However, as discussed, there are times when the project should be continued. Use of the concept of delta P/delta S is a substitute for the intuition or visceral feelings the manager has when he believes that a project will succeed and must be continued.

SUMMARY

Experiments provide information that leads to conclusions about the progress of a design project. It is a process of testing ideas to see which ones lead to the design objectives. Imaginary experiments lead to better-planned actual experiments because they are part of a planning process, essential to economy in testing.

Designed experiments are more than well planned. They are part of a statistical approach to assess the effects of different variables on performance. The methodology requires some knowledge of techniques; but when applied, it provides more information per dollar of cost than conventional experimental methods.

Experimental results lead to conclusions about the progress of the project, and whether it should be terminated if it is not going well. Performance, cost, and time criteria for progress are the internal factors against which this decision is made. External factors are changes in the market, company strategy, and loss of key personnel.

Sometimes a project should not be terminated. Either the external environment has changed, or the internal one has. These assessments must be based on effective marketing input relating to time windows and customer requirements, as well as the rate at which technical problems are being solved.

9
Reviews

Reviewing what has been done before moving ahead is an effective method of avoiding mistakes and repeating past wrong actions, as well as making better decisions. The word "review" has many connotations. It can be merely an inspection of information or a thorough re-study of material. When applied to the design process, a review is a careful re-study of past actions and events.

In a design engineering department there are or should be many reviews of each project, aptly named "design reviews." These can be all-encompassing or limited to certain facets of design progress. An all-encompassing review will include a study of the project's design effort to date, covering performance, cost, and time factors. In addition the review could cover reliability aspects of the design, failure modes and effects analysis, producibility of the design, quality control, inspection and quality assurance reviews, value analysis or value engineering reviews, and "loose end" reviews.

Reviews were discussed briefly in Chapter 3, as they related to the design process. In this chapter, major emphasis will be on the structure and substance of design reviews, dealing with performance, cost, and time. The aspects of producibility, reliability, and so on, will be mentioned within the overall context. Subsequent portions of the chapter will expand on these additional areas for clarification and greater detail. In some companies, these reviews are held separately, to place greater emphasis on a particular area.

DESIGN REVIEW

A design review is a retrospective study of the design to that point in time. It is a formal or informal pause in the activities to look back and to look ahead. The look back determines how far the design has come, whether it meets the objectives set for it. The look ahead evaluates whether the project has been moving in the right direction, and if not, what needs to be changed to bring it back on course.

Design reviews are a natural part of the design process. Every so often, the engineering manager asks, "How's it going?" The informal replies of "Fine," "Having a problem or two," or "Terrible" are no more than a comment, certainly not an informal design review; but the exchange does point out that information feedback is important to a project.

A manager does not assign a project to an engineer, give him a specification sheet, timetable, and budget, and then sit back and wait for him to deliver a packet of drawings, specifications, and test reports on the project due date. Such a course of action would cause most projects to fail, or to go on forever. Engineers cannot operate in a vacuum; they need guidance and feedback to perform the best job possible.

Design reviews are necessary for this and other reasons. Products are becoming more complex, and many engineers are not competent generalists but specialize in one or two technologies of a major project. Some engineers are younger and are not sufficiently seasoned to handle a complex project without some guidance. And management, including the engineering manager, tends to be nervous in a high-risk, expensive project. It wants to know how well the project is going, and how the project problems are being handled.

By definition, a design review is a detailed, retrospective study of the design that:

1. Provides a systematic method for identifying design problems.
2. Aids in determining possible alternatives.
3. Initiates problem solution efforts by appropriate personnel.

Benefits from Design Reviews

When done well, each design review contributes cost- and time-saving methods to meet requirements, make improvements, and prevent problems.

A company that develops new products has a group of conflicting motivations:

- Design engineering wants the maximum possible development time to make sure the design is complete.
- Marketing wants to begin product promotion immediately, with early introduction to stay ahead of competition.
- Manufacturing wants to make sure the design can be produced at low cost before committing to dates.
- Quality control wants to know what areas or dimensions are critical so that sampling and inspection can be planned.
- Reliability wants to know failure modes and weaknesses of the design for reliability enhancement.

- Purchasing wants to know what unusual or single vendor materials must be planned for.
- Finance wants to know if the ROI is favorable.

These conflicting and legitimate wants are part of the normal product design process. The urgency to get a profitable product on the market ahead of the competition provides some benefits to the company. At the same time, it involves some risk that too early completion of the design and release to production will result in a product with some serious shortcomings. This risk must be balanced against loss of market share or missing a major segment of the market with a product that arrives too late.

The design review provides a means of assessing these risks and reduces them to a manageable level. It brings together specialists from the major functional areas of the company, to optimize the design from the standpoints of cost, function, reliability, and appearance.

To accomplish this, the following groups should be represented in a design review:

Design	Quality control
Manufacturing	Reliability
Marketing	Value analysis
Purchasing	Field service

Many engineering groups believe they have an adequate design review system in place. If it is too informal and unstructured, such design review meetings are little more than rubber stamps on the design. At worst, they can be meetings where the product is redesigned and the design engineer castigated as incompetent. Such meetings are counterproductive and in the long run would cause design review meetings to be omitted; or if they were held with relatively balanced groups of opposing forces, confrontation and arguments would lead to a standoff, with nothing accomplished. However, with management support and a problem-solving attitude fostered by both management and the review chairman, much good can be accomplished in a design review.

The major benefit of a design review is to bring questions and concerns about the project into a problem-solving situation where they can be dealt with in a timely, effective way. It can use the experience and expertise of staff members in handling these problems, and to avoid repeating errors in methods and standards.

The key person in a design review is the chairman. This person should not be associated with the project, but should have sufficient breadth and experience to understand the technical issues, run the meeting effectively, and see that assignments are made and carried out. In other words, he must ensure that the right items are discussed and the issues resolved by the right people. The chairman's other duties will be detailed later in this chapter.

When Should a Design Review Be Held?

As a minimum, design review of a development project is necessary prior to prototype approval. At that point, there still is time and opportunity to address problems and provide answers. In some cases, it is done as part of, or in lieu of, the prototype approval meeting. This is a mistake. By then, the project may have little or no slack time and a very tight schedule. If anything must be retested or reexamined, the schedule will slip.

Yet, if manufacturing wants changes to make the parts producible at lower cost, or if reliability engineering is dissatisfied with the failure modes, additional time must be taken to do the work needed to satisfy these legitimate requests, and the schedule may slip. Strong marketing concerns about missing a customer bidding deadline or the imminence of competitor announcements may be strong enough to put off meeting these needs. The net result could be a product that is too expensive to produce, or that may develop field problems, versus one that does not reach the market at the right time. This is a no-win situation.

A better plan is to schedule more than one design review, and time them early enough in the design process that needed changes can be handled without disastrous effects on the schedule. The appropriate number of design reviews is one, two, or three, depending largely on the project. If the end product will have a significant sales and profit impact, or if it is a complex product and the design timetable is tight, then several design reviews are necessary.

Timing of Design Reviews.   Appropriate times for design reviews are as follows:

1. *When concept feasibility has been established.*   At this time the project is ready to be moved into prototype development, and it is appropriate to review what the feasibility study proved and what it did not. The needs of the marketplace and the initial specs must be reviewed and factored into the plan. If an advanced development or R&D group has done the feasibility study, a design review is of particular importance because the engineering group will pick up the project, and must have a thorough understanding of what the device can do.

2. *After layouts have been completed.*   Quite often the first design samples do not resemble the device to be prototyped. When issues of pro-

ducibility, appearance, and features are factored in by the design group, the new product beings to take on its own shape. The design is still fluid enough that concerns about "a little larger here" or "slimmed down there" can be addressed if the product will look better, be more easily maintained, or be lifted and moved more easily.

3. *After prototype testing.*   The prototype test series (one or more) was discussed in Chapter 4. These tests are crucial to the successful development of the product. A design review at this point is critical to ensure that the number of loose ends is as close to zero as possible. Also, it is critical to make sure that the concerns of the other functional groups were addressed before the design was frozen. If the pace of development testing has been hectic and demanding, the design engineers, having reached their goal, are anxious to finish up and get the product into production. Thus, their enthusiasm for a thorough design review, with the possibility of additional testing, is low. They prefer going directly to prototype approval, which may be the wrong thing to do if performance has not been thoroughly reviewed.

Thus, multiple design reviews are necessary to study the progress of the project formally and make sure that problems were identified and solved. Informal reviews are important at the working level, to keep track of day-to-day actions and problem solving; but the comprehensive scope of a well-planned review is more likely to cover all the essential areas than the informal review between engineers, going over test information or drawings.

Project Design Review Information

Well before the meeting, at least two weeks, a packet of information should be sent out to the participants. This enables them to study the background and current-status information and be prepared for the actual meeting. The chairman is responsible for sending out the information, which the appropriate design engineers prepare.

The information packet can include the meeting agenda, as a reminder of the time, date, and location. Significant drawings and product specifications, test summaries, performance reports, and cost data are examples of information to be included. In most cases it is not desirable to send out an information packet that is substantially the same as for a previous review. It is difficult to decide what is new and what is old, and it is preferable to send out only new information, making reference to material sent out for a prior review. The chairman should contact all the recipients a few days later to make sure each has received the information and realizes that a study of the contents is necessary before the meeting. He may suggest (at least the first time) that questions and comments are needed from the participants.

Preparation for a Design Review

One other important duty of the chairman is to make sure that the presenters are prepared for the meeting. He must also reserve a room and make sure it is adequately appointed (with chairs, writing pads, etc., and with visual aids—projectors, screens, operators). More important, the speakers must have the significant points of their presentation ready, reduce the trivia to zero, and be certain of presenting the key information in about half of the allotted time. The other half allows for questions and discussion.

An effective, well-structured meeting does not just happen. It is developed by careful attention to the elements that are important. This is best accomplished if the chairman schedules a rehearsal, or dry run, two or three days in advance of the design review. The participants will, without doubt, complain about lack of time, other commitments, incomplete preparation, and so forth. By insisting on a dry run, the chairman will learn how well the participants have prepared, how well the visuals have been done, and whether or not they illustrate the key areas adequately. Also, there will be a few days left to gather missing pieces of information, recheck important data, and become better prepared for the actual review.

If the chairman insists on a rehearsal before each design review, word gets around that he wants a well-prepared review, and the participants will adjust to it with a minimum of hassle.

One point should be emphasized: the purpose of a rehearsal is not to stage a design review, or to make a slick, flowing production that impresses the "audience" and has them applaud rather than ask probing questions; rather, the purpose of a rehearsal is to ensure that the participants have their data and numbers handy for the questions that are certain to arise. A design review is too important a milestone in the project to be done haphazardly, with preparedness left to chance.

The Design Review Meeting

The chairman keeps the meeting going, prevents major shouting matches, but allows enough give and take to make sure key issues are addressed. Besides keeping the meeting under control, he must maintain a climate of free interchange where problems are being attacked, not people or personalities. If this climate is maintained, the meeting will not degenerate into a "trial" of the design and design by a kangaroo court.

A typical presentation by the project design engineer would include the following:

1. An introduction with background information to bring the listeners to the present situation.

2. Description of the product design's major requirements, usually from items on the product specification.
3. Further description of the design by discussion of the selected approach and alternatives that were also considered.
4. Identification of current problems or expected problems, depending on the project status.
5. Problem-solving approaches being used on current problems.
6. Supporting evidence that the design is meeting the specification requirements and that problems are being solved.

Each speaker in turn follows a similar format to cover the portion of the design review to which he or she was assigned. If a particular project was at the final design review point, the following agenda could be appropriate:

Activity	*Responsibility*
1. Review of design status and major problems, both solved and unsolved.	Design engineering
2. Review of recent test results.	Reliability
3. Review of drawings and specifications.	Design engineering
4. Review of vendors and their capabilities.	Material control or purchasing
5. Producibility factors in the new design.	Manufacturing engineering
6. Planning for inspection and test equipment.	Quality assurance
7. Failure mode and effect analysis (FMEA)—product design.	Design engineering
8. FMEA—product processes.	Manufacturing engineering
9. Assembly and installation instructions.	Marketing
10. Advanced planning for pilot run manufacturing.	Material control and manufacturing engineering

A meeting of this scope is a one-day or a day-and-a-half meeting. Ample time needs to be allowed for adequate coverage of the areas, both in presentation and in response. If each topic requires an hour, with an overall introduction at the beginning and a summary at the end, 12 hours would be needed. On the other hand, if 30 to 40 minutes is enough for each topic, less than a day is involved.

The design reviews prior to prototype approval and after pilot run testing are crucial to the eventual technical success of the product. Allowing adequate time to pursue and resolve key questions here means fewer problems when the product is used by customers.

Action Areas

During many reviews, especially those early in the design process, questions will be raised that the design group did not expect. These should be recorded and investigated by the appropriate group. Whenever possible, they should be answered by a given date.

To facilitate documenting the results of the meeting, a member of perhaps quality assurance, quality control, or reliability engineering should be secretary for the design review. He or she records questions, agreements, action items (and assignments), and pertinent comments.

Shortly after the design review, preferably within one week, the chairman publishes minutes of the review meeting. Timing is important. The longer the delay in publishing minutes, the more solidified the recollections and viewpoints of the attendees are, especially in areas of disagreement.

The minutes should not be a blow-by-blow or word-for-word statement of what went on. As in any good report, they should contain an introductory summary of key findings and decisions for management awareness and comment.

Then, in detail, the key findings of the design review are discussed: the status of questions referred to design or others for investigation, and any significant agreements and points of disagreement. If, after reconsideration, a particular recommendation will not be followed, the decision and the rationale for rejection should be included.

The minutes of a design review should be available for all subsequent design reviews, and some comments on previous action items should be made by the chairman in his introduction for that subsequent review. In that way the minutes are an integral part, not only of each review, but of the total design process.

Design reviews, done in a formal or semiformal way, are important to the ultimate success of the project. They are as structured as the complexity, potential, and risks of the project require. One item of structure that is an important aid in both the evaluation of a design for a design review and the selection of a design approach is the design checklist. While every designer has a mental list of the requirements, and specifications do tie down the most important functional requirements, a number of them can be overlooked without the use of a list.

Table 9-1 is an example of a checklist, which is compiled from several

Table 9-1. Design checklist.

	YES	NO	REMARKS
1. DESIGN REQUIREMENTS—Does it meet:			
Customer requirements			
Design specification			
Applicable industry standards			
2. FUNCTIONAL REQUIREMENTS—Does it meet:			
Mechanical strength			
Shock loading			
Motion, travel, operating time			
Cost targets			
Size and weight			
Projected life			
Mechanical, electrical, thermal, etc., loads			
3. ENVIRONMENTAL REQUIREMENTS—Does it meet:			
Temperature (operating, storage)			
Humidity extremes			
Vibration (operating, shipping)			
Corrosion (salt water, alkaline or acid)			
Submersion (brackish water, soil, chemicals)			
Seals—water, soil, air, gases, vacuum			
Production—degreasing solvents, steam			
Outdoors—U/V radiation, rain, wind, sand			
Combined environmental factors			
4. PRODUCTION REQUIREMENTS—Does it meet:			
Use of standard parts or subassemblies			
Production tolerances			
Well-defined materials			
Familiar processes			
Ease of assembly			
Value analysis considerations			
Inspection/testing standards			
Finishes—protective			
Component interchangeability			
5. OPERATIONAL REQUIREMENTS—Does it meet:			
Field installation assessment			
Instruction sheets			
User maintenance manuals			

Table 9-1. (*cont.*)

	YES	NO	REMARKS
Field service assessment			
Special tools for servicing			
Prototype device experience			
6. RELIABILITY-RELATED REQUIREMENTS— Does it meet:			
Hazardous operation			
Personnel safety			
Failure modes and effects analysis			
Product liability considerations			

sources. Any published checklist can be only a starting point for one's product design, since very few products can use a general list. The major criteria that apply in almost all cases are these:

1. Design requirements
2. Functional requirements
3. Environmental capabilities
4. Producibility
5. Serviceability
6. Special requirements

During the design feasibility study, the list is developed, and key requirements are written down. Each time a review is held, the design engineer or the reviewer will ask, ''Is this requirement met?'' The answer is either yes or no. If no, the degree of departure from a yes answer must be ascertained.

Quite often a design checklist will be used during the review meeting to determine if a particular requirement has been met, or if a new item should be added. As mentioned before, this example checklist does not exhaust the possibilities of what may or should be listed. It is a starting point for developing a unique list for a product.

Other Reviews

While the design review is the backbone review series for a product, other reviews are frequently necessary, either as a subset of the design review or as a separate review on a segment of a project. Some of these are value analysis or value engineering reviews, producibility reviews, reliability reviews, quality assurance reviews, performance reviews, and a loose end review to cover areas often missed. These reviews will be discussed in the remainder of the chapter.

VALUE ANALYSIS AND VALUE ANALYSIS REVIEWS

Analyzing a design for value is a process that is frequently done on an existing product line. If it is done on a new product prior to release, it is termed value engineering; but the correct term is not so important as the discipline and methodology that value analysis (VA) brings to a product.

In simplest terms, value analysis is the process of providing the basic product functions at the lowest possible cost. This cost minimization must be done without reducing the quality or reliability of the product. Ordinary cost reduction attempts to reduce the cost of a part by redesign, or change of material, or vendor. Value analysis looks at the various functions an assembly or device performs and creatively determines ways to minimize those costs.

Value analysis goes still further. It recognizes that several types of value other than cost affect the perception of value. These are use value, esteem value, and exchange value. Since value is not the same thing to everyone, all of the value factors enter in. For example, one purchases a car for many other reasons than just to transport persons from one location to another. Considerations of safety, comfort, appearance, economy, and reliability enter into the evaluation as strongly as sticker price does.

How to Value-Analyze a Product

There are many excellent books on value analysis, and the topic is frequently taught in courses, in seminars, and by VA consultants. Thus, this discussion of value analysis and VA reviews is more of an introduction and overview of the approaches involved than a detailed exposition.

Before beginning a value analysis, one must select a target product to analyze. A good candidate product for VA would have these characteristics:

- High annual sales volume—one of the top items.
- Low or declining profit margins.
- A significant remaining lifetime.

If there are several such products, it will be necessary to restrict the number of candidates consistently with the number of people available to work on them. A balanced value analysis team consists of three to five persons, from different disciplines in the company including engineering, marketing, purchasing, manufacturing, and finance. If shop foremen can be included, so much the better.

It is important not to overload the group with design engineers and equally important not to have the original designer of the device as a team member.

Having too many technical members tends to stifle creative efforts by the rest of the group because they tend to be considered "authorities" on what will work, or will not, by the less technical members.

Steps in Value Analysis

The three steps of value analysis are:

1. Identify the function.
2. Evaluate the function by comparison.
3. Develop value alternatives.

Identify the Functions. The functional approach is the cornerstone of value analysis, with a function being something that makes a device work or sell. Any useful product, device, or service has a primary function, such as to provide transportation, pump water, or indicate time. It will usually have secondary functions as well. A car must resist road shocks, a pump in the home must operate quietly, a watch, especially a digital used to sound alarms, must also be attractive, and so on.

Thus, there are two classes of functions, basic and secondary. The basic function is the purpose for which the item was designed. It is usually associated with cost or use value. The basic function of a pencil is not to write words or write numbers, but to make marks.

The secondary functions enhance the design, or contribute to the esteem or attractiveness values, and affect the exchange value. When an eraser is added to the pencil, a secondary function is generated—to remove marks.

This functional analysis uses only two words per function, a verb and a noun. If a function cannot be described in two words, chances are that more than one function exists, and each must be determined. The two-word approach is a forceful technique that imposes a strict discipline on the participants. It also is a strong aid in determining the appropriate functions.

After the functions have been determined, the device is broken down into its several functional areas. A refrigerator has the basic function of preserving food, and secondary functions of firming butter, freezing food, making ice cubes, and so forth. Today's refrigerators are all assumed to be electrically powered. In earlier times gas refrigerators were used, and ice boxes before them. But in a refrigerator example, the functional areas would include the food storage area, the compressor–heat exchangers, and the containment or housing area. Each in turn is analyzed separately to determine its functions and how it contributes to the basic and secondary functions of the refrigerator.

The costs of each functional area must be determined, usually from cost data in-house. The costs may be standard costs obtained from cost accounting, or may be actual costs determined from analysis of the labor, material, and overhead costs required. It is not important which of the two is used, as long as all of the costs are reasonably accurate and the same type of costs is used consistently throughout the value analysis study. When realistic costs can be associated with functions, the process of value analysis actually begins.

Evaluate the Function by Comparison.   To establish the lowest cost value of a function, it is necessary to begin with its present cost, then probe further. In this phase, the experiences of the group come into play. Evaluating by comparison with other alternatives opens one's eyes to different ways of accomplishing the function. It often stimulates good alternatives.

Consider this: in a refrigerator, the door is part of the provide seal function, as well as part of the provide entry basic function. Years ago, refrigerators had door latches and gasket seals; today they do not have them. The latch is the seal, an elastomeric material, filled with magnetic particles. When the door is closed against the steel frame, the rubber surface seals the cool air in the refrigerator, and the magnetic portion holds the door firmly in place.

Refrigerator manufacturers periodically value-analyze their product to bring costs down and maintain profits, their evaluation by comparison suggesting insights that their engineers could have developed. In this comparison the door latch of the refrigerator has two basic functions—provide entry and provide handle—and it is part of the provide seal function.

As an example of value analysis, Table 9-2 compares costs for latching, as obtained from vendors and from a trip to the hardware store. While costs may vary, depending on the actual items compared, this table shows the kind of cost variance that can be expected from a free-wheeling comparison of ways to secure a door. Ordinarily one does not seriously consider using a bent nail as a method for securing a door; but if low-cost, far-out ideas are never considered, the major cost reductions possible with value analysis will not be achieved.

Table 9-2. Comparison of latching costs.

Military latch—door	$25.00
Refrigerator door handle	$10.00
Car door handle	$9.00
Outside door handle—home	$5.00-$10.00
Window latch—home	$2.25
Magnet	$1.00
Screen door hook and eye	$0.59
Bent nail	$0.01

Normal cost analysis or cost reduction can typically reduce price costs by 5 to 15%, depending on methods employed. Value analysis typically reduces the cost of the function by 25 to 40%. In part such drastic reductions are possible because of the creative, nonstandard ideas that are considered in unrestricted analysis.

While no one would seriously recommend using a bent nail to secure a refrigerator door, or any door, the one-cent cost target for providing the function forces a value analyst to concentrate harder on the problem than he or she might otherwise. Instead of being satisfied with reducing the cost 10%, from $10.00 to $9.00, perhaps a 50% or greater reduction will be possible. The joking suggestion "use a bent nail" could well be a catch phrase in the analysis to encourage a broader exploration of lowest-cost ideas.

Develop Value Alternatives. After the evaluation by comparison is complete, the group is ready to determine alternatives for providing equal value at substantially lower cost. One concept of value relates to benefits and costs. Consider this equation:

$$\text{Value} = \frac{\text{Benefits}}{\text{Costs}}$$

The equation predicts that value is increased if the same benefits are retained at a lower cost, and dramatically increased when there are increased benefits at less cost. Value, however, is decreased when costs increase or benefits are reduced and the other term remains constant.

With this concept of value as a starting point, value alternatives can be sought by a combination of creativity, vendor contacts, literature searching, and brainstorming. An open mind is the most important prerequisite of the value alternative search. Many comments, such as "it won't work" or "there is no other way to do it," are roadblocks to effective idea generation. These and other roadblocks must be overcome to maintain a good level of progress.

The search for alternatives must continually focus on function, not on a part. It is too easy to be sidetracked into using a unique material for a part that complicates accomplishing the needed function at lowest cost. In the refrigerator door example, the idea of using a magnet may have been greeted with some disbelief by the rest of the group. It separated the door-latching function into the two functions of securing the door and providing a handle, and magnets are often either too strong or too weak to use.

Carrying this speculation further, the magnet idea might have been rejected on the basis outlined above. On the other hand, it was appealing because it eliminated the design of the latch mechanism, avoided penetration of the body of the door by the latch mechanism, and thus simplified and improved insulation of the door.

When the rubber gasket and magnetic seal could be combined by availability of suitable "magnetic rubber," the functions of door latch and door seal were also combined.

Although this scenario has been speculative, these events could have occurred as described. It illustrates how functions in a device often are interrelated. It is usually more cost-effective, that is, of greater value, to combine functions. However, in this case, separating functions and recombining them differently produced a better result. This will only happen if the team members keep an open mind in searching for alternatives. All of their creativity and ingenuity will be necessary to develop more "what-ifs," which are the speculative ideas leading to effective value analysis.

The Value Analysis Review

Review of a value-analyzed assembly or design is usually a presentation to the design group or to a management review group, or it may be a segment of the design review of a product being redesigned. The following will assume the review is a separate presentation to a functional management group. The purpose of the review would be to request funds for implementation. A modified version could be used for reporting progress.

The review sequence would follow this series of steps, presented with transparencies or slides for clarity:

1. *Background.* Describe the reasons for selecting the particular device or subassembly for value analysis review.
2. *Problem description.* Describe the specific areas of high cost and poor value that were selected for analysis.
3. *Proposed solution.* Describe the solution proposed. List its advantages, not only from a cost but also from a functional standpoint. Also list disadvantages, which could include the expenses associated with proving the idea, in both design and testing.
4. *Costs and savings.* List the expected costs of implementation, tooling, facilities, and so on. List the savings in the first two years, especially the additions to gross profit margin. Provide an implementation timetable.
5. *Request approval.* If the presentation is to management requesting approval, the team chairman must request approval to proceed with the implementation of the VA program. Approval signifies confidence in the approach and commitment to its achievement.

If the value analysis study has resulted in any models, sophisticated or not, of the proposed changes, include them as part of the presentation. They add to the credibility of the approach and help others visualize the ideas better than any sketches or drawings.

Closing Comments on Value Analysis

In 1946, after World War II, value analysis was begun in General Electric's purchasing department. Since then its use has spread nationwide through a combination of seminars, consultants, and do-it-yourself activities. When projects are undertaken with vigor and determination, savings of 25% are common, and up to 50% are not unusual. Return on the effort required to achieve results is typically 5:1, and it can be higher.

However, value analysis can fail if it receives token or halfhearted support from management; it can fail if it cheapens the product, and does not maintain the basic functions at the target reliability; and it can be ineffective if it is misconstrued as cost reduction. Reducing the cost of a part is not value-analyzing it. To summarize, value analysis is a valuable tool and as such must be used often and with skill, in order to obtain the desired benefits.

RELIABILITY AND RELIABILITY REVIEWS

As a concept, reliability is a statistical estimate. It is the probability that a device will perform its design functions adequately for the period of time intended under the operating conditions encountered in its application. Because reliability must be evaluated by its opposite—failure—it follows that reliability is the science of estimating, controlling, and managing the probability of failure.

Reliability has always been considered as an important factor in design, but more on a qualitative than a quantitative basis. However, when the space programs took shape and the aerospace industry began to grow, reliability was applied widely. With growing concern about the reliability of a great deal of equipment from high-speed transportation to nuclear power plants, the need to estimate and control failures has become vitally important. When growing concern about product liability cases and increases in insurance rates to cover them is added to this, the need for reliable, well-designed products becomes clearer.

Many companies have one or more reliability engineers, sometimes part of the engineering department, sometimes part of quality control. Neither of these attachments is as effective as having a separate reliability group. Quality control is concerned with maintaining established levels of quality and reliability; it is not specifically responsible for improving them. On the other hand, there are advantages to having reliability engineers as part of design engineering. The major one is being in close contact with the persons responsible for designing reliability into the product or system; reliability concepts can be brought in early. Yet this closeness can also be a disadvantage if it affects the objectivity and criticality of the reliability group.

Reliability Budgets

The appropriate budget for a reliability group, or the project budget share for reliability, is often controversial. The initial cost of reliability-related effort is high. It requires organized planning, thorough testing, and objective, critical reporting, without a clear demonstration of value (i.e., benefits vs. costs).

Later, the improved reliability that is built in will save money because the product being tested will not fail, requiring redesign, rebuilding, and retesting. But since one can only hypothesize about what did not happen, it is nearly impossible to quantify the saving due to improved reliability. The major demonstration comes later, with improved designs in service, which have a longer, trouble-free life.

Importance of Reliability Engineering

While many companies doubt that they can afford an extensive reliability program, they understand the importance of one. The complexity of product designs as well as the urgency to complete them and tight budget constraints increases the potential for unreliability.

As the complexity increases, the sheer number of individual parts tends toward system unreliability. If components in a device are statistically independent items, then the probability of successful operation (reliability) is the product of the individual probabilities.

As an example, consider a simple device with only seven components, each with a 90% (0.9) probability of working properly. On an individual basis, 90% is not overly impressive, but it illustrates the point. If the seven components were statistically independent, the system reliability would be the product of $0.9 \times 0.9 \times 0.9 \times 0.9 \times 0.9 \times 0.9 \times 0.9 = 0.48$ (0.9 to the seventh power), which is lower than the toss-of-a-coin probability.

In a larger system, of, say, 70 independent components, a 0.90 probability would be unthinkable, as the probability of successful operation would be only 0.0006. (0.9 to the 70th power) In fact, even if the probability of successful performance were improved to 0.99, the larger system's probability would again only equal the toss of a coin—0.49.

The foregoing illustrates how rapidly reliability drops as the number of components increases. Actual systems are further complicated by interactions. In the simple case, the independency meant there were no interactions between components; but, again in an actual situation, the failure of one component could cause the next to fail—from overload, physical damage, fire, and so on. Or, the vibration of a motor whose bearing was damaged could cause lead failure on an electronic component. Interactions could cascade failures and make the system reliability substantially lower than "independent" assumptions would indicate.

KISS Again

One familiar, basic rule applies to improved reliability as much as to communications and many other areas of technology—Keep It Super Simple. By avoiding complexity, reliability is improved. Keep it as simple as possible, consistent with performance requirements. Other reliability enhancers are based on the redundancy concept, including not only the use of backup components, or circuits that can take over another function, but overdesign of the device. Derating is another type of redundancy, but it is analogous to overdesign. Both concepts mean that the device is capable of functioning at a higher load level than specified.

Reliability Functions

Reliability is just one important facet of a new or improved product. It is as important as performance and price, and occasionally as features and appearance. However, as mentioned earlier, reliability is not as definable as these other facets because reliability is mostly invisible if it is "doing its job." Reliability introduces conflict in design between the apparent high cost of an extensive reliability program and the consequences—both commercial and legal—of producing a product with unsatisfactory reliability; but in its essence, reliability or reliability engineering includes a set of techniques (some of them statistical) generated in anticipation of unreliability and recognizing preplanned problem elimination.

These are some of the major functions of a reliability group:

1. Reliability evaluation
2. Design review
3. Specification, process, material review
4. Test planning, operation, analysis
5. Reliability and failure reporting systems
6. Reliability training and education

The design review mentioned above was discussed earlier in this chapter. A few additional comments are appropriate, if redundant, here. The design review is an essential activity and in no way is a duplication of the design engineer's effort, nor is it an attempt to look over his or her shoulder every day. Unfortunately, most engineers are not probability-oriented, either in design concepts or in testing approaches. Hence, a design approach that appears technically feasible may not have an adequate reliability level built in. In other words, feasibility does not equate to reliability.

Reliability Checklist

Earlier in the chapter, a design review checklist was developed. The following is a more detailed, reliability-oriented checklist of questions applicable for a design review:*

1. Is the item on off-the-shelf device or one developed for this device?
2. Does it perform one function, or more than one?
3. How many critical parts does it have?
4. Can an existing off-the-shelf item be substituted. If not, why not?
5. Is the item to be used at its design limits?
6. If it is off-the-shelf, is this application at similar stress and environmental levels as previous experience?
7. What have been previous problems on similar designs? Are they likely to occur in this design? How can they be eliminated?
8. What are previous hypothesized failure modes? What are new failure modes? How are they different? How is it being tested for failure modes?
9. What is this item's failure history? Is it a critical item?
10. What steps have been taken to eliminate this type of failure? New types of failure?
11. What is the expected numerical value of reliability? Is it sufficient?
12. Have redundancy and derating been introduced?
13. Is the part completely interchangeable with another manufacturer's?
14. If this is a new design, what are its critical weaknesses? How can these be tested early?
15. Has it been designed as simply as possible? Have human error factors been considered?
16. Is it physically and functionally compatible with its neighboring components? Does its physical location affect reliability? Performance?
17. Has it been sufficiently stressed considering vibration, temperature, humidity, dust? Does it have sharp corners to damage adjacent parts or wiring?

---

* Reprinted with permission of D. K. Lloyd and M. Lipow from p. 16 and pp. 28–30 of *Reliability: Management, Methods, and Mathematics,* 2nd edition Lloyd and Lipow, Redondo Beach CA 90277 (1977).

18. What can be done to improve reliability? What are the tradeoffs in per-formace, cost, weight, timing, schedules?
19. Has it been designed for ease of production, assembly, maintenance?
20. What unusual quality control or vendor problems are expected?

Such lists are valuable tools for assessing the costs of reliability and the penalties of unreliability as the design progresses. These and similar questions need to be asked more than once to make sure negative changes are not introduced by accident.

Space Program Reliability Experiences

Reliability concepts have played a key role in the U.S. space program, from the very first rockets, which had reliability problems during lift-off, to the multi-mission space shuttles. The space program designs are not similar to high-production commercial designs, but two valuable concepts have emerged and are key to reliable design whether the output is two or two million. These concepts are, first, the critical importance of design and, second, failure mode analysis.

Spacecraft fly in hostile environments, from the warm, humid air of Florida to the chilling cold and intense heat of the space vacuum. To a large extent, the devices are not repairable; the sample size is small, and extremely expensive to test. The reliability approach is clearly different from that used for large-scale-production items. On the other hand, the key factors are surprisingly similar.

Good design is the most critical determining factor of reliable spacecraft performance. Of component failures that could be analyzed in a NASA study, 42% were errors in design. This was the predominant cause. Defective parts, faulty workmanship, and other causes were less significant. In design the KISS principle is continually emphasized. Also, the use of flight-proven hardware and design ideas improves reliability. Redundancy by derating or backup systems is employed extensively.

Designs are thoroughly checked through the mechanism of design reviews. They are used extensively throughout a program, and are done at NASA field centers, with as many as seven or eight formal reviews on a particular program. They begin during the study phases of a project and continue until the spacecraft is launched.

Failure Mode Analysis

FMECA (Failure Mode Effects and Criticality Analysis) is a second valuable tool. It is a disciplined investigation of possible failure modes and how they affect end performance of the space mission. The purpose is a search for better design with fewer failures.

The analysis is normally carried out by the design engineer with assistance from a reliability engineer. Possible failures for each component are identified. The effects of these failures are specified at the component, subassembly, assembly, and system levels. Then, they are classified by worst-case effects, and ease or difficulty of identifying these failures by test and inspection. This determines the criticality of the failure. Worst cases, of course, can vary from reduced performance to loss of function, to explosion or loss of life. The FMECA studies are frequently done semigraphically by a fault "tree" that allows tracing of possible faults back to a component and its particular failure mode. Reliability texts, handbooks, and courses delve thoroughly into the fault tree and other reliability approaches.

Summary

Good design, design reviews, and failure mode analysis are keys to sound design practices, whether involving a space mission or a commercial product. The importance of design reviews to achieve this end is clearly obvious, whether a design group is responsible for reliability, or there is a separate reliability group.

QUALITY AND QUALITY REVIEWS

In most companies, quality control or quality assurance groups do not report to the design engineering manager. In a few companies they do, on the basis that quality, as well as reliability, begins in the design of the product. The statement that quality cannot be tested into a product but must be designed in, is true.

Quality control sets up inspections and tests to verify that the designed-in quality levels are maintained on a day-to-day production basis; and quality assurance sets up checks, balances, and reviews to anticipate potential problems and prevent them from occurring. Thus, quality assurance sets up the policies, standards, methods, and specifications for monitoring the quality of production; and quality control monitors production to these specifications on a daily basis.

What is quality? Frequently, quality is considered the absence of defects, or conformance to a set of specifications, the "goodness" of a product. Television advertising of beer shows a glass of foaming product, while an announcer's voice describes how one can taste the quality of the product. Because of its sometimes dubious methods of quantification, quality may be difficult to describe, whether in beer or some other product.

However, quality has recently been defined as "conformance to requirements." This precise wording eliminates many imprecise definitions of what constitutes an acceptable level of quality. It insists on the sometimes difficult task of determining what the requirements are. If these requirements are identified, clearly stated, then understanding is clear. The device either conforms to the requirements, which means good quality, or does not conform, which means poor quality.

While this definition of quality may not be universally acceptable, it underscores the reality that acceptable quality requires conformance to some set of requirements. If the customer's requirements are not known or cannot be identified, an arbitrary internal set of requirements or standards is set, against which the device and its component parts are judged. If these standards are too loose, the customer will not accept the product, or will not obtain satisfactory service from it. On the other hand, if the internal standards are too tight, costs will be too high, and the product will not be economical to produce. Parts costs too high, yield levels too low, scrap rates too high—these are all effects that can result from unrealistically high standards.

Quality and reliability tend to go hand in hand, but one must not mistakenly believe that all the quality attributes automatically foster high reliability. For example, excessively close fits and tight machining tolerances may give the appearance of high quality, but if in the "use" environment these close fits cause parts to stick or bind, reliability of the product is seriously affected.

Pre-review Quality Actions

Since quality control actions have their greatest relevance when the device is in production, the development phase of a project is the time when quality control prepares. During design reviews the criticality of parts and dimensions is frequently discussed. Key tolerances, critical dimensions spelled out on prototype sketches, become the basis for inspection fixtures and procedures, to ensure that those areas that are important to the function of the product are checked.

Vendors' plants are often audited prior to production to establish their capability to meet production needs. The vendor's quality control procedures are reviewed, and in some cases a serious discussion of procedures and technique follows. This is to ensure that both quality control groups are concerned about the same areas. Some suppliers are not overly pleased with such an arrangement because it is more stringent than they wish to be. On the other hand, such reluctance is sometimes advance warning of future quality problems with a vendor.

During each of the several design reviews, quality control representatives develop an awareness of the critical areas of design, and often are asked to provide inspection services to assist in pre-production evaluation of vendor parts. Later, at the final prototype reviews, the specifications are being developed jointly between quality control and design engineering. At this time, major decisions are made relative to inspection level, sampling plans, and rejection criteria.

While many of these areas are crucial to the successful use of quality control methodology during production, they can only be qualitatively assessed because the volume of parts has not developed. Thus, the first quality control specs are analogous to the preliminary design specs set at the beginning of the project. They are subject to change and revision as greater quantities of material are received and inspected, as new vendors are qualified and brought on stream.

LOOSE END REVIEWS

When a product is ready for production, the implication is that all documentation is similarly ready. However, in some areas, notably specifications, drawings, and reports, the work often is not up to date. It requires a determined effort by the manager of design engineering to see that his staff finishes these documents on time and completely. Often, this requires special reviews, to cover the loose ends. The three areas will be discussed in turn.

Specifications

The role of specifications in the establishment of mutually consistent criteria is sometimes unclear, but a specification, like a drawing, is a means of communication. Specifications define the key requirements of a design, the materials and processes of manufacturing, performance characteristics of finished goods, testing and evaluation methods, and more.

Engineering may treat specification writing as a necessary evil, and the temptation often exists to write general or vague specifications just to simplify the task. Unfortunately, the vendor will not understand them, and will interpret them to his advantage in the unclear areas. For that reason especially, the design engineering manager must insist that specifications be done thoroughly and well. They must be revised and updated as needed to reflect any necessary changes that occur.

In the same context, specifications, like quality, must reflect the need for conformance to requirements—not conformance to an arbitrary standard, or to someone's desires, but to the real requirements of the product. If not, specifications will not be treated as "law" but will be looked on as "guidelines." Once this occurs, interpretation of specifications will be variable rather than consistent, deviations from specification will be frequent rather than infrequent, the end product will be nonuniform, and quality will be inconsistent.

The overall effect will be that specifications, as guidelines, will be subverted at every opportunity. Individual interpretation will frequently occur, as could this scenario. A shipment of parts is received and is slightly out of specification. Quality control has rejected them, but manufacturing needs the parts. So purchasing approaches product engineering (which has the last word on specification deviations). If the first engineer refuses to sign the deviation, a second is tried, and sometimes a third, in order to find a way around the spec. The third engineer may not be aware of why two other engineers have refused to sign the deviation, so he signs it, to keep production going.

The result could be insignificant or chaotic. If the part is actually usable, production continues with no problem. If the part proves troublesome, production is halted, and after the fingerpointing stops, it is clear that everyone is "clean" because engineering okayed the deviation, and the part should have been okay to use.

This example is not intended to criticize any group in a company for taking steps to keep production moving. Rather it is a criticism of an attitude where specifications are treated as guidelines rather than laws. As guidelines, they are subject to wide and various interpretations without penalty; but as laws, they entail responsibilities, and penalties if they are broken. Because specifications are vital to purchasing items from the outside as well as defining products and items on the inside, they must be realistic and enforceable.

Realistic Specifications.   A realistic specification considers these factors:

1. Specifications need input from more than one source. While there always is concern about anything developed by a committee, specification development requires a team concept so that serious constraints are brought out into the open before the specification is released.
2. Requirements of customers and limitations of processes must be determined. "Getting the key facts" is another way to phrase it. Tolerances, for example, should not be arbitrarily set by a design engineer without inputs from industrial engineering, quality control, and sometimes purchasing. There is a cost penalty associated with unnecessarily tight tolerances.
3. Critical characteristics and dimensions must be identified. Not all dimensions, tolerances, and characteristics are equally important or critical to the device's performance. Those that are must be identified and classified as to criticality. When product integrity must be protected, it is vital to know what is critical, what is important, and what is merely desirable.

Considering these factors, a realistic set of specifications can be written. Once written, they need to be reviewed and approved. Once approved, they are "the law." This requirement must be fostered by the engineering manager, his peer group of managers, and the division general manager.

Adhering to this tough requirement will result in a number of benefits over time. Vendors will find that they must either meet the specifications or have their product returned. Internal production is similarly constrained on parts made in-house. While enforcement is tough and occasionally painful, it is done with the knowledge that the specs are realistic. The minimums are minimums. Deviations are likely to result in assembly or customer problems.

Specification review must be more than a tally of various specifications to make sure they have been written. It must be a last check that realistic specifications have been generated, there is reasonable consensus, and they have been approved.

Drawings and Reports

Documentation includes specifications in the broadest sense, but in this chapter specifications are described in a more narrow sense that allows separate consideration of the other major areas of necessary documentation, namely, engineering drawing and reports.

Engineering Drawings. These drawings must be complete and unambiguous before the final design review; so a thorough study and check are needed before release. All mistakes, errors, and omissions must be eliminated. While it is not always achieved, this goal must be attempted. Errors of any kind built into the drawings will cause future problems.

In checking drawings, the first rule is that the checker be someone other than the one who made the drawings. The person who made the drawings will likely miss errors because he or she knows the parts too well. The checker, however, must be familiar with the field of work involved and must have enough experience to understand what has been drawn.

All calculations must be checked, not just dimensions and tolerances. Calculations should be checked by a second method. One such approach is the inverse or backward check. One takes the answer and calculates backward to see if the original figure can be obtained.

The checker needs to look for two general things—correctness of the information on the drawing and necessary information that has been omitted. Experience enables a checker to determine the second of these, which can be as important as the first. While a drawing should not be covered with auxiliary views, explanatory notes, and references, enough of these must be used to make the drawing complete.

Dimensions must be checked, not only for accuracy but for usefulness, by the machinist or fabricator making the part. If a series of additional calculations must be made to determine a dimension to be measured, the drawing is inadequate, and additional dimensions must be added.

After the drawings have been checked, and errors and omissions noted, the marked prints are returned to the drafter for correction. A competent drafter will recheck the checker, rather than blindly make corrections. In some instances, the checker may be wrong because of unfamiliarity with the part. The differences must be reconciled before checking is complete. The initials of the checker on the print indicate that the drawings are ready to be released. If the drawing is made with fabrication in mind, and checked with the same premise, a good shop drawing will result.

Reports.   It is important that reports be completed, reviewed, and routed in a timely manner so that technical information can be received, reviewed, understood, and challenged by the other members of the engineering department. The design engineer manager must insist that the reports be complete and up to date. If not, his engineers will get started on other projects, and despite good intentions, the reports will go unwritten after the product is in production.

SUMMARY

The review process is one of retrospection, looking back at what was done to see how it met the requirements and how it will impact the future. Design reviews are the backbone of project success, provided they are done well and effectively. The space programs have emphasized the key role that good design plays in the development of space "products." It is no less important in commercial product development. Reliability engineering and quality play important roles in bringing a product to market that meets customer requirements for usefulness and value. Specifications and other documentation are the vehicle by which engineering ideas can be translated into the product for the customer. They also must be done completely and concisely to make a viable end product.

10
Communications

INTRODUCTION

Effective communication is the goal of every human and living creature interaction. Much has been written about communication problems because many attempts have been made to improve communications where they are faulty or inadequate. Fostering, developing, and maintaining effective communication with his subordinates is an important leadership function of an engineering manager. This chapter deals with the essentials of communications—oral and written—for the design engineering manager, and includes ideas for better listening.

Communication is an essential part of a functioning engineering design group. Though it comes in many forms, a communication is not established by the sender transmitting a message. The receiver must receive the message, understand it, and in some way feed back information that the message was not only received, but was understood.

Feedback in communication can be in the form of a question, a request for clarification, a restatement of the message, agreement with everything, or disagreement. Without some type of feedback, the information sender has no idea whether the message has been understood, or even received.

In oral communications between manager and subordinate, there is always the danger that the message will not be understood; but the subordinate will not wish to appear stupid, and so will nod or give some other indication that suggests understanding. Sometimes a perceptive manager can pick this up, sometimes not. One test is to ask the subordinate to rephrase the instruction in his or her own words. Another is to wait for results and see just how well the message was understood.

After a time, the manager will become aware of which employees grasp messages quickly and which do not. The perceptive manager will also analyze his own communication techniques and learn whether he is making it easy or difficult to understand the messages he sends.

EMPLOYEE CONTACTS— REGULAR AND RANDOM

In order to know what is going on in his department, the manager must have regular communications with each of his subordinates. The regular and special reports needed are discussed later in the chapter. This section refers to meetings with the manager and subordinates to discuss various aspects of all the ongoing project activities.

The manager should allow about one hour each week to meet with each of the employees reporting directly to him. This can be broken up into two or perhaps three meetings during the course of the week. The meetings have several key purposes. One of them is to see how well each subordinate is planning and executing his work. Another is to learn what progress or problems are occurring on the various projects.

A third, very important purpose is to demonstrate the interest of the manager in his subordinates. Doing this is a bona fide morale booster, for the manager frequently does not realize how important the interest he demonstrates by his visits is to subordinates. A typical complaint about managers is that they are so tied up reading reports and going to meetings that they never get out into their departments to learn what is really happening.

One hour per week may seem like a considerable amount of time. If a manager has eight or nine subordinates reporting to him, that means about 20% of his work week is spent in key employee contacts; but this is important, key usage of time. (Uncontrolled, of course, it could also be wasted, "shoot the breeze" time.) If on a weekly basis, preferably on the same day each week, the manager and each subordinate meet for business discussions, much can be accomplished.

For example, a review of the past week's activities is helpful, and a preview of the next week's major efforts is very important, particularly if project priorities change. Correspondence sent and received can be reviewed, and portions of test reports can be studied for clarification. To aid in keeping these discussions focused on the key areas, the manager should keep a notebook showing principal items to be reviewed, with space for a brief note as to their status and whether they should be reviewed again or not.

Many engineers have an aversion or at least a dislike for planning and reporting; but focusing on a list of items reminds them of the areas the manager believes important, and subtly suggests they take a similar approach with their subordinates.

Weekly meetings with key subordinates can also provide the manager with early warnings of areas that are not going well, perhaps in the engineering area, another product section, or another department. Sometimes awareness of inventory shortages or manufacturing problems has not surfaced. The manager who becomes aware of potential problems before he is faced with them has an opportunity to think about them and prepare contingency plans to handle them.

Regular, communications-oriented meetings with individual subordinates have a strong positive influence on morale. The interest of the manager in specific projects or parts of them shows the importance and priority of the work. The manager's concern that the engineering work be done well and correctly puts emphasis on the quality of output. Whether the employee needs correction or the majority of the work is well done, the positive influence of the manager's coaching will make the subordinate a better engineer.

Another benefit of the regular meetings is that the manager learns firsthand which are his better subordinates. The person who writes the best, most lucid report may not have the best technical handle on a project, or may not fully understand the key issues, the major problems. By the same token, the subordinate who is the best face-to-face communicator may not be the best project engineer.

Questions and Answers

The manager must express his questions and comments carefully to understand what his subordinates mean when they give him certain information. Often a subordinate will hate to admit he is having problems, and will go to great lengths to avoid telling his manager the unpleasant truth. The experienced manager can usually tell if an employee is telling the whole truth, skirting the edges, sugar-coating bad news, or lying. Some persons cannot disguise their nervousness when their words are not truthful. For example, if an engineer is asked, "How's the project going?" and he answers "Just fine," but his voice sounds weak or off key, or his mannerisms are nervous, his answer needs investigation.

At other times a subordinate will use sweeping generalizations such as "The project would be fine, but costs are up." In inflationary times, one might nod his head sympathetically as costs tend to rise rather than fall; but effective correction cannot be discussed without a better understanding of what the answer means.

The manager knows that statement could relate to higher project costs than planned, higher costs for outside test labs, higher model shop costs, or even a higher cost estimate for the new product than planned. So, to get at the real question he first asks, "Which costs, specifically, are up?" When he finds out that it is the cost estimate for the new product, he asks, "How much higher are they than planned?" With that answered, he asks, "Which parts specifically are high, and by how much?" Then he may ask, "Is it material, labor, or overhead that is responsible?'' In this process, the manager is asking for more detailed information.

This may seem like pulling teeth to get to the core of the problem, but it is necessary to continue to probe to sharpen up a subordinate's loose thinking process. If he does not know the answers to some of the questions, he must find them out; otherwise accurate assessment of the situation is impossible.

Besides enabling a manager to get at the truth, this questioning technique provides a learning experience for the subordinate. Gradually, he or she understands that the manager wants information to be precise and accurate. An employee will then self-monitor oral and written reports. He will train himself to report in the most accurate way, and will do so because his manager is really listening.

Listening to Subordinates

A key managerial communications skill is listening. A manager may spend as much as half a day in listening, but his time may be either completely wasted or unbelievably useful in finding out what is going on. By listening, he may learn answers to questions he has not yet asked, or pick up solutions to problems.

Before some ways to become a better listener are discussed, a short commentary on poor listening is in order. Listening is not the pause between making effective comments to a subordinate. Too often a manager "knows" what the answer is and cannot wait for the subordinate to stop talking, so he interrupts and passes on the information. Other times, what the manager has to say is so important that, impatient to speak again, he drums his fingers nervously as the subordinate finishes.

Another listening defect occurs when the manager listens with just one ear and shuffles papers or signs letters at the same time. The manager may truly be hearing every word the subordinate is saying, but he is not giving full attention to listening. He may miss at least half of the verbal and all of the nonverbal clues to a deeper meaning that the subordinate is trying to convey. The effect on the subordinate can be devastating. He doubts that the manager is really listening because it appears that signing the letters and reviewing them is a much more important activity.

While there are other ways to demonstrate lack of listening, the worst two have just been described—making one's own point without responding to the other's thoughts, and giving only partial attention to the speaker while doing other things. On the other hand, there are some positive things a manager can do to become a much better listener.

These listening techniques include concentrating on the message being delivered, evaluating the content of the message and not overreacting to the emotion-provoking words, avoiding distractions, and listening between the lines.

Concentrating on the Message.   There is an old story about the king who killed the messenger who brought him the bad news of a battle lost by his army. It is a natural tendency to be somewhat irritated with a person bringing unpleasant news and pleased by one bringing good news. However, quite often the bearer of bad news may also provide an explanation for, or perhaps a partial solution to the problem. Other times, the subordinate's message is not clear-cut and up front; it may be near the end, or in the middle of his comments. So it is up to the manager to concentrate on getting the message and asking appropriate questions if he does not get it.

Not Overreacting to Emotion-Provoking Words.   Sometimes a subordinate will use inappropriate words or phrases, such as "stupid mistake," "management goofed," or a slur on someone's ethnic origins. The manager's emotional reaction may come so close to triggering a spontaneous expletive of his own that his efforts to "keep his cool" cause him to completely miss the information transmitted. A good listener must refuse to be distracted by these emotional triggers.

If the manager writes down the word or phrase that bothers him (as if jotting a note), he can later review the phrase more calmly and decide why it makes him react as he did. The act of writing it down is in itself a "detachment" from the emotional impact of the words. At a later time the manager may elect to counsel the subordinate on his poor choice of words, but certainly not at the time of delivery. If he does it then, the message may be lost.

Avoiding Distractions.   It is difficult for a manager to avoid being distracted when listening to a subordinate, particularly if he is concerned about several major problems, and the subordinate's concerns are about a smaller problem than his own. Unfortunately some of the subordinate's words may trigger the manager's thinking process so that he begins to review his own problems and completely misses what the subordinate is trying to say.

At the end of the subordinate's comments the manager may have to ask questions to get back on track, or terminate the meeting abruptly to concentrate on his own problem situation. Incidentally, it would be better to do this than waste the subordinate's time, if the manager could not concentrate on the message.

The telephone is one of the most common distractions faced by a manager and a subordinate. Its very ringing may break one's train of thought. If the manager answers it, the entire meeting may be wasted. The caller could be anyone from the company president to a salesman who has a serious gripe or even just a humorous story to share with the manager. That "It'll be just a moment" phone call can easily be ten minutes or longer.

Telephone distraction can be avoided only by planning. The manager needs to instruct his secretary to take all his calls and say that he will call back. The company president or vice-president may insist on getting through, but most others will not. He can also have the phone company turn down the sound level of his bell, so that it is a soft buzz instead of a loud jingle. Since the phone call is both an interruption and a distraction, its negative aspects must be factored into a meeting where listening is important.

Speedy listening is still another possible distraction. People listen at a speed of nearly 500 words per minute, but talk at only one-fourth that rate. In fact, some people talk even more slowly as they try to emphasize their point by accenting every word. To avoid the strong temptation to let his mind wander, the manager should mentally summarize what the subordinate has said, reviewing the points that have been made; and he can ask himself what point is going to be made by the trend of the conversation.

He can also evaluate the speaker's evidence by mentally questioning it. Are the points accurate? Is this the full picture with both pros and cons? Or is he providing only the information that will prove a point?

Listening between the Lines. This questioning of meanings, using the difference between listening times and speaking rates, allows the manager to do more than merely listen to the words; he can watch the speaker and, by observing him, keep his mind open to evaluate meanings. In that way he is "listening between the lines." Do the speaker's inflections or tone of voice suggest something? Perhaps he seems open about the project in his choice of words, yet is using arm and hand actions that suggest the opposite. On the other hand, a speaker who is quite open and honest will often support his points with sketches on paper that illustrate the point more clearly than mere words do.

The need to have regular meetings with each key subordinate is amplified by the need to make these meetings meaningful. By skillful questioning the manager can learn what he needs to, and by skillful listening he can often obtain insights and answers to questions he has not yet asked.

Coaching Subordinates

Another aspect of regular meetings is that of coaching subordinates. Just as in sports, coaching is much more than telling a subordinate what to do under a specific set of circumstances. Coaching is a form of training that sometimes is given as a directive—simple and explicit. Other times it occurs in the manager's questioning of a recommendation to determine whether it is sound and appropriate.

Coaching is not accomplished by solving a subordinate's problems, even if they happen to be in an area where the manager has technical expertise. Often a subordinate will come in with a problem, perhaps a fresh one he has not evaluated, or an older one that he has been unable to solve. The manager should not attempt to solve the problem but should instead insist that the subordinate analyze the problem situation, attempt to determine its real cause, and then bring in several possible solutions.

In a follow-up meeting, the manager should review the proposed solutions and have the subordinate select the one most likely to solve the problem. Then the subordinate carries out and implements the solution. The subordinate may make some mistakes, and the manager may let him make them—up to a point—because mistakes also are a part of employee development.

The objective in subordinate coaching is to improve the employee's skills in the interrelated areas of problem analysis and decision making. Thus, if a subordinate says he has a problem, the manager listens, responds, and directs the subordinate to study and evaluate the problem. If a subordinate brings in data on three or four design variations (or problem solutions), the manager listens to them, and then tells the subordinate to rank them according to their value to the company—and to select the one with the highest value, unless it is clear that some other choice is better.

The manager must continually fight his own tendency to come up with rapid-fire solutions to his subordinates' problems. Good coaching requires that he use his skills and knowledge to help the subordinate develop strengths in these same areas. It is directed "nondirection" that forces the subordinate to generate answers instead of questions, that allows growth and development. Occasional mistakes must be recognized and dealt with as important learning steps to better performance.

These three facets of managing—questioning, listening, and coaching—must be a part of the regular weekly meetings between manager and subordinate.

INTERFACING WITH OTHER DEPARTMENTS

Interfacing is a key form of communication and interaction, whether it be between people or between departments. The interface between two people, groups, or departments is the boundary between them. For effective interfacing, the boundary should be a flexible zone, allowing information flow in both directions. Interfacing is discussed elsewhere in this book (see Chapter 1), but additional comments are included here because of its importance to effective communication.

In science, the interface is the boundary layer, which may be thick or thin, solid or flexible, permeable or impermeable. Chemical adsorption of a substance, for example, occurs if there is an adequate interface. Carbolac, one of the active carbon blacks, has a surface area of 1000 square meters per gram—close to a half acre of interface per gram of material. It is essentially an "all-interface" material.

In semiconductor electronics, crystals of silicon are doped with tiny, precise amounts of additive elements so that the electrical conductive properties at the crystal boundaries are enhanced and controlled. In mechanics, an interface between two metal blocks may be a thin sheet of rubber that compensates for the irregularities of the two materials and enables them to transfer loading effectively. This is especially true if one of the hard materials is ceramic instead of metal. Then, the flexible interface allows the brittle ceramic to absorb shock loads without being fractured.

These references to interfaces in science and mechanics are a background for considering similar interfaces between design engineering and the other segments of the business in which it operates. Interfacing is so important that the editor of this series of books has planned for a complete book on the subject. This section will highlight the importance of good interfacing between departments.

Interface Problems

A frequent problem that exists between engineering and other departments is that information is never transferred with enough time allowed for an effective job to be done. This is particularly true when drawings and bills of material are transmitted to manufacturing engineering for a new design. In a rigid, structured interface situation, the packet of drawings and bills is transmitted to the chief manufacturing engineer via the company mail. This may be the first time he has heard of the project, or seen the drawings. He may not have been previously advised of the deadlines, and may have his group heavily loaded with production problems and tooling improvements.

The results are predictable. Numerous large and small errors will be found in many of the drawings, and discrepancies in the bills of material. The work packet will be sent back to the engineering manager, with a terse summary of the mistakes and perhaps a few editorial comments. By the time the errors are cleared up and the packet is sent back, considerable project time will have been lost.

Improving the Interface

A much better interface, more flexible and less sharply defined, occurs when a member of the design engineering group has been designated as the interfacer. After an initial discussion by the two managers regarding the importance of the project and its timing, the interfacing begins. The design engineer meets on a regular basis with a designated manufacturing engineer well before the drawings and bills of materials are completed. They review the key parts and assemblies, from both a functional and a manufacturing point of view. Changes for ease of manufacturing are relatively easy to make.

Tolerances, manufacturing capabilities, assembly problems are discussed beforehand; prototype parts are used wherever possible. The manufacturing engineer is invited to the design reviews and progress meetings to better understand the device. Then, when the packet of drawings and bills of materials are hand-delivered to the manufacturing engineer, or his manager, the group is ready to move ahead with them. Errors and discrepancies are found, and are dealt with on a priority basis. As one would expect, the more flexible interface is much more effective. It recognizes the needs of both groups, factors in the project details before they are "cast in concrete," and allows for change and interactions that ensure the effective participation of both groups.

The Interfacer

Interfacing is facilitated by having the right person as an interfacer, one who works effectively with others, who can be both flexible and stubborn as the situation requires. However, good results can be obtained even with inter-facers who are less skilled and less people-oriented than this. Conscientious application of interfacing techniques and follow-through make the job effective, even though the individual interfacer may not be the best possible choice. With a superior interfacer, superior results are likely to be obtained.

REPORTS, DOCUMENTATION, PAPERWORK

The major output of the engineering design department is not new ideas, new products, improved products, or lower-cost products. The output of the engineering design department is paper! It is the many pieces of paper that comprise information on drawings, letters, reports, bills of material. While computerization in many of these areas is changing the amount of paper, information on how to make this, correct that, improve other areas is the output. It is true that much information is transmitted orally, in meetings large and small, and in telephone conversations; but the bulk of key information needed to run a design department and to keep track of how well it is being run is communicated in writing.

The proliferation of written information that can accumulate is staggering; the presence of the ubiquitous paper copier has simplified sending a copy of a memo or report to anyone who may have a vague interest in its subject matter, instead of routing one or two copies to a few individuals who need to see and react to the information. A paper explosion due to overuse of copy machines has occurred. Unfortunately it tends to retard information transfer, as people tend to balk at the inch or more of incoming paper that may be delivered each day.

Short of placing actual restrictions on the use of the copier, or the promulgation of copies, there seems to be no effective means of controlling the amount of paper copies generated. Individuals would rather be nonselective than to target copies to those who benefit most from the information.

One possible answer will be found in the computerized office, where letters will be filed in the computer's memory and will be retrievable on a screen rather than on a piece of paper. If correspondence is stored as electronic mail in a computer, files and desk drawers will be less likely to contain reams of paper that are read only once. The use of electronic mail has already begun in many companies, and it will expand.

An Engineering Information System

Since the design department's major output is information on paper, it is important that the information be clear and concise. The old acronym KISS could be updated slightly to "Keep It Simple, Succinct." The general types of documentation included are correspondence (letters), test abstracts and test reports, progress and status reports, summaries of business and technical situations, and more. In total, they comprise the engineering information system.

Such a system may be relatively structured or relatively unstructured. Structure includes numerical or alphanumeric systems for files of project correspondence for general or specific engineering areas such as product engineering and application engineering and files for patents, patent disclosures, and legal correspondence. Within a filing structure, the placement and retrieval of important pieces of information are easily accomplished; but without a system and structure, individual engineers will maintain their own files, which are much less accessible than the structured ones.

An engineering file system can be quite simple to structure. It can be based on project category, or only on the next available number. It can contain codes that enable one to predetermine where to search. The following listing shows such a simple system.

File no.	Purpose
G 84001.00 to G 84099.00	General files for miscellaneous correspondence in the areas of product engineering and application engineering by product line, patent disclosures, general product information. The decimal point divides the identification portion of the number from the subcategories used in projects.

The "G" stands for "General" category; 84 is 1984, the year the file was opened. Each year a new file folder is established and maintained for each category. Ninety-nine options may be too many general categories, with not all used; but since these are simply reserved numbers, that is not an important concern.

The next series of file numbers is used for projects.

Proj. no.	*Purpose*
P 84101.00–199.00	Product Line #1 Projects
P 84201.00–299.00	Product Line #2 Projects
etc.	etc.

This sequence uses a "P" for "Project"; 84 is again the year the project starts, and 101 is the first project in product line #1. Similarly, product line #2 projects begin with 201, and so on. The project number is maintained throughout the project's life, for several years if necessary. Thus, in 1987 or 1988, a scan of the incompleted projects quickly flags those that are older and have not been completed or are closed for some reason. Allowing 99 numbers for each product line assures that a number of years will go by before the 99th project is opened and the numbering system must cycle back to zero. If this is necessary, it can be done without causing confusion. The same number is, of course, used for correspondence files as for time tracking.

The system as shown is suitable for nine product lines, which should be adequate. However, if there are more than nine, two alternatives are possible. One, use a two-digit field for product line, which allows for 99. Two, use an "alpha" or letter designation for each product line, which allows for 26 lines. Either way should work out satisfactorily.

The system is well suited for computer monitoring of engineering time, or for manual monitoring. The decimal point may not be. If not, drop the decimal point, and separate the two numbers by a space, or an "X." These last two digits are useful in project cost tracking.

For example, if the project is broken down into experimental, prototype, pilot run, and early production phases, the numbers 01, 02, 03, 04 can be used, respectively, to define and track how much effort is going into each phase. This is relatively simple but is valuable for establishing both a track record and a data base for future project estimating.

A better use for the two digits is to keep track of the costs on each work package of a project. Since the work packages are well-defined segments of a project with a target cost, timing, and man-hour estimate, the two digits provide a way to track. Few if any projects contain more than 75 work packages, so it should be a simple matter to charge time to the appropriate work package, provided the individuals involved understand the whys and hows.

Thus, for project P8410100 (omitting the decimal point) the effort for work package #12 would be charged to P8410112. When the computer printed out the detail, it would also be programmed to print out the summaries for P84101nn, summing all of the work package costs from 8410101 to the last one, perhaps P8410199.

This simple numbering system provides a method for handling and tracking a large number of projects of an engineering department in a systematic and simple fashion. Only a small amount of training is required to understand and apply it, and relatively little effort by a clerk or other timekeeper to monitor the information going to the computer. If the department has no filing system, or an inadequate one, the method just described can be a significant improvement in information handling.

However, setting up a filing system for engineering correspondence is only a basic step in the process of managing engineering paperwork in a usable information system.

Engineering documentation includes progress reports, test reports, general correspondence, patent-related correspondence, cost information, and new product ideas. While many documents can be generated for review and comment, only the vital data on the key projects should be automatically generated and distributed frequently. All others should be developed as needed.

However, it is important that general reviews and brief summaries of all projects be written on at least a quarterly basis, to determine what has been done and what is planned for the next period.

Writing Engineering Reports

Technical reports are the major vehicle by which the results of substantial engineering effort are correlated and presented. Sometimes engineers balk at writing them because of the work it takes to bring all the necessary information together, check all the results, and summarize everything in a cogent report that is both well organized and readable. They may complain that writing reports takes away from their time to do engineering and keep their projects on target. This is a writing cop-out; such views are incorrect. A well-written engineering report enhances the image of the engineer-author, and, whether it represents good news or bad, demonstrates the performance of the engineer on the project as well as the performance of the device.

Many engineering reports are difficult to read because they are dull, filled with ponderous detail, and written in stilted, passive-voice prose, where each sentence resembles the previous one in form. Few reports contain short sentences for variety. Too many are collections of long sentences, containing qualifying thoughts, hedges against taking a strong positive or negative reaction to the data presented. In addition, their structure and orientation are designed to discourage the reader as much as the writer.

For example, test reports are often organized under the following headings*:

Background and Introduction
Purpose of Test
Description of Equipment and Test Samples
Discussion of Test Procedures
Statement of Results
Discussion of Results
Conclusions and Recommendations

Although this structure is recommended in many colleges and companies, it does not take the reader into consideration. Using it, the writer will normally follow the format as he or she writes each part, and will get so involved in describing equipment and samples that these areas tend to be overemphasized and make unrewarding reading. The test results may then be briefly stated and the discussion even more briefly, with the conclusions and recommendations almost nonexistent monosyllables.

The reader will often jump to the conclusions and recommendations first, rather than wade through a mass of uninteresting information. Turning through pages of data and graphs may not appeal to him, and he may ignore much of the rest of the report after reading the end.

Reverse Sequence. One way to avoid this structural problem is to reverse the order of presentation completely. The report would follow this sequence of topics:

Conclusions and Recommendations
Discussion of Results
Statement of Results
Discussion of Test Procedures
Description of Equipment and Test Samples
Purpose of Test
Background and Introduction

This resembles the layout of front-page stories in a newspaper, which puts the most important material first, to get the reader's attention quickly.

---

* Reprinted with permission from *Effective Writing for Engineers–Managers–Scientists* by H. J. Tichy. Copyright © 1967 by John Wiley & Sons, Inc.

Reading a technical report with this format, one can continue reading until he has covered the main points of interest. He will know that the remaining material is less important. The main disadvantage of this structure is that the reader must know and understand the purpose of the test in order to understand the conclusions and recommendations. If these sections are difficult and complex, the reader may ignore the entire report.

A Combined Approach.   By using a combination method, the writer can develop a sound structure and organization as well as maintain reader interest in the whole report. Such an approach incorporates the interest-catching method of putting conclusions first by placing a summary at the beginning, and then follows the conventional format. Structurally, it looks like this:

Summary
Background and Introduction
Purpose of Test
Description of Equipment and Test Samples
Discussion of Test Procedures
Statement of Results
Discussion of Results
Conclusions and Recommendations

The summary is necessarily brief compared to the contents of the report. Depending on the type of report, the writer may summarize the significant ideas in it and the results. Or the summary may be results, conclusions, and recommendations. A third option is to summarize only the key recommendations, conclusion, or result.

A summary "lead" allows a busy reader to study only the summary, and also lays the groundwork for a careful study of the report by those who need more information. If the summary is well written, the reader knows he has obtained the essential ideas and can read the entire report at his convenience. Placing important material at both the beginning and the end of the report catches the reader's interest early; it is then developed anew as he reads through the entire report. This type of report writing demands creativity and effort by the writer, but it is much appreciated by the reader.

Report Abstracts

An executive summary at the beginning of a major report helps top management to be aware of the major topics covered without reading each page. In much the same way, a test report abstract, routed or distributed to each engineering department member allows them all to be aware of the test programs and results without reading each report.

A test report abstract (see Fig. 10–1) is a one-page document that forces the author to be clear and concise. It consists of four parts: background, purpose of the testing, significant test results, and the next action to be taken. It contains the report title and the appropriate file and identification numbers.

Care must be taken in preparing an abstract, with its space limitations, not to be so general that specific results and their implications are left out. It is better to feature one major result that triggers action than to try to squeeze in a dozen results.

The abstract provides many advantages to the engineering department using it, including these:

- It is a communication medium for prompt sharing of technical information throughout a whole department.

TEST REPORT ABSTRACT

TEST REQUEST #	DATE	AUTHOR	FILE NO.
REPORT #			REF:
REPORT TITLE:			

BACKGROUND:

PURPOSE OF TEST:

SIGNIFICANT TEST RESULTS:

NEXT ACTIONS TO BE TAKEN:

Fig. 10–1. Test report abstract form.

- It is a vehicle for generating questions or comments relative to the project or the testing.
- It is a useful introduction to the test report for those who need to study it completely.
- For short test segments of a large program, it can substitute for a formal report.

The major disadvantage of the abstract in that it may be used too often as a substitute for a formal report. Then most of the details on test results and procedures may be lost in some engineer's file. This possibility is likely if abstracts are allowed to be two or more pages in length. Second, if abstracts are not reviewed by a supervisor before being distributed, some may be poorly written and difficult to understand. This reduces the value of all abstracts, so a measure of control is required.

Despite a few disadvantages, the test abstract is an excellent communication means to disseminate technical information promptly after the testing has been completed. Abstracts can be an important part of an engineering information system.

Reports to Top Management

Occasionally the engineering manager must submit a semitechnical report to upper-level management. While the manager may consider delegating the work to one of his key subordinates, this approach is risky. Too often the report will be overly technical and will not provide the information in a form that management needs. It is better if the manager prepares it himself.

When writing reports for upper management, it is important to remember that many upper-level executives do not have an engineering background, or have not used it for many years. Moreover, management is always keenly interested in ways to save time and money, and, of course, to increase profitability. Taking all of these factors into account helps one to better structure the report and its contents.

Structure of a Report to Management

As executives are extremely busy people, the report should be structured to provide them with useful information and capture their interest. Its structure is relatively simple:

Conclusion and Recommended Action
Supporting Data
Reinforcement of Recommendations
Ideas for Future Investigations

The report might begin with, ''This report shows how RST Company can increase profitability by 10% to 14% by redesign of the electronic sensing product line.''

After stating such an attention-getting conclusion with the action needed to achieve it, the writer introduces the supporting data. He then reviews the results of studies and work completed, using subtitles that consolidate each thought and emphasize points of interest. To enhance readability, he uses many subtitled sections rather than a few. Illustrations, diagrams, and drawings can be used to reinforce major points effectively. They need to be kept uncluttered, clear, and to the point.

The conclusion may be only a paragraph or two, recapping what was stated earlier and tying everything together logically. Quite often additional studies are needed, either to further improve on what has been started or to chart a course toward a different new product or process, perhaps triggered by the study just completed. These ideas for further study belong in the concluding section.

Although the report needs to be relatively nontechnical, sometimes technical information must be included, particularly if management asks other managers for comments on the report. The writer places all detailed technical data in the back of the report, as an appendix. It is a mistake to put it in the middle; there it may only confuse the reader and cloud the major issues being addressed. Placed in an appendix, the technical information is available as needed, but does not intrude on the main message.

Writing and Editing the Report

Several drafts may be necessary to clear up and properly edit the report. In an early stage, one may get carried away with words and use ten words to describe vaguely what five can handle crisply. Long, ponderous phrases are not used in speaking, so they do not belong in writing. One effective way to cut out excess, unclear words is to read each draft out loud, and listen to the words being spoken. More often than not, the ponderous prose, the lengthy phrase, becomes obvious and can be deleted without loss.

Both long and short sentences should be used, to increase readability. Using both active and passive voice adds life. Particular points are emphasized by posing them as questions at the beginning of a paragraph and answering them in its body.

The first draft frequently is too long, too wordy, too vague. However, if the major points are stated, editing pares them down. It is worthwhile to set a report aside for a day or two before re-editing it. Otherwise familiarity with phrases and concepts blocks one's critical ability, and mistakes that should be caught are missed.

A competent secretary can also be a help in writing a good report, if the manager is willing to ask for comments, and to use them where appropriate. Some secretaries are more attuned to listening and thinking about what was dictated than others. One secretary will listen to the words, phrase by phrase, and type then exactly as dictated. Another will listen to each sentence and relate its structure and content to the subject of the report, and thus will be able to offer constructive criticism of the report.

Whether reports are made clearer and more concise by reading them aloud, editing, and redrafting, or by secretarial input, it is important that they be written with the reader in mind. Logic must be sound, conclusions backed by fact, recommendations based on conclusions.

SUMMARY

Departmental communications involve transfer of information and feedback from each person receiving the information. Frequent, planned contacts with each subordinate are necessary for the manager to know what is going on in his department and where the problems are. These exchanges require a search for truth, whether it be pleasant or not, and listening, to evaluate everything that is being said. The overlap of one department's output into another department requires development and maintenance of good interfaces. An informal, early, "mushy" interface is a much better information-transferring method than a formal, fully structured one.

Since information is the main product of an engineering department, internal systems for sorting, filing, and retrieving information are important. It is of no value to have written thorough reports on a key topic and be unable to retrieve them from the filing system.

Good engineering reports must be readable, interesting, and accurate. The traditional way of writing reports can be improved by providing a summary in the beginning to remove the mystery from the report and prevent the reader's tendency to read only the last page to find out what happened. Report abstracts contain essential information about a test and can be routed throughout a department to provide information without the risk of mislaying a bulky and valuable report. For short tests, abstracts effectively replace a long test report.

Reports to upper management are less technical, more bottom-line-oriented than engineering reports. They must be factual, interesting, and easy to read by a busy top manager. This requires thorough preparation by the engineering design manager. It is best that he prepare these reports because he knows the needs of his manager for the information being transmitted. Relating the financial benefits to the technical work accomplished adds to the managerial interest level.

The acronym KISS originally meant, "Keep It Simple, Stupid!" A variation for improved communications is "Keep It Simple, Succinct."

11
Meetings

Meetings are the best means of communication between people. They occur face to face, where reactions can be observed and assessed. Everyone receives the same information at the same time. Yet many meetings are a waste of time. They bore; they obfuscate; they demotivate. They are unproductive, tending toward trivia, bickering, side issues, daydreaming, and napping.

How often does one call an associate only to find that he or she is "at a meeting"? The meeting will end "sometime this afternoon," or "I don't know when." This situation occurs all too often in just about every company. It is not restricted to engineering; long meetings also take place in marketing, manufacturing, quality control, and purchasing, without exception.

There is no single aspect of running a business that gets more criticism than business meetings. Some of these criticisms are:

- Meetings begin at 8.00 A.M. sharp and end at 5:00 P.M. dull.
- A meeting is where people go to listen rather than to work.
- A meeting is where people talk about the work they should be doing.

Unfortunately, most of these criticisms are true. Too many meetings are ineffective, disorganized, and poorly run, and the cost of an unprepared meeting is excessive.

COSTS

Poorly run meetings are costly. They are expensive from both a direct and an indirect standpoint. Consider one meeting with six participants, each making a salary of $30,000 per year. The average hourly salary is $14.50 per person; fringes and overhead of 25% or more will increase the cost to at least $18.00 per hour per person.

For six participants, the meeting costs the company $108 per hour, higher than the cost of many consultants. If the meeting lasts five hours, it has cost the company $540. Relating this to business conditions, if the company makes a 5% after-tax return on sales, sales of $10,800 would be required to generate enough profit to cover the costs of this one meeting. If this group meets for five hours every month, the annual cost is nearly $130,000 in terms of product sales. Multiply this by the number of groups that meet for various purposes, and the costs quickly become astronomical.

High as these costs may be, there is a second indirect cost factor—the time of key people that is wasted in meetings when they could be solving problems. Pareto's Law (see Chapter 7) reminds us that 20% of the people do 80% of the work. If too many of this 20% of the design group attend too many meetings for too many hours, engineering productivity will drop too much.

A well-run meeting requires careful preparation, effective execution, and thorough follow-up. With these three characteristics, the meeting can be as short, or as long, as needed to cover the essentials and reach the necessary conclusions.

EFFECTIVE MEETING PREPARATION

Poor preparation is one of the major causes of poor meetings. Conversely, good preparation not only causes good meetings; it may prevent unnecessary meetings from being called in the first place. The best way to prepare for a meeting is to consider the six key questions—why, what, who, how, when, where—and add a seventh—what if (what if we don't have a meeting at all?).

Why

Why are we having a meeting? Here are ten possible reasons:

1. To receive reports from participants.
2. To reach a consensus prior to a decision.
3. To discover, analyze, or solve a problem.
4. To gain acceptance of a program, idea, or decision.
5. To achieve training objectives.
6. To reconcile conflicting views, or clear the air.
7. To provide essential information for project guidance.
8. To assure equal understanding by the staff members.
9. To obtain immediate feedback when rapid response is essential.
10. To take up a matter stalled in normal channels.

What

The specific subject matter and the objective of the meeting comprise the "what" of a meeting. The objective should be clear-cut, and focus on an essential few items. It could be a problem that needs creative inputs or review. Only specific aspects should be covered. Attempting too many topics in a meeting leads to confusion and frustration.

For a problem-solving meeting, this could be the general "what" plan:

1. Problem statement.
2. Statement and review of the facts.
3. Review of objectives of project.
4. Review of possible solutions in view of facts and objectives.
5. Selection of potential solutions for implementation.

Whether this is done in depth, or on the back of an envelope, defining the "what" serves two useful purposes. It develops a pre-agenda for the meeting and crystallizes either the need for a meeting or the awareness that a meeting is not necessary.

Who

After the agenda has been set, the task of determining who should attend the meeting follows. Without a doubt, the list should include only those who can best contribute to, or benefit from, the objectives of the meeting. If at all possible, they should all be of equal status. Attendees should be the individuals who:

1. Must implement the decisions reached.
2. Have good ideas.
3. Can approve or disapprove the results.
4. Represent differing views.
5. Can attend the meeting.

If the right people are unable to attend, the meeting will be a failure. Also, if a meeting discusses a complex subject, there should be fewer people present to better understand the issues. As the number of attendees increases, the probability of reaching a consensus decreases.

As a general rule, the number of people should be as small as possible, consistent with the objectives of the meeting. Few is good; many is bad. With too large an attendance, individual participation will not occur; there will instead be inattention, subvocal side discussions, and wandering away from the subjects. If attendees cannot participate, many will feel their opinions are unimportant and not worth presenting.

How

The agenda determines how the meeting objectives are presented to the participants. A written agenda is important to planning and running a good meeting, and equally important to the participants, who know in advance what is going to be covered.

All too often, people go to a meeting with only a general idea of the agenda, uncertain whether they need to bring along any reports, prepare any comments, and so on. The meeting leader glances occasionally at a few notes scribbled on a piece of paper, or on the back of an envelope. Nobody knows when the meeting is over until the leader makes a brief summary, or says "That's it" and rises from the table. An agendaless meeting is a voyage into unknown waters, a walk along a foggy path, an ill-planned venture that is usually unsuccessful.

The meeting leader needs to prepare and distribute an agenda prior to the meeting. The agenda should be as complete as possible, but even an incomplete agenda is far better than none at all. The members have some idea of what will happen and prepare for it accordingly. It is also a good idea to send the agenda out prior to the meeting so there can be some feedback from the attendees. It should be distributed only a few days before the meeting—with time enough to review and understand it, but not so early that it is forgotten and mislaid.

An agenda is a preview of the meeting, so the meeting chairman needs to lay it out with a logical flow of ideas in mind. If the meeting will cover a specific topic, that objective should be clearly stated, at the beginning. Following that should be listed as sequence of steps to define and explore the objective, and reach a conclusion or recommend actions for attaining it. Figure 11-1 is a sample agenda that illustrates this approach. Note that several individuals have been selected for brief presentations on various aspects of the subject.

It is important not only that these participants be forewarned of their assignment, but that the meeting leader (in this case the engineering manager) contact each to discuss the program and find out each person's thoughts about the topic. It is of little value to have an opponent of training programs discuss their advantages—unless, of course, there is some reason to believe that in studying the advantages the "opponent" may change his or her views.

There are several advantages to preparing and sending out an advance agenda. It generates thinking, improves understanding, and gives participants a chance to prepare for the meeting. Coming to the meeting, they are aware of the purpose of the meeting, are motivated to participate, and will derive more satisfaction from the meeting itself.

There are few disadvantages. One is the possibility that opinions can crystallize before general discussions occur and an impasse will be generated

4 June 1985

TO: DISTRIBUTION LIST

FM: P. GARDYK

RE: TRAINING PROGRAM PLANNING MEETING

WED. 12 JUNE 1985, CONF. RM #4

9:00–10:30 A.M.

SUBJECT: Shall we establish a formal training program for new engineers?

AGENDA ITEMS:

1. Past experiences with training programs (ABC)

2. Advantages of division-wide training group (CRT)

3. Disadvantages

4. Alternative approaches (RJB, JRM)

5. Costs expected—2-year period (MAC)

6. Recommended action

Distribution: A. B. Cassel   J. R. Miski   R. J. Bronars

C. R. Thomas   M. A. Connors   H. G. Aspen

Fig. 11–1. Meeting agenda letter.

instead of progress toward a discussion. However, this is a small disadvantage compared to the real advantages of an agenda.

Where, When

The agenda defines the location and time of the meeting, but a few more comments are necessary. The location should be convenient to all or most of the participants, and should of a size appropriate for the number of participants involved. Using a huge room for a group of six is a poor idea, as is crowding twelve people into a conference room designed for six. The room should have facilities appropriate for the discussions—flip charts or white board for illustrating points, transparency or opaque projectors with a screen for visuals, adequate lighting and ventilation.

The time of the meeting is important. It should be set by a consensus of the participants. Not only does this indicate that their presence is important to the meeting; it also ensures that they will not plan business trips then.

The time of the week is somewhat important. Monday morning meetings, or meetings held first thing any morning, are not the best idea. Frequently a supervisor has been thinking about a problem over the weekend, or driving to work in the morning. He may want to confer with a subordinate, either to give some directions, to share a new insight, or to learn how things are going. "Locking him up" in a meeting before he can make his calls will mean an inattentive, preoccupied participant.

Late afternoon meetings also are not the best idea because many people are tired and will not give their best attention then.

The worst time for a meeting with long discussions is after lunch. This is particularly ill-advised if the room is darkened for the presentation of slides. It is highly likely that someone will get so relaxed that he will fall asleep. Sometimes this is humorous, sometimes embarrassing; but always it means a loss of attention to what might be an important part of the meeting. Avoid meetings after lunch unless the discussion will be lively and spirited.

The best time for meetings are generally midmorning and midafternoon. Early morning meetings held in the middle of the week are usually good also.

The meeting duration should be as short as possible. The manager should not hold a 90-minute meeting to cover 30 minutes of material—unless the meeting will end in 35 minutes and people will feel exhilarated at accomplishing something important in such a short time. Too often, Parkinson's Law prevails, and discussions expand to fill the time available.

Long meetings are sometimes very necessary—reviews, for example. If well structured and paced, the meeting will move along, and the time will be well spent. If not, the meeting will drag, people will excuse themselves to make important phone calls, and the meeting may get out of control.

Allowing the right amount of time for a meeting takes some practice. It is better to err on the short time side in order to keep the meeting moving.

What If

What if, after doing all the preparation and planning, the meeting chairman telephones a few participants and finds that a controversial issue was settled once they reviewed the agenda and thought about it? Maybe no meeting is necessary. Wonderful! If a few more phone calls show that a consensus does indeed exist, there is no reason to have a meeting just to validate that conclusion.

The chairman writes a memo to the appropriate people stating the consensus reached and the direction to be taken. He cancels the meeting room reservation and goes on to bigger and better things. It is seldom possible to avoid a meeting in this way, but it is a happy circumstance when it occurs.

These basic questions provide the planning framework for an effective meeting. Figure 11–2 summarizes these points in checklist format. A simple "Let's get together a few people and discuss this" meeting may not require all the planning effort of a larger, more structured, formal meeting. However, time can be wasted and people frustrated just as quickly in a short meeting where the key people are somewhere else, the objective is unclear, and the agenda is haphazard, as in a longer meeting.

Planning for a meeting can be done with the six or seven key questions approach, even on a 3 × 5 file card or a napkin in the cafeteria. It is important to plan for each meeting, and not only because of the high costs of the time spent in meetings, or the tying up of people who could be doing their regular assignments. Plan each meeting because the participants will be so amazed at attending well-structured, focused (instead of pointless) meetings that they will look forward to them and participate well. These meetings could even be an aid to career advancement.

CONDUCTING A MEETING

Conducting an effective meeting is much easier if it has been planned. The effective meeting leader can be compared to the orchestra conductor who brings out the best capabilities of the total group, with everyone participating.

Most meetings have similar structures. Generally the meeting leader or chairman opens the meeting (on time), provides some background discussion, and quickly gets to the objective of the meeting. Various participants are called on, or volunteer, to present a set of facts and opinions relevant to the central issue or objective. Discussions usually follow, and the group is asked to vote, or in some other way to arrive at a consensus.

While most meetings are run this way, some get out of control and run away on a tangent or run amok on an inconsequential point that no two participants can agree on. When this happens, the meeting leader must be blamed because he or she did not keep the meeting running on course.

As in almost all planning, there are a number of things the meeting leader

1. Why—Is there a good reason to have this meeting?

2. What—What are the major topics to be covered? Are they limited in number? Are they clearly defined?

3. Who—Who are the key people necessary to make this meeting successful? Can they attend?

4. How—How are the objectives presented in the agenda? Did the participants help develop it?

5. Where—Has the meeting room been reserved? Is it appropriate for this meeting?

6. When—Is the meeting limited in time? Is there adequate time for discussion?

7. What if—Is is still important to have this meeting?

Fig. 11-2. Planning checklist for effective meetings.

must and should do during the meeting. The most important one is to keep the meeting focused on the agenda and bring it back into focus each time it starts to wander. At the same time, the leader's role must be minimal. The more he or she talks, the less others will express their views. The meeting participants must do the talking and the discussing. Occasionally, though, someone will state his points in a confused manner that no one understands. Rather than ask if there are any questions, the leader should clarify the obscure points by asking that member a few key questions, or rephrasing his statements clearly, without condescending to the member.

All contributions must be accepted and acknowledged, but not with flowery or extravagant praise. Differences of opinion will and should be encountered; that is the purpose of having a meeting. However, hostility and name calling must be prevented, or stopped assertively, if they occur. The individual must be called on to cease and desist immediately.

In this context, each participant must be allowed to speak without interruption. Questions may arise, but if the speaker is distracted by an impromptu question from the floor, his train of thought may be lost. It is far better to lay down a no-interruption ground rule at the beginning of the meeting, than to try to proclaim one as the meeting goes out of control.

Another form of interruption is the private conversation that two members may have while someone else is speaking. Out of courtesy, this should not occur. If it does, the meeting leader must insist on quiet, or more specifically must ask those offending members to stop their conversation.

The meeting leader generally has an idea of what ought to be discussed in the meeting. To make sure pertinent points are raised, he should have a list of cue comments or reminder questions jotted down to use appropriately. Sometimes a participant forgets to make a point, and this reminder will enable him to complete the discussion.

Someone should take notes during the meeting. If the leader can do it without losing his effectiveness in running the meeting, fine. If not, he should assign someone to perform this important function; and the leader will then jot down a few key ideas as they occur.

If a matter needs to be voted on—by either group consensus, show of hands, or secret ballot—the leader makes sure all the members have enough information to enable their intelligent vote. A consensus is preferred, but if the matter clearly is controversial, a vote is indicated. When the meeting leader is satisfied that all of the important points have been raised, and people's views have been heard, he makes a verbal summary of the discussion, clarifies the principal ideas and differences developed, and puts the question at issue to a vote. Once that has been completed, he and the group recommend how best to implement the decision reached.

To summarize, the meeting leader must be a leader, a conductor, a controller. He must state the issues, follow the agenda, and keep the meeting moving in a timely manner. He must let everyone comment or discuss the subject, but must bring the group back to the issue if they wander. He summarizes all the discussions, takes a vote if one is needed, and ends the meeting on time.

Conducting an effective meeting is a learned experience. Some groups will be docile, others quite headstrong and demanding. Skill in conducting meetings can be learned only by conducting meetings according to a plan and adapting the plan to the realities of the situation. The ideas developed in these pages should make the learning period shorter.

12 JUNE 1985

TO: DISTRIBUTION LIST

FM: GARDYK

MEETING MINUTES—SHALL AN ENGINEER TRAINING PROGRAM BE ESTABLISHED?

RE: 11 JUNE 1985

SUMMARY OF DISCUSSION

1. The four major advantages of a division-wide training program are:
   (1)   , (2)   , (3)   , (4)

2. The three disadvantages are: (1)   , (2)   , (3)

3. Cost estimate for first two years and staff requirements are detailed on attached supplement.

Votes taken: (1) To set up training unit   5 Yes   1 No

   (2) Personnel to interview applicants immediately

   4 Yes   2 No

Assignments: RJ Bronars will obtain . . . .

   CR Thomas will assist . . . .

Distribution: A. B. Cossel   J. R. Miske   R. J. Bronars

   C. R. Thomas   M. A. Connors   H. G. Aspen

Fig. 11-3. Brief summary of a meeting.

POST-MEETING FOLLOW-UP

A record of the proceedings of most meetings is a necessity, particularly if a decision results and assignments are given for its implementation. Minutes should not be a comment-by-comment recap of the meeting: "He said, then she said, etc." Rather, the minutes summarize the results of the meeting in a manner similar to the meeting leader's summary that he gives before taking a vote.

Figure 11-3 illustrates a brief meeting summary. The leader should write the minutes himself and make sure they accurately cover the meeting discussions and decision. The format of minutes will vary between organizations and departments, but everyone appreciates receiving a concise meeting report.

SUMMARY

A business meeting can be an effective tool for face-to-face transactions in key business areas. Many meetings fall short of this realistic goal because they are planned inadequately, if at all, and conducted poorly. Spending time in initial planning of a meeting allows the manager or meeting chairman to focus on the objectives and major areas to be covered. Sometimes by pre-planning, the leader determines that a meeting is not necessary. The agenda resulting from pre-planning is itself the plan for the meeting. Successful meetings are conducted by "working the plan" and being responsive to what actually occurs during the meeting.

Conducting a good meeting is a learning experience, reinforced by good experiences and modified by bad ones. Most meetings run overly long because the meeting leader does not know how to, or chooses not to, curtail discussions or keep them focused only on the agenda. To be effective, decision-making meetings require a decision, either by consensus or vote. In either way, an action is agreed upon by the members.

The post-meeting minutes provide a record of what was accomplished and what remained to be done. Any incompleted actions require follow-up by the meeting leader.

Meetings get a bad reputation in many companies because they are not done well. That reputation can only be changed by meeting leaders who do a better job of planning, conducting, and follow-up.

12
Publications and Presentations by Staff Members

INTRODUCTION

Communication of project status, problems, and progress is accomplished internally by written presentations such as memos, reports, and reviews, and by oral presentations. Communication to the outside world—to users, technical experts, even competitors—is accomplished via technical papers, articles in technical magazines, presentations to industry groups. In addition, technical writing may be needed to explain certain application approaches to the engineers of companies purchasing and using equipment. To serve all these needs, a certain amount of writing by staff members is both necessary and desirable.

However, most engineers are doers and do not want to be writers. A very few are both. Some who are doers can be encouraged to spend part of their time being writers. Readers of technical literature prefer that it be written by practicing engineers such as themselves, rather than by technical writers, unless they are expert at their craft.

The manager who can develop some of his engineers into occasional or even prolific writers will do well for the company and for the engineers. In doing so, he will enhance his own reputation as the manager of an effective design group that does its job well and lets others know about it.

This chapter discusses concepts of effective writing; the types of written communications that go to the "outside" world, namely technical papers and magazine articles; presentation tips for those giving talks or presenting papers; and what the design manager can do to promote and encourage writing by his engineering staff.

WRITING FOR ENGINEERS AND MANAGERS

Whether one writes a technical paper reporting R&D results, an article for a company journal or trade magazine, or any other publication, the fundamentals of good engineering writing are the same. Good engineering writing must be readable and well organized. Suggestions for readability are given in this section, while organization is discussed in the following one.

Readability

Above all, good writing is readable writing. That means it is clear, concise, and directed toward the reader. The writer must aim for his audience and neither overestimate nor underestimate its technical level. This is not an easy task, but is a necessary one that improves with practice. If the engineer is writing about new areas of technology, he must begin with the technology levels with which his readers are familiar, and work into the nonfamiliar areas. In that way, he makes the unfamiliar familiar, starting with knowns and working into unknowns.

The writing engineer should be willing to go into enough technical detail to support his findings. Readers who are engineers usually want enough detail that they can agree or disagree with the conclusions. Giving a superficial treatment of a technical subject, without much detail, is worse than not writing about it at all.

One effective way to enhance the readability of a technical article is to complete all of the illustrations before writing the paper or article. Each drawing, diagram, or circuit layout should be prepared before the words are written. The figures do not have to be in finished or final form but must be reasonably close to it. Any items of interest in the illustrations should be labeled for highlighting in the article.

The reason for doing the artwork first may not be obvious; but ask any engineer how a device or a circuit works, and the first thing he will do is reach for a pencil and something to write on—even a napkin or the back of an envelope. As he sketches the diagram, his descriptive commentary will flow, with his pencil moving ahead of his words. Thus, many technical papers can almost be the equivalent of a written (rather than oral) explanation of the functions of a design, or circuit, and what results occur under different test conditions.

The important relationship between graphics and good technical writing is usually overlooked in courses on writing, and especially so in the writing courses engineers must take in their undergraduate curriculum. Such courses tend to focus only on the form and structure of sentences and paragraphs, instead of emphasizing the interrelationship of written and visual information, a subject just as important as grammar.

ORGANIZING AND WRITING

A good paper needs organization. Initial preparation of the visual material is an important step in organization because it focuses the writer's attention on the main topic to be covered. Relating each graph or diagram to the whole, as discussed above, is a good start in organizing the paper. Two other steps are important also: developing an abstract or synopsis of the paper, and developing an outline.

Abstract

An abstract is a one- or two-paragraph summary of what the writer is going to say. It is a summary that includes the topic material and the writer's attitude toward it. In a less formal approach, the abstract is an introduction to the paper, where the reader gets his first insight into what the paper is about.

Outline

Any paper or article consists of three major parts:

- The introduction or abstract, which establishes the tone of the subject and the author's view (see above).
- The body, which is the meat of the paper, containing supporting evidence, test results, and findings.
- The conclusion, which summarizes the findings, draws some inferences from them, and includes the most important ideas the author wants the reader to remember.

However, the outline must be more than just the three broad parts—introduction, body, and conclusion. The body portion must be outlined thoroughly because there is a great deal that must be said clearly and in an evolving, developing fashion. There are many ways to outline, which can be found in books on writing. Some of them are good, others difficult to follow.

How to Outline. The following is a simple method, suggested by E. T. Thompson, editor of *Reader's Digest:*

- On 3″ × 5″ file cards, write all the points that need to be made, using one point per card.
- Divide the cards into piles, using one pile for each group of points that are closely related to each other. (If the paper were about a series of tests, all points about electrical tests would be in one pile, mechanical tests in another, and so on.)
- Arrange the piles of points in a sequence. Which are the most important? These points should be presented first or, in some cases, saved until last. Some points must be presented earlier so that later points will be understandable.

- Next, within each pile, do the same thing—arrange the points in a logical, understandable order.

Once this is finished, number the cards—in pencil—so that if they are disturbed, the sequence will not be lost. This is a practical and flexible outline. One can readily add, delete, or change the location of points. The location of each figure or illustration should either be marked on a separate card, together with its number (Fig. 1, etc.) and title, or it should be entered on the card that covers the particular topic.

Combining the illustration sequence and the card file outline is a powerful and accurate way to organize a paper or article. All that is left, before the actual writing, is to jot down some notes for the conclusion.

Conclusion.   The conclusion is a summary that wraps up the ideas presented. It includes the ideas the writer wants the reader to remember. It should not be overly long, nor should it be a retelling of what was developed in the body. Let the reader see the value of the points which he now understands. Then stop.

WRITING A PAPER OR AN ARTICLE

Writing is hard work that requires concentration—but so do many other jobs of an engineer. Just as one develops a plan for a project, he develops an outline for a paper. The engineer plans his work, then works his plan. In a similar way, the writer outlines his writing, then writes to his outline.

Writing the first draft may be fairly easy. One follows the outline and writes one paragraph after another, developing ideas, clarifying points, explaining the illustrations. Some will write the first draft in longhand; others will dictate it; a few will type it. In today's modern office, the aspiring engineer-writer may type it himself into his desktop computer using a word processing program.

In defense of a handwritten draft, one can say it is a time-honored method that allows one to think, ponder, and scratch out words as one goes along. The truth is that handwritten drafts are time-consuming and cumbersome. In addition they need to be typed by someone, who may introduce some errors because of illegible words, miss a line of text, or, in some other way, confuse an important point. Also, the typing will often be done on a time-available basis, after important letters and memos are finished. It may take a number of days to get the typed manuscript back. So, if at all possible, the writer should learn to type and put his drafts into a word processor, rather than prepare a longhand draft or dictate his ideas.

Revision

Most writers hate to revise what they have written. Revising is usually much more difficult than setting down one's ideas in the first draft. Revision consists of subjecting one's "perfect prose" to close scrutiny. It involves reading and rereading, checking for clarity, continuity, and aptness of thought. The mechanics of paragraph structure, sentence structure, transitions between paragraphs are very important and need to be studied.

The writer must ask himself such questions as:

- Is the topic clearly defined?
- Is it supported by facts?
- Are there irrelevant facts or extraneous material?
- Does the writing flow, or is it jerky?
- Does punctuation serve its purpose?
- Is there variety in sentence structure and paragraphs?

Another self aid in revising a paper is to read it aloud, preferably into a tape recorder. Even if it is not recorded, the act of reading it aloud, to one's self, is powerful. Using either approach, the writer can hear his own thoughts, and catch repetitions, omissions of key points, words that are disjointed, or ideas out of context. As he reads, he considers these questions:

- Are the sentences too wordy?
- Do the ideas connect to one another?
- Is it difficult to read?

Reading out loud, one finds sentences that are too long. They show up glaringly. Sentences with difficult word constructions or foggy meanings can be spotted quickly. Revision usually means a complete or substantial rewrite into a second or even a third draft, to make sure it is right. The aspiring writer/engineer will do this willingly to get his message across as clearly as possible.

Additional Helps in Revising.   The total process of revision also includes examination of paragraphs for type of development and length, and sentences for clarity, variety, length, and use of the correct words. The author also reviews the complete paper to make sure it is both readable and understandable—and that it is also interesting.

To improve both writing and revising, two books should be available and used, a dictionary and a thesaurus. A dictionary helps one determine correct word meanings; the thesaurus locates synonyms to avoid overuse of words or to provide new words that better express one's shades of meaning. A third basic book is *The Elements of Style* by Strunk and White. This thin paperback, less than 80 pages long, distills the essence of good writing. It covers rules of usage and composition, matters of form, and commonly misused words and expressions, and makes some choice comments on style. It is worth reading and using.

Various other books provide valuable insights into writing, especially technical writing and revising. One runs the risk of omitting equally valuable references by naming any. However, these two have benefited the author and may help the reader. The first is *Effective Writing for Engineers-Managers-Scientists* by Tichy, published by Wiley in 1967. It delivers the product its title describes. The technical person using this book will write better papers, reports, and business letters. The second is *Revising Prose* by Lonham, published by Charles Scribners' Sons in 1979. It focuses on revising one's writing, whether technical or nontechnical. Written in a light, tutorial style, it packs writing punch in its brief, 126-page content.

A book on writing is valuable if it enables one to write with greater fluency, directness, clarity, and brevity. Any prospective writer will do himself well by searching and sampling until he finds one or more books that help him do this.

Submitting the Article or Paper

When the paper is completed, what does one do with it? Depending on the paper, the company policy, and the internal technical writing department, the author or his manager may still have a lot to do, or very little. The following paragraphs will cover general concepts and guidelines that some companies have followed successfully. If one's company has no such guidelines, these can be used as a basis for establishing them.

R&D Reports, Technical Papers.   These two categories are described together because of their similar writing form and style. The R&D report is usually written to present findings to other technical people within the company to keep them apprised of new developments or directives in R&D. It frequently begins with a management overview to condense the information and highlight any significant findings. It has limited circulation, generally to the R&D group and to other engineering groups that may have an interest in the ongoing work.

Usually the R&D report must be approved by the engineer's or scientist's department head, in some cases by the R&D manager, before distribution and routing. The technical paper, on the other hand, does not stay within the company. Instead it is sent to the "outside world" where it will be read by potential customers and scrutinized by competitors—the former, to gain insights into what the new product can do for them; the latter, to glean information about technical progress achieved, by reading between the lines.

Thus, before the paper is written the subject matter needs to be cleared by the appropriate level of technical management. Any sensitive information, in areas where the publication of significant research or development work might be of greater value to a competitor than to a customer, must not be published. It could be deferred until later, when potential patents or new products are far enough along that the knowledge transmitted is timely for the company's image but not for the competitor's education.

A management review should be held, even on a paper that has been cleared for writing, once it has been written in final form. While a "hold" or an edict to delete certain portions of the paper will not be welcomed by the writer, it may be necessary, to avoid the risk of teaching competitors the results of research that may have cost the company hundreds of thousands of dollars.

One final thought on technical papers: many associations—the IEEE (Institute of Electrical and Electronic Engineers), for example—have a special form on which the paper must be typed and submitted. The forms can be obtained before one writes the final draft. If the writer is not certain of the need for special forms or formats, an inquiry to the association will save the trouble and irritation of last-minute retyping to meet a deadline.

Trade Magazine Articles.   Many magazines accept articles on new products, process improvements, design tips, calculation aids, management concepts, and so on. In some cases a magazine will accept a published technical paper and excerpt from it certain areas of interest to its readers.

The design engineering manager himself may write articles for technical magazines, or may encourage his staff members to write articles in certain areas that would interest technical readers. A clear policy needs to be established regarding who needs to clear the subject matter to be written about, and whether any working time can be spent on writing the article, preparing illustrations, or having the final drafts typed.

Some companies allow an engineer wide latitude on "nonsensitive" article writing, provided all the work is done on the employee's own time, without the use of the company facilities (word processors or typewriters). Others will allow a certain amount of time spent, or allow manuscripts to be typed by a secretary on a time-available basis.

If company facilities are used, should payment for an article be split between the company and the engineer, or should it all go to the engineer? The latter arrangement is preferred, since article payments are more like honoraria than substantial payments. A typical article of 12 to 15 double-spaced typed pages will be published as a 4-page article, for which the author may be paid about $150 by many trade magazines. Thus the payment is not so substantial that the question of who gets paid what is important.

The company benefits because its name is mentioned; the author benefits because he gets a byline as well as some extra dollars as a reward for his work. Because of this dual benefit, technical magazine writing should be encouraged.

Most technical magazine editors are willing to work with writer-engineers. A simple query letter to the magazine's general editor, outlining an article idea and how the writer intends to develop it, will receive a reply. The answer may be negative, if the magazine has published similar articles in the recent past; but if the idea is relatively new or not recently published, the editor will respond affirmatively and encourage the engineer to write and submit the article for probable publication.

In some cases, the final draft or manuscript must be substantially revised to bring out the essential ideas, but the editor still welcomes these articles. The writer, seeing his handiwork in print, can compare his typed manuscript to the printed version and find in which areas the editor "tightened" the writing, removed unnecessary words, sentences, and so on. This is a valuable learning experience for the engineer who wants to become a writer as well.

Articles for Company Magazines, and Technical Product Literature. Many companies support a technical writing group either in a marketing department or in advertising. This group writes and publishes descriptive literature on new products, product releases for trade magazines, catalog sheets and application guides for product use, and a company magazine. Technical writers generate the volume of words needed to support all of the documents to be published.

Quite often one or more writers will work with the design engineering manager to get facts and data for an information or application article the writer is planning. Many of these situations can be turned into a writing opportunity for the engineer, particularly if the subject is complex or detailed. Technical writers may not understand the intricacies of the design well enough to document it clearly without considerable review and editing by the design manager or design engineer.

Actually this is a desirable type of role reversal, in which it is preferable for the engineer to write the material and the technical writer to edit and polish it. Should this happen, the beginnings of a writer–editor relationship can be established within the company, and more writing by the engineer probably will be requested, since most technical writing groups are overloaded and understaffed. A more direct approach is for the design engineering manager to maintain contact with the editor of the company magazine or the head of technical writing services—often the same person.

Writing good technical articles is more complex than it seems. Often a solid technical explanation that does not read well can be trimmed and edited into a readable, factual article, or application guide, by the technical writers, using a combination of technical and writing skills. The manager who "volunteers" that he or one of his staff will prepare an article for the company magazine, or a first draft of a product application guide, is providing a service for all concerned—the company, the editor, himself, and the engineer who will do the writing.

MAKING PRESENTATIONS

A paper accepted for publication often must be presented at a meeting. The prospect of presenting one's findings to a group of technical peers and experts can be frightening and devastating to contemplate. But it need not be. Stagefright and butterflies in the stomach are natural reactions before one gives a paper. They are signs that the body is "cranking up" with a flow of adrenaline. Recognized and properly channeled, they can be useful reminders that a little (or a lot) of nervousness beforehand will result in a more focused, interesting presentation from the podium.

The keys to a successful presentation of any type of information to an audience are preparation, pictures, and practice. The extent to which all three are utilized prior to presentation will have a major effect on how well the presentation is done in front of the audience.

Preparation

Preparing what to say takes a lot of work. Many speakers can ramble for an hour and a half about a subject they know well, without much preparation. But a lot of effort must go into preparing an interesting, fact-filled 30-minute talk.

If the presentation involves a paper the engineer has presented, all the better. The paper must be reread, the references studied, and state-of-the-art information researched. Telephone contacts with experts on some facet of the subject may be necessary, to bone up on facts that may be needed to answer questions from the floor. Enough work must go into a presentation to cover what the audience wants to know and will ask questions about.

Keying the presentation to what the audience needs to know is important for a successful presentation. This is done to focus the presentation and prevent it from being an assortment of information on the history, growth, and details of a subject. Instead it becomes a clear exposition of a handful of key points that have value to the audience. If the speaker presents too much information, he will lose the audience's attention.

Pictures

Just as visuals are vital to a good paper, they are the key to a good presentation. However, visuals that can be handed out a piece of paper will usually be too detailed and cramped for an audience to follow and understand. More visuals are needed in presenting key information than in illustrating a paper. Visuals can be done via chalkboards, flip charts, slides, overhead projectors, movies, and working models. Several different types can be used in the same presentation, but must be planned for, budgeted and prepared in advance.

Audiences expect visuals, and visuals enhance the value of the presentation. Visuals can emphasize key points and show comparisons and relationships, to channel thinking. In general, a good visual aid *is* better than a thousand words! The advertising or technical writing department can assist in developing good visuals, as can outside services. Some companies sell products for the preparation of more effective visuals. This allows the engineer–writer–speaker to prepare visuals himself and create inexpensive effective aids

Practice

For a presentation to go smoothly, cover the main points, and be effective, it must be well practiced. Sometimes a talk can be memorized, but it is better if the main points are memorized, and the talk is allowed to flow. A presentation has an introduction, a body, and a summation. The old advice on presentations still applies: "Tell the audience what you're going to tell them (introduction), tell them (body), then tell them what you've told them (summation)."

The introduction sets the tone and sharpens the audience's listening level. The body covers the major points, complete with visuals and anecdotes. The summation summarizes, and leaves the audience with something to remember.

A presentation must be practiced several times in order to be good. The key points for the introduction, body, and summation should be set down—written or typed—on numbered index cards, $3'' \times 5''$ or $4'' \times 6''$. Then the presentation is given to an audience of one—the speaker in front of a large mirror. This is effective because the speaker by glancing at himself will see his facial expressions, his stance, his gestures or lack of them.

Another way to practice is to give the presentation, record it on a tape, and play it back several times to better remember the flow of the talk. It takes time to prepare a good talk. Time each reading, and strive for a 20-minute presentation. This is ideal. One hour is the maximum because an audience gets restless beyond that limit.

The Actual Presentation

The presentation itself can be much less traumatic if a lot of good preparation goes into it. But, just prior to the presentation, the combination of a lump in the throat and butterflies in the stomach often makes the speaker wish that the earth would quake and he would be mercifully dropped into an abyss.

As we know, disasters rarely occur, and almost all speakers manage to start their presentations, some smoothly, some not so smoothly. And once into the presentation, the speaker finds his voice becomes firmer, his butterfly stomach quiets down, and he has an interesting message to give his audience, who have come for just that purpose. The talk goes smoothly; the visuals illustrate according to plan. The speaker is actually disappointed to find that his talk is ending—he could have gone on longer. The applause afterward is a pleasant accolade.

Pre-presentation jitters are a natural sign that the body is gearing up. Being keyed up usually means the speaker will do a better job. WIth thorough preparation and a touch of self-confidence, the first-time speaker, as well as the more seasoned veteran of several talks, will be pleasantly surprised with his performance. He had something of interest to say to the audience, and they were interested in hearing it.

Using Jokes in a Presentations.   Most public-speaking instructors stress the value of humor as part of every presentation. It helps hold audience attention and adds emphasis to otherwise obscure points. However, unless the presenter has a natural gift for joke telling, he may feel insecure about his ability to deliver a punch line well—and he may do it poorly. His "best" gag may be followed by an embarrassing silence.

It is best if the jokes are relevant to the topic, and brief. If it is a talk to economists, one can deliver a one-liner such as, "A meeting with economists is where everyone says there is no such thing as a free lunch, while eating one." Sprinkling a few short humorous comments throughout a talk can earn a few chuckles and renew audience interest; this is much better than aiming for a huge guffaw or belly laugh at the beginning.

Stories usually should not be current ones; listeners who have heard them may not laugh. Should this happen, the speaker must just smile, shrug his shoulders, and proceed.

Because the element of surprise is basic to provoking laughter, let it be a surprise. The speaker should never alert his audience by beginning, "That reminds me of a story . . . ." Before delivering any humorous remark, the speaker ought to check it for sexism, racism, and personal prejudice, especially toward lower-level employees. After the laughter fades, the audience may resent a display of disregard for others.

THE MANAGER'S RESPONSIBILITIES
TOWARD WRITING AND PRESENTATIONS

Encouragement and support of engineers who take on the additional role of writer or speaker are the two main responsibilities of the manager. Not everyone can create a new product, invent something patentable, write an article, or give a talk. But with encouragement and support, the talented engineer who gives it his best effort will usually succeed. These creative contributions can only be encouraged and fostered. They cannot be made a bona fide requirement, or many will fail at it who would otherwise succeed.

Some engineers will attempt them on their own because they recognize that publication of technical articles and technical presentations are two of the best means for an engineer to advance professionally. Others need extra encouragement from the manager, plus some advice and counsel from the advertising or technical writers, or from other engineers who have already published an article or paper. Those who have published are almost always ready to help someone who aspires to do it. If the manager himself has an article or paper published, he can be sympathetic, understanding, and sufficiently demanding to bring out the writing potential in his staff.

SUMMARY

Writing technical articles in an unwelcome chore to many engineers (and their managers). But the design engineering manager who is interested in developing his promising engineers will look for such opportunities and encourage his people to write about their work, and to make technical presentations at industry meetings. As mentioned earlier, everyone benefits—the company, the manager, and, most important the engineer.

PART III
EVERYDAY WORKLOAD
OF THE DESIGN
ENGINEERING MANAGER
AND OTHER KEY AREAS

13
Everyday Workload
of the Design Manager

INTRODUCTION

Each day in the career of an engineering design manager is filled with activities, meetings, unplanned visitors, impromptu conferences with staff members. Each day is busy. The necessary busy-ness and the unnecessary busy-ness must be balanced. Sometimes the "unnecessary" or impromptu bits of business are more significant and far-reaching than is immediately apparent. And some of the planned meetings are little more than get-togethers to communicate what already has been accomplished.

The everyday workload of the design manager covers the gamut of everything involved in staying on top of the job. It includes handling priority shifts, motivating employees, looking for signs of obsolescence and turnover, and deciding the who, when, and how of promoting and assigning, and hiring and firing. It includes planning for merit reviews or appraisals, and financial increases. At the level of his peer managers and superiors, it includes some negotiations and politics as part of the everyday requirements of getting things done and putting one's self in a good light. It also includes searching for, but not always finding, a portion of quiet time to do some thinking and planning for tomorrow and the day after.

These are the major stimulations and pressures that comprise a design engineering manager's daily workload. It is an interesting, fast-paced day, which often passes so quickly that one wonders where the time went. Since no day could possibly include the entire variety of incidents that occur and must be dealt with, no attempt will be made to discuss them in a chronological sequence. Rather they will be discussed as a series of equally important factors that the design engineering manager must address on any given day.

STAYING ON TOP OF THE JOB

One of the responsibilities of a manager is to stay on top of the jobs in his department and at the same time effectively delegate responsibility and authority to his subordinates. This seeming paradox causes different managers to address the situation differently. One may delegate the job completely and lose track of what is going on and what problems are occurring until it is too late to manage them. Another may need to know every detail of the job, and so will tell the project engineer what to do and how to do it.

Both these ways are wrong, of course. The difficulty is in keeping one's hand in whatever one delegates without taking over the project and without losing track of what's actually happening. To amplify, the manager must be able to:

- Spread the project work among his subordinates.
- See that it is done correctly.
- Have enough time to take corrective action should something go wrong.
- Develop the talents and abilities of his subordinates.

While staying on top of the job is neither easy nor automatic, it is doable—and some managers do it better than others. There are several ways in which the manager can keep on top of his projects and successfully delegate. All of them require him to select a style of management that includes those four elements necessary for effective delegation.

In one of several ways, he must periodically audit what is going on. This audit, like a financial audit, is a check point on what is happening, compared to what was planned. The process of auditing enables the manager to establish where the project is, and where it is heading, and enables him to make corrections with a minimum of lost time, efforts, and results.

A number of methods can be used, singly or in combination, to delegate the work yet keep on top of the job. Five of them are: previewing direction, questioning progress, requiring written reports, scheduling conferences, and setting deadlines.

Previewing Direction

The manager requires his project engineer to submit a project or a job plan before he starts the project. In this way, the manager learns how the subordinate interprets the problem, how it will be attacked, and a detailed procedure that can be corrected or adjusted. This is equivalent to the project plan approach described in an earlier chapter.

Questioning Progress

The manager asks the "How's it going?" questions to ascertain progress and problems. It is the most important, yet the most revealing and time-consuming, method of "hands-in" delegation. It requires deft timing and a clear attitude of friendly interest, as well as careful listening, to both what is said and what is left unsaid. With good rapport between manager and subordinate, this method is extremely successful in maintaining effective project delegation and at the same time keeping the manager aware of what is going on.

Requiring Written Reports

This method requires periodic progress reports to be written by the subordinate, with or without specific deadlines. These save the manager's time, as he can read them at his convenience and respond to those areas where he needs clarification or more information. But the method's effectiveness depends on how well the subordinate expresses himself in writing and whether he includes all the pertinent facts needed to judge progress without making the report unwieldy. Deviations from the intended path of progress may not be easily spotted.

Scheduling Conferences

This style calls for formal oral reports at specific times in the future. It is intended to produce the detail of a written report but with greater clarity and directness. It can also have the positive features of informal questioning. When a project tracks on schedule, the scheduled conferences can be very useful. But if conformance to scheduled conferences is adhered to without deviation, it may be that the plan is being followed without any additional creative experimentation. This approach can eliminate innovative results because the subordinate is calendar-constrained.

Setting Deadlines

This can be the best of methods. It sets a time limit for successfully completing a delegated assignment. The only pressure it imposes is that of time. It allows the subordinate freedom of thought and actions, gives the subordinate authority to act on his own, and reduces interruptions of the manager's work. The disadvantages are the loss of time and material if the solution does not work, the loss of interim help the subordinate might have received from the manager, and the setback in subordinate morale if the solution does not work.

The length of time is crucial in this method because the manager has allowed the project to be out of his control until the deadline date. Thus, the shorter, the better.

No one of these methods will be fully effective for keeping on top of a job or project that is delegated. However, a combination of them can be. Setting deadlines with interim progress reporting is one such combination. Adding a weekly or biweekly "How's it going?" session adds another dimension. Varying the combination with each project engineer is probably the most effective method. Some enjoy the oral progress report; others get tense or forgetful and prefer to submit a written report with all the details. Regardless of the combination used, the manager must, on a day-to-day basis, keep on top of the projects he has delegated by figuring out how to "keep his hand in" without destroying the effectiveness of his delegation.

PRIORITY SHIFTS

Some businesses have engineering problems that come up quickly and must be fixed quickly. These range from shop floor problems where the design engineer must assess and correct a problem condition, to customer problems where urgency is the rule, and prompt response is necessary. Under these circumstances the manager is tempted to throw priorities out the window and have his key people respond to the challenge.

While this approach sometimes must be taken because of the urgency of the situation, it should be carefully thought through, even if only for a half hour, because of the long-term effects of such short-term responses. These problem situations pop up like brush fires that must be put out before they grow. The manager needs to ask, is it an engineer who needs to solve the problem, or can someone else do it? Must the problem be solved immediately, or can it possibly be set aside for a few days? Who has all the facts about the problem? And so on.

Handling Priority Shifts

Priority shifts can cause disruption of the necessary functions of the design group. A manager cannot have a group of firefighters and still get his fire prevention projects done at the same time. While it is true that urgent problems must be dealt with promptly, it is equally true that not all problems are urgent, and that urgent problems do not always require the skills of a project engineer to solve them. The design manager needs to sort these conflicting requirements out, and keep them in perspective.

Priorities do need to be shifted at times. Such shifts must be made with care and forethought. Some long-term projects must be escalated because of competitive actions that were not anticipated. Others need to be deemphasized because market needs have changed.

It is usually relatively easy to escalate project effort because that means greater effort, increased activity; but deemphasizing a project, or even setting it aside for a few months, must be done carefully because more than momentum is lost—documentation can be lost as well. Frequently the project engineer has a number of project notes, ideas, and hunches in his head or scribbled in his project notebook. With a lapse of time, these notes lose their freshness and significance. They become obscure and unintelligible. The incomplete test reports become less valuable and are eventually forgotten. Over a period of several months, previously acquired data and knowledge are practically useless.

When priorities must be shifted, these potential problems must be recognized and time allowed for recovery and getting up to speed once the project is restarted.

MANAGING UNDER PRESSURE

Only in an idealistic world does a manager find no pressure. In the real world, there are pressures each day upon the design manager, and on his staff members. When economic times are bad, or when they are good, there is pressure to do things better, faster, less expensively, with fewer mistakes, with fewer people. Deadlines, unfavorable business levels, rising costs, production quotas, tyrannical bosses, confused priorities, personal problems or setbacks, organizational changes, changes in assignments, all contribute to pressures on the job.

It is easy to accept the fact that pressure exists, and impacts the design engineering manager in many of the areas that were mentioned, and in other ways that were not. Pressure builds stress, and, more than anything else, stress is what disturbs and causes depression or other negative effects in individuals. The side-effects of stress can be compared to the physical stress that one endures, say, from carrying a weight over a distance or up a flight of stairs.

For a short distance and a short time, the stress is enjoyable. Carrying the load, one moves ahead confidently at a normal pace or perhaps faster than normal, to show that he can handle it. Then, after a while, fatigue begins to set in. The weight seems to get heavier. Beads of perspiration form on the brow and in the armpits. One's breath gets shorter. If there is an end in sight, a limit to the distance, one will press on, to accomplish the goal, reach the destination.

But if there is no end in sight, if the extra load is to be carried "forever," it begins to be wearing and exhausting. The constant stress becomes too much, and the person has to stop and rest. Or he may be so fatigued from the constant, unremitting stress, that he stops trying.

Types Of Stress

Hans Selye, the originator of theories of stress and life, did research on the effects of stress on laboratory rats. He learned that sometimes rats died young with no discoverable disease, only the prolonged exposure to stressful situations in the laboratory. Extending this concept to humans, he speculated and later proved that serious physical illness and death can result from prolonged periods of stress.

Other psychologists later determined that not all stress is to be avoided; some stress is essential to a useful enjoyable life. In fact, many retired people die early because there is no stress, no challenge to their life. It was determined there are two types of stress, dys-stress and eu-stress.

Dys-stress covers the noxious pressures of work that, when unchanged, lead to loss of vigor, emotional disturbance, and serious illness. It is the prolonged stressd that Selye found earlier. Eu-stresses are the interesting, fun challenges of work (or sport), the pressures that are enjoyable and necessary. These generally come in short bursts.

Managing under Stress.   Obviously most job pressures will be of the dys-stress type, which initially may be challenging. But as they continue to grow as assignments are added, frustrations and tension grow. One's effectiveness and energy are sapped by the continual pressure.

To manage under stress, the manager must think, communicate, plan, respond to the total situation. Then he must focus on the results needed and prioritize the efforts. Only then can he begin to reduce the dys-stress and blend in some eu-stress to balance the stresses on himself and his staff.

Stress on the Job—A Case in Point

Consider this situation as one way in which stress develops and continues. In the XYZ Company, the general business situation was poor. Sales dropped, and the economy was down; so the company initiated a series of layoffs in all key departments, reducing design engineering to a low staffing level, for example. Some new products with long-term potential were being developed, and a number of short-term cost reduction projects were being vigorously pursued to improve profitability at the present business level.

Then business increased about 10%. With this slight increase, there was significant additional work to be done. There were incoming orders for nonstandard parts that had not been made for some time. The nonstandard part drawings were outdated and needed changes so the parts could be manufactured correctly. Then a piece of old production equipment broke down and needed repairs. As there were no drawings of the parts available, engineering was directed to develop drawings immediately so that the repair parts could be built.

The major cost reduction project had reached a point where engineering design and testing were needed to determine which of the lower-cost ideas was acceptable and could be phased into production in the shortest possible time. And, because of a creative design idea, it was clear that a new product, which had the potential for good sales in a poor market, was ready for final engineering evaluation.

At the same time, the general manager restated his interest in several long-range design projects that were being worked on. He told the engineering manager not to lose sight of the strategic importance these new products would have. In other words, none of those projects could be allowed to slip—not any of them.

The manager was well aware of the importance of all of these projects—the shop floor support, the short-range cost reduction, the new product ready for release, and the important long-range products. The need to do all of them was clear, but it was difficult to handle all the projects simultaneously. Everyone had more than his share of work to do, and each mistake meant something had to be done over. Each idea that failed on test needed to be reviewed and analyzed, then changed and tested again.

The pressure on the engineering manager and his small group of design engineers, technicians, and drafters was increasing. They had too much of a good thing. They had a tiger by the tail and could not let go. All the projects were key; all were A-1's.

Then, quite unexpectedly, one of the design engineers went home sick one day, with abdominal cramps. The manager hoped it was only the flu, that the engineer would return in a few days, rested and ready to go back to work. The following morning the engineer's wife telephoned to say that her husband's pains had become worse, and the physician had diagnosed a gall bladder problem, requiring immediate surgery. The engineer would be away from the job for at least three to four weeks, depending on how quickly he recuperated.

Relieved that the health problem was diagnosed but disappointed that his engineer was out for a month, the design manager felt the pressures of the job nearly double. He still had everything to do, but had even fewer resources to do them with. The job was starting to get to him. No more eu-stress each time a success was achieved; each day brought more dys-stress as the jobs fell further behind. His staff was feeling the pressure as well. And with one staff member out, the "almost" balance of job requirements versus available staff to do them was completely out of balance and getting worse.

Driving home that evening, the manager decided that the pressure of the job was getting to him. He either had to come to grips with the impossible situation of having all projects be top priority, or his group would falter and fail. After dinner, he sat down with a pad and pen, and reviewed once again the major areas:

1. The shop floor problems: Designing the required parts was a complicated job, but his people were handling it well until the design engineer went to the hospital.
2. The cost reduction programs: One of them was right on schedule, but two others were beginning to slip because of the shop floor workload.
3. The product development: This was ahead of schedule. A small amount of confirmatory testing had to be run, but otherwise the project was ready for final prototype approval and release for pilot run.
4. Two long-range projects: These were in early development. One was on schedule; one had slipped badly.
5. Miscellaneous projects: These included a group of small projects—one man-week or less, each—that were used as fill-in projects.

As the design manager thought more about the project load, and his people, he began to organize his thoughts about what was doable and what was not. He developed a tentative overall plan and priority approach that would utilize the people he had, and get the work out. The next morning he called his project engineers together for an overview meeting on their programs, to test the reasonableness of his initial analysis. He solicited their opinions on who could fill in for the hospitalized engineer and what the trade-offs would be.

Then, he refined his priority listing, which was clearer to him now. He conferred with the marketing and manufacturing managers to get their input, and then went in to the general manager with his revised priority listing. It consisted of continued emphasis on the shop floor support and key cost reduction programs. The project ready for prototype approval would continue, but the other two long-range projects would be held up for three months or until the heavy load of urgent, current projects had subsided.

The general manager, a "do it all now" person, was not delighted with the revised priorities. But after some in-depth discussion, he realized that the manager's proposal was realistic, and he reluctantly accepted it. Because of the staff shortage he could do little else.

The engineering group accepted the changes willingly because their new workload, while still heavy, represented reality instead of wishful thinking. The job pressures were much less intense because the short-term efforts included their own high points of accomplishment, and the long-term efforts would be started soon. In fact, some of them would be continued, unofficially, as the engineers could squeeze in a little time here and there.

The design engineering manager's response to the pressure was better late than not at all, but he should not have waited for the illness of a key subordinate to trigger his thinking and actions. Had he known them, he could have used the following keys to managing under pressure:

- *Think.* Grab or steal some time from the day to get away from the pressures of the day to think through the situation. Get away from the tangle of trees and look at the forest.
- *Plan.* Think in some detail of how to get from the present "mess" to a better-functioning operation.
- *Focus.* Planning naturally leads to prioritizing, which is focusing on the essentials. Not all jobs can be first. Otherwise all of them are also last.
- *Respond.* The general manager must be told what can be done, and what can't. He may not like it, and may order or negotiate more promised output. But the design manager must assert his position and reinforce it with fact and opinion.

It is through this process, or one similar to it, that the manager can manage under pressure. He needs to manage the pressure, rather than have it manage him and then disorganize him and his staff. He will be able to reduce the dysstress of steady, increasing pressure with bursts of eu-stress from challenging segments that end on a successful note.

THE DESIGN MANAGER'S QUIET TIME

A manager will not have any quiet time to think about tomorrow's actions, or how to improve his operations, unless he grabs a chunk of time to do his thinking. Quiet time is a conference with only one person attending, where concentration instead of conversation takes place.

In the preceding example, the manager who was under pressure had to do his thinking at home one night because it was impossible to think clearly under the day's pressure. In the same vein, the manager must occasionally spend some time reviewing where everything is and where it is going, even if he is not under intense pressure.

Quiet time can be found at work if a manager has:

- A private office.
- A competent secretary to intercept his calls.
- An understanding with his staff that he is not "in" unless a matter is so

urgent that it cannot be settled immediately without his personal intervention.

This works for perhaps an hour at a time, which is enough if the manager takes his quiet hour two or three times a week.

Conference Rooms for Quiet Times

Another way to find a quiet place is for the manager to reserve a conference room for a specific time period and alert his secretary to his "meeting" at that time and when it will be over. The manager can have a quiet hour or two in this way, and be relatively certain that he will not be interrupted.

Other departments in the same building or building complex usually have conference rooms that can be reserved when the design manager needs one. Sometimes managers of other groups are willing to cooperate in this fashion. Why? Why, one might ask, must a manager employ such ruses, to carve out reasonably large chunks of time to be alone and think? The time should be there automatically, but it is not.

Activities, responses, questions, and answers will fill most—if not all—of the day. A succession of weeks can slip by, filled with busy moments, without any reflection on how things are going—or in what direction they are moving. Most, but not all, of the manager's day must be spent in such activity.

If none of these ideas for finding a quiet time works, the manager has a few more options. One is to rise an hour earlier in the morning, have a cup of coffee to clear away the cobwebs, and use that hour to plan activities and change direction where needed. Depending on where he functions best, he may do this in his den, at his kitchen table, or at the office, arriving before anyone else. Or, in the evening, he may sit in an easy chair at home and do some thinking instead of watching television.

Quiet time is essential, but the manager must take it when he needs it and for the length of time necessary. Too short a time is as bad as none.

MOTIVATING ENGINEERS

"Motivation" is defined in the dictionary as a need or desire that causes a person to act. The word "motive" has deeper meaning than a "reason" for acting a certain way. It recognizes an emotional, rather than logical, set of forces at work within the individual. The ending "ation" suggests the word "action." Thus "motivation" suggests the inner needs or desires—or perhaps drives—that cause engineers and others to do things.

All Motivation Is Self-Motivation

It should be clear from the start, then, that a manager cannot "motivate" his people in the direct sense. No one can motivate anyone else. The best any manager can do is to learn what motivates his people to action, to find challenges in the work that are consistent with their motivations, and to provide an environment or climate that allows the motivated person to "do his thing" and at the same time "do the company's thing."

Our awareness of motivating needs comes from work done by Abraham Maslow in the 1940s. In his psychological studies, he established that man has five levels of needs, structured in a needs hierarchy.

These needs, beginning with the basic level and ending with fulfillment, are:

1. Physiological: A person wants to stay alive and be alive.
2. Safety: He wants to feel safe and secure.
3. Social: He wants to socialize with others.
4. Self-esteem: He wants to feel worthy and accepted.
5. Self-actualization: He wants to do the kind of work he likes.

In this description, the words "wants" and "needs" can be used interchangeably. If a person's physiological or safety needs are not being met, he will strive to have them satisfied before considering his social or self-esteem needs. After a brief review of this hierarchy of needs, it is clear why being fired can be devastating to a person. Almost every level of need is challenged, from self-actualization to safety and security.

In an article on motivating engineers, Dr. Frank Sterner developed this list of problem areas regarding engineers and their motivation:*

1. Engineers expect individual treatment. They resent being treated as a group, or as employee numbers.
2. Engineers have different development paths. Some are theoretical; others are hands-on people with good mechanical skills.
3. Engineers are trained in logical thinking. Their years of study have not prepared them for the subjectiveness of many situations in a complex organization.
4. Engineers are technology-oriented. They tend to be less concerned about costs and schedules than about technology.
5. Management is subject to a jaundiced evaluation by engineers. Being in-

---

* Excerpted with permission from *Professional Engineer,* July 1969 issue. From "Managing and Motivating Engineers" by F. Sterner.

dividual thinkers, they often are turned off by seemingly illogical actions of management.

6. Engineers are answer-oriented. They do not accept explanations per se; they want to be involved, to be communicated with.

Sterner also perceives problems that engineering managers have in the motivational area:

1. Background: The engineering manager's background is strong in engineering; he is usually weak in behavioral science.
2. Logic: Like his engineer subordinates, an engineering manager is not well prepared for subjectivity and unpredictability.
3. Professional attitude: Managers often face a conflict between what makes sense to them as engineers and what is best for the company.
4. Career: Despite many protests to the contrary, engineering managers have chosen management for the higher income and status that result.

Listing these problem areas does not suggest that there are immediate solutions, but, rather, highlights areas that the manager must be aware of when he deals with the lack of self-motivation frequently found in engineering subordinates.

An Environment To Foster High Self-Motivation

Since the manager and his superiors cannot motivate individuals per se, but can only develop an environment, it is important that this environment be one where motivation is fostered and grows rather than withers and dies.

Raudsepp observed that the effective manager, recognizing self-motivation as the driver, tries to make each employee more self-directed, more responsible, more capable of using his or her own talents and skills.* This requires an enlightened management approach, toward participative management and away from the traditional, autocratic type. Traditional management fosters dependency and a reluctance to make decisions, both of which are likely to demotivate employees.

What motivates each person is a blend of background and experience. The key elements in an individual's motivational makeup are his or her current image, achievement expectations, and ability. The first two relate to the Maslow hierarchy; the third is a natural gift. All three factors influence how a person reacts to an environment, and how well he will do.

---

* Excerpted with permission from *Machine Design,* September 9, 1982 issue. From "Reducing Engineering Turnover" by E. Raudsepp.

1. Current image is how the individual perceives himself. His personal image can be high or low, true or false, subjective much more than objective.
2. Achievement expectation is not necessarily consistent with current image. One person may expect to become engineering vice-president, another engineering manager, still another a technical expert. A competent, talented engineer may have quite modest expectations because of a poor self-image.
3. Ability is both natural and learned. Again people have different levels of ability and use it differently. A person with a high level of ability may use it to the fullest for superlative performance; or a person with little ability, using it well, can surpass a person of greater ability who does not use his gift.

Managerial Responsibilities. One of the more difficult tasks of the manager is to understand his subordinates as they are, and to learn each subordinate's value system. He needs to realize that each person will act and react within the framework of his own value system, not anyone else's. Another important concept is that people do not change very much. They may give the appearance of changing, but they do not change. However, the manager can help a subordinate gain insight into his shortcomings, and from this, the manager can either temper or ignite his subordinate's aspirations.

How the Manager Can Foster Self-Motivation

The foregoing discussion covered generalized approaches to engineer–subordinate motivation. The following are several specific actions a manager should take to foster an employee's self-motivation and hopefully to ignite his aspirations.

1. Establish clear objectives and standards for each subordinate.
2. Frequently evaluate the subordinate's programs and direction toward accomplishing them. Help him make necessary adjustments.
3. Take corrective actions promptly, including discipline.
4. Provide rewards promptly and amply when results are good.
5. Do not tolerate incompetence. Encourage competence and excellence from each subordinate.
6. Consider expectations for each person in terms of his capabilities.
7. Assign intermediate goals—small successes to build confidence and expand horizons.
8. Give the employee the opportunity to fail and to learn from doing so, in less critical areas, whenever possible.

9. Give each employee an understanding of his part in the organization's total goals. This provides a sense of involvement and allows him to tie his goals to the company's.

Work Factors That Improve Self-Motivation

Nothing thus far has been said about the working conditions of the environment. What about lighting, temperature control, desk and office size, color coordination, or benevolent personnel policies, vacations, and so on? There is a school of thought that places great importance on these factors, but most authorities do not. One authority, Emery, is relatively outspoken. He states that human relations concepts can be forgotten if one's work meets these five criteria:*

1. Work that is important to the organization and recognized as such.
2. Work that is challenging. There is a real possibility of failure within the limitations or resources available.
3. Work that is doable from the start. Success in the project is quite likely, given a solid and consistent effort, sparked by occasional all-out bursts.
4. Work that is fitting to the individual's particular capabilities.
5. Work that is valuable. One can talk about it with pride to his children.

The work itself, in Emery's view, provides the real challenge that brings out the capabilities of the employee. While the role of the manager is not specifically defined, it is subtly woven into all of the five conditions. Another writer, Jacobs, takes a somewhat different, yet similar, approach, with four keys to motivating engineers.† The keys he focuses on are Pride, Identification, Professionalism, and Recognition (PIPR), in this manner:

- Pride
  - Generated by management.
  - Based on high standards of performance.
  - Based on individual competence.
- Identification
  - As engineer identifies with his project.
  - As reinforcement to pride.
  - Fostered by contacts with customers.

---

* Reprinted with permission from *The Compleat Manager* by D. Emery. Copyright 1970 by McGraw-Hill, Inc. All rights reserved.
† Reprinted with permission from *Chemical Engineering,* August 14, 1978 issue. From "Four Keys to Motivating Engineers" by J. J. Jacobs. Copyright © 1978, by McGraw-Hill, Inc., New York, N.Y. 10020.

- Professionalism
  - Attitude based on pride, competence, identification.
  - Ethical decisions based not on what the customer wants to hear, but on what is right.
  - Company willingness to stake its reputation on the engineer's conclusion.
- Recognition
  - Good performance noted and rewarded.
  - Bad performance constructively critized.
  - Participation in professional society activities.
  - Presentation of technical papers.
  - Patent and patent disclosure awards.
  - Job advancement

All of these PIPR keys are impacted by the actions of the manager. An observant manager will recognize that behind the logic and intellectual honesty of a competent engineer lies a desire for approval, an ego needing support and expression, and the personal need to excel. The degree to which an engineer is motivated by these forces determines his productive output.

There is nothing more rewarding to a manager than to provide a climate where engineers can feel recognized for having contributed productively to the success of an important project. In such a circumstance the engineer will outperform his own estimate of his capabilities.

STAFF OBSOLESCENCE AND TURNOVER

Obsolescence begins as soon as an engineer receives his diploma. He may then be current in some aspects of the technology field he enters, but technical obsolescence will set in rapidly unless the engineer does something about it. The company and the engineering manager can play a role in preventing it, but much depends on the engineer himself.

Turnover is also somewhat related to the professional climate, which can enhance self-motivation or retard it. The problem of turnover has causes that can be identified and often corrected. The following section covers both subjects, and ways in which they are related.

Obsolescence

Engineering is a profession that requires one to learn continually with frequent updating, in order to keep current on advances in technology. Otherwise obsolescence sets in. An exponential process, obsolescence begins slowly and is almost unnoticed; then it picks up speed, and when visible, it almost can no longer be corrected.

An engineer who is now in his fifties probably graduated carrying a slide rule, and may have heard about transistors but did not see one. Today, semiconductor chips power the tiny calculator in his shirt pocket that provides number-crunching possibilities beyond his needs; he gets digital accuracy greater than the accuracy of his measuring equipment. Yet if he is not becoming familiar with computers and their capability to make his computational work easier and more productive, he is becoming obsolete, if he is not obsolete already. Five years after graduation, an engineer working in electronics or microprocessors is probably seriously deficient in his knowledge, compared to today's graduate, unless he has made strong efforts to stay informed.

Obsolescence has several sources. Indifference or laziness on the part of the engineer and extremely rapid advances in many technologies are two of its important causes, related to the individual. Also, there is at least one major internal cause of obsolescence: engineer specialization in a company or technology.

Through specialization, an engineer can come up to speed quickly and become more competent in a given technology area in a short time. Once competent in an area, the engineer may tend to resist new assignments that would require acquiring substantial new knowledge and experience. The longer this specialization persists, the more difficult it becomes for the engineer to take the personal risk of change and to justify novice performance at higher pay levels. Knowledge ages quickly; products lose market share, and technologies become unprofitable and are dropped. In any of these situations, a company can end up with demotivated, aging engineers who are performing substantially below their potential. They have become a group of obsolete engineers, of greatly decreased value to their company.

Alternatives to Specialization.   One alternative to specialization is generalization. Instead of being extremely competent, even expert in one narrow area, an engineer becomes adequately competent in several areas. To become competent in several technology areas is not easy, but it can be done. It takes perhaps a year to become more than conversant in the new field; but in two to three years, an engineer can become competent, and in five to eight years, an expert. The time can be shorter if the two fields are related.

For example, an engineer familiar with the chemical reactions and properties of epoxy resins could branch out into elastomers (rubber compounds) in a comparatively short time. Instead of being an expert in epoxies, the engineer could become competent in several areas of engineering materials, especially plastics and rubbers.

Obsolescence is not due to age per se. However, if an engineer stays in the same discipline throughout his career, he tends to become an obsolescent engineer as he grows older. He has few if any new challenges, and seems to see the same old problems over and over again—in new disguises. Some persons tend to feel comfortable doing the same job—their routine—every day, but the engineer's talent for solving problems and developing creative ideas dulls if it is unused. An engineer may become obsolete if he does not use all of the "tools of his trade"—all of his skills—for some time; he may find that they have become rusty and outmoded from lack of use.

The manager must vary the assignments given his engineers to make sure they are sufficiently varied and challenging. Self-motivation is a key factor in avoiding obsolescence, but the manager's desire to get people up to speed in a short time can force them into specialized areas. Although in the short run this can be beneficial, in the long run it is a detriment because it leads to obsolescence.

Turnover

Engineers change jobs, and leave a company, for many reasons. When engineers leave after only a year or two, the company's investment in training is largely wasted. Worse than that is the loss of productivity in the engineering department. Projects must be reassigned, new engineers hired and trained; and the important, informal working relationships need to be reestablished.

Turnover tends to occur more frequently with new engineers, in their first and second jobs after college. In a recent study of engineer turnover, Raudsepp found these major reasons why 58 young engineers left their first job.* They are listed in order of importance to those engineers:

1. Inadequate compensation
2. Uncertain future with the company
3. Higher salary offer elsewhere
4. Change of mind about specialization
5. Job was different than expected
6. Work was too routine
7. Nothing more to learn
8. Not enough responsibility
9. Poor supervision
10. Misrepresentation by recruiters
11. Inadequate training program
12. Lack of recognition for performance

---

* Excerpted with permission from *Machine Design,* September 9, 1982 issue. From "Reducing Engineering Turnover" by E. Raudsepp.

This list shows the types of dissatisfaction that engineers encounter on the job and company practices that need improvement. It does not need much explanation. However, a manager may wonder why *he* has a high rate of turnover and is not guilty of the causes just cited. The reason may be a difference in perceptions. Perhaps in the manager's perception the company pays well, individuals have a great future, and so on; the young engineer, however, may see things in just the opposite way. The view of the experienced engineering manager may be less significant than the "reality" perceived by the young engineer.

Alert management can solve the problem of turnover by addressing the issues that cause it. Nothing can eliminate turnover, but improved policies and practices can make a big difference. The following practices can have a beneficial effect.

Salary Administration.   One plan that is being used increasingly is to break the usual six-month evaluation period for new employees in two, with half the available increase distributed each three months. Although the total amount of money involved is only slightly different, the new recruit has an earlier evaluation, and realizes that he is getting close attention and good treatment. This can be a very effective technique.

Recruitment Follow-through.   The recruiter should stay in touch with the new employee by letter or telephone for the first several months. The employee should be encouraged to ask questions about his work, his design group, his supervisor, and should then be put in touch with people who can provide answers. This extra personal touch is valuable because the recruiter probably made quite an impression on the engineer and may have been one of the reasons why he joined the company.

Guided Training.   Some period of orientation should precede any training, to provide information about the company, the engineering department, and job-related subjects. Actual training is usually a direct project assignment to challenge the engineer, or a series of rotated assignments to give the young engineer a broader view of the engineering department. Each method has advantages, the second of the two providing the better opportunity for the engineer to decide what he might like to do best. Either can be effective.

Whichever method is used, the engineer's first year on the job should be sufficiently challenging to minimize dissatisfaction or boredom. If the manager monitors the engineer's progress and attitude toward the job, he can address and deal with any concerns the engineer has.

Interview When the Employee Leaves.   When an employee terminates his employment voluntarily, an interview prior to his departure can be informative and valuable. An employee who quits is likely to disclose his real reasons for leaving if he can remain anonymous. Such an interview should be conducted away from the workplace. This interviewee should not be asked to write or sign anything, and should be assured that his comments will be used to upgrade the policies and work conditions of the company. If these findings are reported back to the supervisors and managers and considered as insights, corrections can be made to the system that will reduce turnover.

PROMOTIONS AND REASSIGNMENTS

One of the happier tasks of the design engineering manager is to promote one of his subordinates to a higher-level position. This may come about because someone has left, creating a void that needs to be filled, by a new engineer who has successfully completed his training period or a more experienced one who can be promoted, say, from engineer to senior engineer. The actual act of promotion is the culmination of observing a person's growth and devlopment, and providing some coaching and guidance along the way.

There are a few guidelines to be considered both prior to and after a person is promoted. They include promoting on results, providing a trial period after the promotion (in some cases), and providing training for supervisors and group leaders.

Promoting on Proven Results

Just about every employee believes he is promotable and should be promoted if a job opening exists. Some will be assertive enough to speak out and ask to be considered for such promotions. This is fine. However, the manager must consider those who have not asked as well.

One of the main criteria is past, proven results. What has the employee done in the recent past to demonstrate that he has made specific important contributions to the company? What has he done on a continuing basis to show that his performance is not timed for evaluation? Is he a consistent performer, doing things that need to be done, volunteering to do jobs, asking for additional assignments?

In focusing on accomplishments, it is important that the manager not let his personal feelings bias the evaluation. The person must have earned the promotion on results, not on being likable, pleasant, or outgoing. Personality characteristics do not normally enter heavily into the judgment. What does enter in is how well one handles his daily assignments. An engineer's development is 90% the result of how well he does his day-to-day work.

Thus, primary emphasis must be on how successfully the employee concentrates on doing "today's job" particularly well. The manager evaluates by measuring the results.

Trial Promotion

If a person is promoted and some unsuspected factor emerges that makes him ineffective in the new position, the promotion cannot be rescinded without a loss of face for all concerned. If the newly promoted supervisor cannot supervise, or if he supervises with an iron hand, the promotion may have been a mistake. It cannot be ignored or glossed over with the bland statement that "It will all work out soon." Key people may quit, projects veer off in the wrong direction, and worse.

One way to avoid this catastrophic problem is to provide a period of trial promotion. It could be called an interim action or a temporary promotion. The promotee is given an opportunity to have the supervisory job for a while, on a trial basis. Perhaps he will not want the promotion permanently because he will not like the shift from technical to nontechnical responsibility. Perhaps he will not like being a supervisor, period. In either case, he should have the opportunity to return to his regular assignments, with no stigma attached.

In some cases, temporary promotions can be appropriate. They can be only for the duration of one phase of a project, with a definite time limitation. In that way, it is clear that the temporary promotion is project-oriented. Using such an approach, the manager can assess the supervisory and/or technical leadership potential of his staff members on a no-fault basis, if things do not work out well.

Supervisory Training

Training for supervision is another way to assess whether a high performer is a potential supervisor or not. Many companies offer presupervisory training courses where the fundamentals of supervision are taught and prospective supervisors can learn what is likely to be expected of them should they be offered a promotion. These courses are valuable, but they must be available to any employee who is interested, not only to those whom a manager believes to be "management material." The manager's positive or negative biases should not enter into this preselection process. Also, some aspiring supervisor may not be an obvious choice to the manager.

Supervisory training is another form of no-fault evaluation. The employee may learn that supervision's responsibilities are not for him, or the firm may learn that some employees are poorly suited to supervisory positions.

Reassignments

A reassignment is not the same as a promotion. A person may be reassigned from one engineering group to another, or to marketing, or to manufacturing. Such reassignment is generally for one of two reasons, both of which relate to the person's value to the company.

In one case, the person may be reassigned for broader experiences with the company, to prepare him for an eventual promotion with greater scope. In another, a valuable employee is working in an environment where personality conflicts, either with peers, subordinates, or supervisors, prevent him from using his talents effectively. Reassignment will give him another opportunity to find a niche in the company where he can excel.

In the first case, the reassignment would be temporary; in the second, hopefully permanent.

PERFORMANCE APPRAISALS AND FINANCIAL INCREASES

Every company has a system for appraising the performance of employees and making salary changes to recognize higher levels of performance. The system may be relatively structured, based on specific evaluations with numerical values assigned to them; or it may be an informal evaluation, reviewing items of accomplishment during the previous year and listing some areas for improvement.

Whichever method is employed, the most important aspects are two: that performance be appraised objectively and that the results be communicated to the employee, giving him an opportunity to respond to the appriasal.

Who Should Appraise?

The immediate supervisor should evaluate and appraise each subordinate's performance. He is closest to the subordinate, and has assigned various projects and set certain objectives, which were or were not attained. He is best able to see personal development, new skills being learned, and old ones being improved.

The manager should review each appraisal in semifinished form to satisfy himself that appraisals are done thoroughly and that the employee is reviewed fairly, as well as to find out how his own impressions compare with those of the reviewer.

When Are Appraisals Done?

Most often, appraisals are done once a year, prior to an increase in pay. For new employees, an appraisal every quarter, whether there is an increase or not, can be a valuable counseling and coaching tool. Some companies plan increases on an anniversary date, so they are done at various times during the year. Some companies time all appraisals for, say, January 1 increases. This approach has one major advantage, to balance the disadvantage of a supervisor doing many appraisals in a short time: with all appraisals done at once, all employees' performances can be better compared. The manager gets a valuable overview of how all of his employees are doing, simultaneously. Seeing them scattered throughout the year, he gets only a fragmented idea of performance and skill building.

The all-at-one-time method is preferable because of the broader overview it provides the manager. Relative employee performance can be compared, the evaluation criteria of supervisors compared, and budgeting for increases both compared and finalized. The workload appears heavier only because it is concentrated. Actually less time is required to do several at once because the evaluator's mind is attuned to the evaluation process, with the need to complete them all.

What Should Be Appraised?

An appraisal is performed for several reasons—to decide on salary increases, to make promotions based on skills improvement, to determine attainment of specific goals. Each factor is important, from the manager's viewpoint and the employee's. The first two factors are more the result of having an appraisal; the last one reflects the results that determine the extent of salary increase.

Goal attainment is easy to measure but more complex than it may seem. Either a goal is achieved or it is not; judging purely on results, this is easily quantified. But a deeper analysis is needed:

- How close was it to completion?
- Was the goal realistic?
- Was it achievable?
- Why wasn't it completed?
- What was actually accomplished?

From a study of the goals assigned and the employee's progress toward achieving them, much can be learned about his performance, attitudes, and so on. Both near-term and long-term goals need assessment. If a system of project management is in place, long-term projects have a series of milestones, each of which is a checkpoint for progress, a subsidiary goal.

Figure 13–1 shows a form that can be used for self-appraisal by the employee prior to his review.* If the supervisor also fills out a form on the

---

* Figures 13–1 and 13–2 excerpted with permission from *Machine Design,* November 9, 1976 issue. From "Getting Extra Mileage from Performance Appraisals" by E. Raudsepp.

PERFORMANCE APPRAISAL

NAME _____   DEPARTMENT _____

PRODUCT GROUP _____   SUPERVISOR _____

SELF-EVALUATION

To prepare for your performance review, it is important that you evaluate your performance during the past year in terms of progress toward goals, and your efforts. On this form, list your most important goals of the previous year and comment on the results achieved. Send the completed form to your supervisor one week before the review, and keep a copy for your own use.

_____

1. _____
    results:

2. _____
    results:

3. _____
    results:

4. _____
    results:

5. _____
    results:

General Comments:

Fig. 13-1. Self-evaluation form.

same objectives, a much broader perspective of performance can be gained when the two are compared. When used as a reference for the next review, the historical perspective is valuable as a tool to enhance performance by showing strengths and weaknesses in attaining goals.

Skills Development

Each engineer has a number of skills that are developed to a greater or lesser degree. These include the technical skills necessary to do the work, of course. In addition, there are the planning skills, leadership skills, and others needed to do the job better and prepare for promotion or new assignments. Figure 13–2 shows a skills inventory, which also is intended for completion by both the employee and the supervisor. The skills inventory, while somewhat subjective, points out areas of strength and weakness. Weak areas often can be strengthened by incorporating one or more of them into the employee's work objectives for the next year.

The inventory helps determine which skills need to be developed if a subordinate is to do his job better or be able to take on a more challenging assignment. The career questions at the end encourage the subordinate to think ahead to future career goals and to assess current efforts to improve skills.

Work objectives for the next year round out the pre-appraisal effort. Figure 13–3 shows a form for listing work objectives. It should include:

- Short-term objectives to be completed in less than one year, project-related.
- Long-term objectives, which will require one year or more for completion, also project-related.
- Skill development objectives, which will vary in the time frame required.

As before, maximum benefit is obtained if the objectives are listed independently by the subordinate and supervisor, then adjusted or combined. The number and type of objectives are difficult to quantify. If there are as many as ten, it will be difficult to show significant progress unless each is easy to work on. The optimum number is probably five to seven. This allows two or three short-term and long-term objectives, plus one or two skill development objectives.

The Appraisal Review

Having the subordinate appraise his own performance, with the supervisor doing it independently also, provides significant insights into how the subordinate perceives his performance in the light of the supervisor's needs. They come together in the appraisal review when supervisor sits down with subordinate and gives him an appraisal of his accomplishments, skills, and next

SKILLS INVENTORY

Your performance last year was a combination of your skills and how you used them. Look over this list of skills and consider how they contributed to your performance. Rate each of your identified skills as:

1. More developed than necessary to do the job.
2. A good match to present assignments.
3. Development needed before you are competent in it.
4. Not needed for present assignments.

It is not necessary to rate yourself in all skills.

PLANNING SKILLS

1 2 3 4

() () () () 1. Goal setting   Developing aggressive but realistic group and personal goals.

() () () () 2. Scheduling   Establishing priorities and time sequences for project work.

() () () () 3. Budgeting   Allocating resources to accomplish goals.

ORGANIZING SKILLS

() () () () 4. Organizing   Using an effective, systematic approach on project and non-project work.

LEADERSHIP SKILLS

() () () () 5. Decision-making   Selecting best alternative course of action.

() () () () 6. Listening   Identifying essential information in verbal exchanges.

() () () () 7. Presentation   Speaking effectively to groups of peers or others.

() () () () 8. Writing   Expressing thoughts effectively on paper.

(continued)

1 2 3 4

() () () () 9. Developing people Improving performance of others through training.

CONTROL SKILLS

() () () () 10. Establish standards Defining criteria for comparing results to goals.

() () () () 11. Measure performance Determining progress rate of ongoing jobs and completed work.

() () () () 12. Evaluate performance Appraising results achieved on work in progress.

GENERAL SKILLS

() () () () 13. Follow-up Ensuring that work begun is completed.

() () () () 14. Time management Self-management of the time needed on projects and non-project work.

() () () () 15. Persistence Pursue valid goals despite obstacles

() () () () 16. Risk taking Pursue valid goals despite potential negative consequences.

() () () () 17. Stress tolerance Perform effectively despite pressure and conflict.

() () () () 18. Impact Control the attention and determine the actions of others.

() () () () 19. Career planning Pursue a personal career plan.

() () () () 20. Interpersonal Getting along well with others.

A. What career goals and questions do you need to discuss with your supervisor to perform more effectively on the job and advance your career? List them on a separate sheet.
B. On another sheet, list your ideas on how to improve those necessary job skills you have checked off as "3."

Fig. 13–2. Skills inventory sheet.

WORK OBJECTIVES FOR NEXT YEAR

NAME _____   DEPARTMENT _____

PROUCT GROUP _____   SUPERVISOR _____

List your primary work objectives for the coming year, from your perspective. Include short-term objectives (less than one year), long term objectives (more than one year), and one or more skills improvement
objectives. Indicate how they can be measured. Use a separate sheet if necessary. Send the completed form to your supervisor one week before the
review, and keep a copy.

1. _____

2. _____

3. _____

4. _____

5. _____

6. _____

7. _____

Supervisor's Comments: _____

Fig. 13-3. Work objectives form.

year's objectives. It should be done by a verbal summary first, where the
supervisor summarizes the subordinate's performance in general and then
with specifics. Next, both review the written copies of the other areas. The
supervisor will comment on where he agrees or disagrees with the subordinate's evaluation and why. They progress from last year's accomplishments
to skills development, to next year's objectives.

   In each case, the supervisor should solicit the subordinate's comments and
reactions. Both should agree on the next year's objectives for skills and proj-

ect goals. This review typically should take about an hour. It is important that the subordinate provide some feedback, to enable any disagreements to be aired and some understanding to be reached about their resolution. In some cases, it may be advisable to schedule a second meeting to enable the subordinate to study the supervisor's comments, particularly when they are critical. This will enable the subordinate to come up with some ways to improve on the shortcomings the supervisor has found.

A second meeting allows a more comprehensive response by the subordinate when there is need to develop a plan of action. With experienced engineers, the second meeting allows a better exchange of views. However, it is optional, and the subordinate should be given the right to say it is unnecessary if that is his view.

When the appraisal interview has been completed, both supervisor and subordinate sign their agreement for the next year's objectives. The subordinate is given a copy of all the sheets. During the year, he can and should refer to the objectives and skills he is working to improve.

Some managers and supervisors think that the appraisal process takes too much time and effort, but a thorough appraisal is appreciated by most subordinates because their performance has been evaluated, and their successes have been noted. Some will be concerned about the negatives, others reinforced by the positives. The appraisal should be no more detailed and comprehensive than one the supervisor would like to receive from his manager.

The value of mutual, consistent, congruent goals cannot be overemphasized. The appraisal process, even though only an annual or semiannual event for most, reinforces congruency of goals and directed progress toward them.

Financial Increases

Performance appraisals and financial increases generally occur sequentially. All too often, however, the increase is not consistent with the performance level during the evaluation period. Many times these raises are referred to as merit increases; but in times of inflation, with living costs increasing, most of the "merit" increase is a recognition of higher wages being paid to starting engineers. In recent years it has often been difficult for a manager to provide a substantial increase for his star performers of the year, and keep his newer engineers' pay above that of new hires.

Engineering managers, recognizing this paradox, will set aside some extra money as a "merit" award, over and above the salary increase that would result from satisfactory service. But, as mentioned in Chapter 7, the concept of close time coupling between achievement and reward has a greater positive effect on behavior modification than does a once-a-year recognition.

Salary administration requires fairness to all. It requires that a subpar performance be penalized with a subpar raise or no raise at all. This should jolt the employee into realizing that his present performance rather than his past successes or years of service alone is the basis for the financial penalty. Similarly, superior performance should be rewarded at a level that will make the employee's eyes pop when he sees the size of his pay increase.

HIRING AND FIRING

These two areas, which are entirely different, represent the extremes of an employee's association with a company. They will be discussed separately, and any similarities in discussion are coincidental rather than intentional. Hiring is the process of bringing new and needed skills into a department; firing is the process of removing inadequate or undeveloped skills from the department. In both cases, individuals are involved, with personalities and needs, as well as skills.

Hiring

Hiring an employee is the end of the process of selection and interviewing that leads to choosing which of the available candidates for a position is the best. The personal interview and evaluation are the most important factors in a sound hiring decision.

Planning and preparation for the interview are extremely important, yet some managers believe that an interview will be better if it is unstructured and spontaneous. That might be more relaxing for the manager, but the interviewee rather than the manager will be in control. After the interview, a number of unanswered questions may remain unanswered. A good candidate may be overlooked, or on unsatisfactory one may be offered the job.

Prior to interviewing anyone, the manager must have a list of the skills and abilities he is looking for, not just one that contains the job description items. The list may be fillable only by a superman, but most managers can bridge the gulf between the theoretical ideal and the practical actuality. Some items on the list can be obtained from the supervisor whom the new person may be working for. They may express a need for skills that the previous employee (whether he left or was fired) did not have, or perhaps necessary skills that he did possess.

Reviewing Resumes.   This activity also should be done before any interviews are planned. There are excellent resumes and poor ones, but the quality of the resume does not reflect the capability of the applicant. Professional resume writing services can package the skills of an applicant very attractively.

Some items to study in a resume are:

- Length of stay and number of previous jobs. A job hopper may be looking for an elusive opportunity rather than trying to develop on each job. Or, he may have made several poor choices.
- Progression in responsibility in past employment. Has it increased, stayed constant, or decreased?
- Nature of previous work experience—duties performed, accomplishments that relate to suitability for this job opening.
- Educational background, special hobbies, activities related to suitability for the job.
- Unexplained breaks in employment that indicate something the candidate wishes to conceal.

The resumes should each be evaluated and ranked, on the basis of very promising to unpromising, so that the interviews can begin with the best candidates for the job.

The Pre-employment Interview.   The manager, assuming that he will do the interview, should call the candidate and make the appointment. This shows his interest in the candidate, who will thus be more interested and cooperative. Time and place should be mutually convenient, in the manager's office or a nearby conference room. There should be no phone interruptions, no drop-ins, and no third party in the interview. All of these can prevent rapport between the manager and interviewee. Enough time should be blocked out that the interview can be unhurried.

The interview itself needs some structure, and the manager should control it. Questions asked must be open-ended, not answerable by a yes or no. The manager needs to restrain himself and not monopolize the conversation, as this again is losing control of the major objective—gaining insights about the candidate's suitability for the job.

Open-ended questions that encourage narrative replies will provide greater information than a mere expression of opinion. Questions should be used, but not overused. Their main purposes are to start and keep the candidate talking, to obtain specific information, and to check conclusions. Asking too few questions can allow the interview to deteriorate to the level of social conversation.

Effective questioning is an art by which the manager can develop insights into the qualifications and suitability of the candidate being interviewed. It takes practice, and reflection afterward.

Questions to be explored during an interview include:

- What are the candidate's major responsibilities in his present job?
- What does he think he has done particularly well?
- What difficulties has he encountered in his present job?

- Why does he want to leave his present job?
- Is he satisfied with his work progress and career advancement?
- What does he think are his superior's strengths and weaknesses?
- What present job frustrations does he want to avoid in a new job?
- What kinds of people does he find the easiest and the most difficult to work with?
- What are his strengths, and where can he improve his performance?
- What are his key motivations and areas of greatest self-confidence?
- What are his career objectives?
- What salary does he expect, and how did he arrive at that figure?

In asking these and other questions, the manager must be careful not to suggest that a certain answer is desirable or expected, or to be avoided. A good interviewer knows when to be silent and for how long. A certain amount of silence can be constructive and will encourage the candidate to elaborate on his answers.

Note taking during the interview can be a good idea, particularly if it helps the interviewer record some fact or impression that might otherwise be missed. Excessive note taking, however, may give the interview a question-and-answer tone. Or, it may be negative, if the candidate tells the interviewer some personal information, possibly unfavorable, and sees that it is written down. He will say no more on that subject. If possible, avoid note taking during the interview. Instead, record impressions and pertinent information after the candidate has left. Key facts can always be double-checked later.

If the candidate is a promising one, it is a good idea for him to be interviewed by the actual supervisor he will be working for. If this supervisor does not conduct the primary interview, he should at least take the candidate on a facilities tour, describe some facets of the job to him, and form some impression about him. These "casual" interviews often elicit key information—either positive or negative—that did not come out of the main interview.

Interview Analysis. If the interview is successful, the manager gains insights about the candidate in these areas:

- Stability
- Work output
- Perseverance
- Loyalty
- Ability to get along with others
- Self-reliance
- Leadership
- Motivation
- Emotional maturity

The candidate learns about the requirements of the job, either by asking about them or by having them described by the manager. He learns something about the people he might be working for, and forms some impressions about the company. If the candidate continues to be promising, the initial interview should be followed by a second. Important factors overlooked in the first interview may be observed in the second. During a second interview the manager will be able to ask the questions he thinks of after the first. More important, he will firm up his evaluation of the candidate.

The decision to hire or not to hire should be made after "sleeping on it" if the manager uses this technique in his important decision making. A phone call to the candidate is the best way to announce the good news (or the bad). A letter confirming salary and other major fringe benefits, job title, and starting date should follow in a day or two.

Role of the Personnel Department. The foregoing assumes that the company's personnel department plays a supportive or minor role in the process of hiring a new employee. This is often true. A personnel department can help prepare and place newspaper ads, contact employment agencies, screen resumes, set up interviews, and handle the necessary mechanics of placing the new employee on the company payroll.

Personnel should play only this supporting role and not a major one in the interview and selection process. The right/wrong decision of hiring a new employee—from evaluation of resume to job offer—must be handled by the group for whom the new employee will work. Also, the personnel department should not check references. The manager needs to check the references himself to better determine the impact the candidate had in his previous job. In doing so, use open-ended questions—as in the candidate's interview—to elicit as much information as needed. Educational credentials need to be checked, also.

By thorough planning, conducting, and follow-up in candidate interviews, the probabilities increase of selecting the best candidate for the job opening. A new employee will be found who likes to work for the engineering department, and the department will be pleased to have him aboard.

Firing

Firing or terminating an employee is an emotional, difficult task for a manager. Some find it so difficult that they ignore the reality of an employee's incompetence and let things go. Eventually the situation deteriorates; other employees are disturbed that nothing is being done about someone not carrying his share of the workload. Perhaps the business situation worsens, so the manager is forced to take action. Firing in bad economic times under the guise of company financial problems is a poor, but frequently used, excuse. It is much better for an employee to be fired in good economic times when a new job can be more easily found.

When a manager is newly appointed, or sometime after he has taken over the job, he may find a subordinate who is not doing his job in a competent manner. Technical work is incorrect; plans are wrong; deadlines are ignored or chronically missed. Clearly the employee is not doing the job competently. Assume that the supervisor has counseled the employee, attempted to help him correct the situation, flagged serious problems during performance appraisals—but no improvement is noted, whether the subordinate tries or not. This is a situation in which the subordinate is not competent to do the work for which he has been hired.

If corrective actions have been attempted, the employee has been counseled about his poor performance, yet nothing has improved, the only appropriate course of action is to terminate his employment and search for a better replacement. The firing decision is never an easy one because the supervisor or manager feels that somehow he, too, has failed. But the longer a decision is delayed, the more demoralizing it is for others in the department who may be shouldering an extra workload to compensate. These employees may resent the fact that they are overloaded because the manager does not recognize and correct the situation with the problem subordinate.

Preparing for Subordinate Termination.   Firing or termination is one of the most difficult of management tasks because of the emotion involved. The first step is to do the preparation carefully and objectively.* Company policy may dictate that poor performance be documented in a certain way to meet legal requirements. This must be checked beforehand with the personnel department. Pre-approval of the manager's superior is necessary also.

The Termination Interview.   When all documentation is in order, with all approvals received, the manager reviews the major points to make during the termination interview. They cover facts about poor performance, corrective actions that were not followed, company procedures that may have been ignored.

Next, the time and place must be set for this interview. Privacy and freedom from interruption must be assured. Then an appointment must be made with the employee. When the two meet, the manager needs to set the tone. Small talk is out of place. The probable beginning is a phrase such as "I have some bad news for you." The seriousness of the situation will be evident to the subordinate.

A review of the problem situation—a brief one—should follow. The manager points out the contributing factors and efforts that were made to

---

* Excerpted with permission from *Chemical Engineering,* September 7, 1981 issue. From "Replacing Poor Performers with Good Ones" by R. Bhasin. Copyright © 1981, by McGraw-Hill, Inc., New York, N.Y. 10020.

deal with them. He points out the principal shortcomings—budget overruns, consistently missed deadlines, and other major factors. Then the manager indicates that these conditions could not be allowed to continue, and states his decision, "I find it necessary to end your employment here, today."

Sometimes the news will be taken calmly because it was expected, other times not. The person may become angry and vent his emotions. It is important that the manager be empathetic and not match anger with anger. Calmly and gently he must close the interview by telling the terminated subordinate what he is expected to do. He may have to turn in credit cards and keys, empty his desk immediately. The activity will give him something to do, as his thoughts are in turmoil. In closing, offer any support you can in good conscience give—references, or job-finding help that may be provided by the company.

Termination is a difficult task for a manager, although in many cases the terminated employee will find a better job than the one he left. This is not immediately apparent, of course. In the long run, the engineering department will benefit by utilizing the best talent it has.

NEGOTIATIONS

Negotiating some type of agreement is an area that a design engineering manager may face in relationships with vendors, or it may occur in day-to-day activities. Almost everything, from daily encounters with individuals to structured meetings with attorneys or other company executives that go on for days, can involve a form of negotiated result.

In simplest terms, negotiation is a process that focuses on gaining something wanted from the person or persons who have it. Each person is a negotiator unless he accepts whatever conditions are given him in every circumstance without questioning or offering alternatives.

Since negotiations can end with one party getting all or part of what he wants from the other, one could say that what one person wins, the other loses, as a result of the negotiations; but such negotiations are unsatisfactory because no one wants to lose. The preferred way of negotiating is to develop a win–win result where both sides believe that they have gained something of value from the negotiations.

In order to achieve a win–win situation, each side must be aware of the other's needs and try to satisfy them, without losing sight of its own. Then both are working together to resolve differences and reach a joint goal.

Negotiations have three elements that are present in almost every situation:

- Information: the needs and wants of one side or both.
- Time: deadlines. One side may have a time constraint on its decision making.
- Power: the capacity to exercise control over people, situations, events.

In a negotiating situation, the negotiator must learn as much as he can of the circumstances of the other party. Does he have enough information, or must he find out more about the other party's needs? Who has a time constraint? Is it the other side, or his? Is the deadline real, or arbitrarily set? A real time deadline can be an advantage to one side, a disadvantage to the other.

Power is sometimes an advantage; at other times the absence of power is. If the decision maker is the negotiator, he has the power to commit to the negotiated outcome, at the negotiating table. The next day, he may have second thoughts about one portion of the negotiated settlement, but it's too late. The negotiator with less power can also agree to the negotiated outcome, but with the proviso that he needs his boss's signature. Getting that signature allows valuable extra time, to review and, if need be, force a renegotiation of terms.

With these three elements common to every negotiation, the beginning negotiator, or the experienced one, can recognize when one of the elements is being used against him. If he has a late-afternoon flight to catch and the negotiator for the other side keeps digressing to other topics, he can suspect his desire to leave on time may result in last-minute deadline terms that are not to his advantage. But, by being willing to stay an additional day and saying so, he can blunt the other side's otherwise effective tactic.

Negotiating for Mutual Benefit

The best negotiating strategy is to strive for mutual benefit, where both sides feel satisfied. Both see acceptable gain for themselves. In order to work for mutual benefit, three important areas must be emphasized:*

- Building trust
- Gaining commitment
- Managing opposition

Building Trust.   This process takes place after one or more meetings, phone calls, and/or written messages have occurred. If both negotiators conclude that the negotiations will be open and aboveboard, and they are striving for a settlement that satisfies both, trust can be developed. This can be done by finding out the other's needs and honestly trying to meet them. Some risk is involved, but the adversary relationship can be transformed to one of collaboration in problem solving.

---

* From *You Can Negotiate Anything.* Copyright © 1980 by Herb Cohen. Published by arrangement with Lyle Stuart.

Gaining Commitment.   Commitment can be obtained by finding out what is really needed by the other side (as one builds trust) and structuring the settlement to incorporate the major needs. One cannot be opposed to a settlement that meets these needs, and will more likely become committed to the negotiated terms.

Managing Opposition.   Opposition is necessary in any negotiations because there will always be some areas of difference. It may be a difference of ideas, on how the negotiation is structured, or on some of the terms. Or it may be a visceral opposition—disagreement, opposition to the person. Accusations of underhandedness, dishonesty, scheming, all are examples of visceral opposition. It is difficult to overcome, once it has developed. It can be prevented by one's attitudes before negotiating, by being noncritical and understanding (keeping one's cool), and by not judging the other's motives and actions.

Compromise

Collaboration in negotiations is not equivalent to compromise. Compromise means each side gives up something it wanted. Compromise is almost a no-win result, not a win–win. Neither side has strong commitment to the agreement because neither receives what it wanted. There may be some solace in saying "Half a loaf is better than none," but there is no satisfaction in such negotiations.

For the manager who will become a negotiator as part of his job, there are training seminars and books that provide further insights into negotiation and skill development. The idea of win–win in negotiation to mutual satisfaction is powerful in many circumstances. It can move people away from emotional confrontation to cooperation and commitment.

POLITICS

In any organization, politics plays an important role in achieving organizational results. Only the naive engineering manager believes otherwise. Political activity goes on at all levels, sometimes consciously, sometimes not. It involves the utilization of tactics and people in order to influence decision making. This does not necessarily mean that people are being "used" in the negative sense of the word, but rather that various people and their actions are a necessary part of the process of self-advancement of an individual, or the promotion of an important project that needs the support of key upper-level management.

The word "politics" has different meanings for different people. For the engineering design manager it is part of the world in which he operates. If he wants to win further promotions and advance into upper-level management, he will have to understand the political climate of his company and use it to his advantage. If he wants to stay in engineering management, he needs to use enough political acumen to make sure he and his department are not stepping-stones for someone else aspiring to the top. However, if an engineering design manager aspires to a top-level position in his company, he should study the backgrounds of the company's top executives.

Unless those individuals have come from engineering operations, the chances of his achieving the top level are slim. Companies and boards of directors tend to look for the same background in their next chief executive that they did in the present or previous one, especially if the executive has been successful.

The Political Process

While each company's political environment is different, and must be assessed, the concept of developing and using politics in organizations is relatively the same. It involves the utilization of effective techniques to influence decision making.

Politics consists of three major processes, neither strange nor unusual, that are used in daily decision making and problem solving: information gathering, analysis, and implementation. Political specifics are somewhat unique to the job situation, however.

Information Gathering. This involves collection of data on the "players" and the "plays." The first step of information gathering is to identify the players. Some hold positions of influence; others are seeking it. Some can provide technical advice on how one maneuvers to gain influence. Identifying and categorizing the individuals involved is a necessary first step.

The second consists of gathering information about each. What are their qualifications, age and family status, strengths and weaknesses, professional and political interests, objectives and past performance records, both actual and perceived?

The third step is understanding the relationships among players. This includes ongoing alliances: family relationships, college ties, friendships, positive ties due to particular issues, and so on. Also one should recognize who are enemies (and why), patron/disciple relationships, and past debt alliances.

Regarding the plays, one needs to understand what the basic rules are within the company. How do people get chosen for a position? For example, must certain lengths of time be spent in particular positions, or must certain age requirements be met? The informal rules are equally important but more difficult to learn than the formal ones.

The plays come in a number of categories: those related to advanced information gathering, those related to information disseminating, those related to positioning a person for consideration regarding key issues, those related to the actual selection process. The process also involves evaluating which of the other players may be doing what to further their own careers.

Analysis. Analysis involves establishment of explicit goals and setting a "campaign" strategy and tactics to meet them. The goals could include a promotion, increased budget, larger staff, new responsibilities, greater compensation. Once the goal or goals have been met, the plays to be implemented must be selected, so that as little as possible is left to chance.

Implementation. This is carrying into action the overall strategy and the tactics that will make it happen. It requires not only tactical action but contigency thinking and timing. Knowing when to use a certain play is as critical as deciding to use it. If event X occurs, then play Y is implemented. X may be the success or failure of a particular play, or the occurrence of a particular event. Implementation should also be based on a timetable where various plays will build on one another.

Implementation also requires interactive thinking. As new events occur, new choices may be necessary or new tactics implemented. The players may change; the formal rules may change; alliances may change. The making of a play is likely to change some of the information gathered. If it becomes known that the person is seeking a particular promotion, that fact may alter existing alliances or cause a new candidate to emerge.

Using political acumen to win a promotion in a company is similar to running for political office. A similar series of goals, information gathering, analysis, and implementation is involved in each. The manager seeking to use the politics within his corporation to advance his career aspirations needs to study politics as a prelude to planning and making his own moves. Politics is neither good nor bad by itself; it is neutral. Playing politics requires a manager to set goals and use methods that are morally neutral, or good. If not, there is a high risk that unscrupulous methods will not work, and his personal career aspirations will suffer.

SUMMARY

The manager must be effective in many areas to handle his daily workload. Keeping on top of the job is probably the most important of them. If he does this effectively, he will have a good perspective on the motivational climate, whether any of his staff members are becoming obsolescent, and how he can turn this problem situation around. Reducing staff turnover can be accomplished by essentially the same means as preventing staff obsolescence, by assignments that are meaningful and challenging.

High performers with upward potential are promotable, but sometimes cannot handle the shift from technical excellence to supervisory excellence. The matter of promotion and its benefits/problems for the individual needs to be considered carefully.

Hiring a new employee must be done with care and planning. Sloppy hiring interviews can lead to poor fits between the person hired and the job requirements. In the short run this increases employee turnover; in the long run, it may mean termination or firing or an employee.

Pay raises are always welcome, but the performance appraisal that precedes them each year is of greater long-term significance. With subordinate and supervisor participation it can become effective in all of the key areas: self-motivation, preventing obsolescence, employee turnover, staff member termination.

Negotiations can occur informally on a daily basis, and on a formal basis between two persons or representatives of two companies. When the design engineering manager becomes a negotiator, his use of power, information, and time in negotiations will help him be effective and avoid being manipulated by others. An effectively negotiated agreement is one in which both sides are satisfied because they worked together toward a joint solution of the differences between their positions.

Politics is a natural factor in human relations, but many profess that they do not want to have any part in the politics of an office. However, a manager aspiring to a higher-level job will not achieve it merely by superior performance. He needs to understand and use the tools of the political process and apply them to his business situation. Using company politics effectively for personal gain is something like running for political office. While the political methods used by some may be wrong, playing politics in a company situation is neither right nor wrong per se.

14
Creativity and Innovation

Creativity and innovation are the twin forces of progress in industry. Creativity is the series of mental processes that brings something into being that did not exist before. It is either devising something completely new, such as inventing a light bulb, or combining existing things in new ways, such as using electronic components to build the first electronic calculator. Innovation is the process of introducing change, of bringing creative ideas into commercial reality in the marketplace. The two work well together.

The two concepts are related just as brain and brawn are related. One generates ideas and brings them into being—creativity. The other seizes a particular idea and makes it into something usable—innovation. Both are important aspects of a company's growth and development. In fact, when creativity and innovation cease, a company stops growing.

Creativity is perceived as part of new product development. It is. But in a broader context, it should also be part of improvements to existing products. Likewise, innovation is not restricted to the changes associated with new product introduction; it can be a new package for an old product, or the use of a larger-size freight car to transport a product to market at a lower cost.

In a company, there are certain factors that either enhance or inhibit creativity and innovation. These include the general environment, either in the company or in the design engineering department, and resistance to change. The manager seeking to improve the creativity of his staff must find it in his subordinates and develop it to a high level; and he must understand his creative employees and learn how to manage them. This chapter deals with these issues, as well as two others: the myths and realities of the innovation process, which is change-making, and how to prepare for change and implement it effectively.

COMPANY ENVIRONMENT

Most articles and books relating to creativity and innovation talk about the "environment" or the "climate." Those are interesting word choices because they compare creativity to growing things in a garden. The analogy is appropriate. A gardener prepares his soil, tills it, and loads it with nutrients and fertilizers. After preparing his garden, planning where his various crops will be, he plants seeds.

The soil here is people, specifically, people's minds. Seeds are problems, and opportunities. Sometimes they are idea "sprouts" that are planted in people's minds, with the expectation they will germinate, grow, and flourish.

The environment has a major effect on what happens to the seeds planted in the soil. If there is too much rain, the seeds will rot, and they may wash away if planted too near the surface. If there is no rain at all, the seeds will lie dormant, and some will shrink, wither, and die. If there is a little rain here and there, the seeds will probably sprout, but the resulting plants will wilt from lack of moisture.

Similarly, temperature, humidity, and sunshine are important factors. Temperature changes—a sudden spell of hot weather followed by a late spring frost can devastate an entire crop. Too much humidity can encourage the growth of molds and fungi, yet some plants require a humid environment to flourish.

Sunshine is another key environmental factor, the source of both energy and warmth. Too little sunshine causes many plants to grow more and more leaves to capture as much sunlight as possible; their fruit often suffers as a result. Too much sunshine is generally not a problem unless one of the other environmental factors is lacking. Sunshine without adequate rainfall leads to drought. Cold weather is sometimes modified only slightly by sunshine.

Thus, the environmental factors individually and in combination are vitally important to the growth and development of plants and crops of all kinds.

The above comments are not intended to be a primer for gardening, but to suggest, by analogy, that the company environment similarly can stimulate or inhibit creativity in its staff members. While it is true that one must start with good people (soil in the farming example), people can be developed more fully in the right environment. The company's environmental factors are not quite as simple as temperature, humidity, sunshine, and rain; but nevertheless they are, to a large extent, identifiable.

Sometimes the company environment for creativity and innovation can be deduced from its published information. For example, one company, a technology leader, listed these three long-range objectives in its annual report:

1. Creating technically complex products that solve important customer problems in unique ways.
2. Developing, manufacturing, and selling these products ourselves.

3. Insisting that all products be fundamentally new and proprietary in nature.

The company annually spends over 8% of each sales dollar on research, engineering, and development. It is, by design, a company committed to creativity and innovation. A continuing stream of new and unique products verifies its commitment to creativity over its 25-year history (Raychem Corp. 1982 Annual Report).

Innovation-oriented companies have a number of characteristics that, taken together, determine an environment that fosters creative ideas. Each must provide products and services that satisfy customer needs at prices that are "profitable," both to the customer and to the company. Without customers, a company cannot survive. Without sufficient profits, a company cannot long endure and maintain growth by investing some of its profits back into the company. The product must be important enough for the customer to use, and priced at a level the user believes is fair to him.

Product development is the focal point of company effort. It is impossible to be creative if there is nothing to create. Creativity cannot exist in a vacuum, without needs. In almost all instances, there is a need to develop new products, and many innovative companies actively pursue such needs, recognizing the value of the old saying that "Necessity is the mother of invention." Still other companies develop broad competence in basic technology areas, and can quickly develop spinoff products, which are only modifications and adaptations of the basic technology. Although both types of functional approaches are viable, the latter has better long-range innovative potential, particularly if the technology is complex and cannot be quickly learned or copied by a competitor.

These are some other characteristics of various types of creative and innovative organizations:

1. Open channels of communication. The company has suggestion systems, brainstorming approaches, small groups whose job it is to develop ideas. The organization encourages and fosters contact with outside sources for information and ideas. Ideas are sought, stimulated, received, acted upon. This positive approach causes more ideas to be generated.
2. Positive approach to ideas. Ideas are evaluated on an objective, factual basis. The merits of an idea are important, not the originator's status. New ideas are tested experimentally instead of being prejudged on rational grounds. Everything gets a chance. Ideas must be proved wrong; their originators do not have to struggle to prove them right.
3. Risk-taking perspective. The organization tolerates risk and expects

people to take chances. It allows them freedom to choose and pursue problems; its employees have fun. Failures are viewed as necessary learning experience often leading to greater success.

4. Heterogeneous personnel policies. Such policies as flexible working hours allow for the idiosyncracies of creative staff members. There is room for unusual, nonconformist employees.

5. Security of routine that allows innovation. In this type of company, production-oriented staff handle routine, standard actions, allowing creative individuals to "roam," thus separating creative from productive functions. The company has separate units or methods for generating versus evaluating new ideas.

6. Unique organization. The company has its own original and different objectives. It does not try to emulate its major large competitor, or any other company. It wants to be and become "itself."

Not every company will possess all of these traits that identify the environment in creative organizations, but creativity is seldom found in a company that provides none of them. Rigid operations, with clear structure and definite policies for everything and everyone, discourage creativity. Creative people either will leave to join other companies, or creative instincts will be stifled. The more assertive, self-confident individuals may leave to form companies of their own.

The design engineering manager must do everything in his power to develop an environment that fosters creativity. He needs to encourage his manager to recognize the need for a climate that enhances creativity, not one that discourages it. Since a marketing department needs to be as innovative as an engineering department, his ally may be the marketing manager. Together, their team of two can promote and initiate many of the environmental factors that improve the creativity of both staff groups.

The manufacturing manager similarly needs to be involved. Like the engineering manager, he needs to carry on routine functions and at the same time provide an environment conducive to change.

RESISTANCE TO CHANGE

Some organizations, such as the one whose strategies were described earlier, have a built-in creativity orientation, from top management on down. In such cases, a design engineering manager carries on and embellishes those policies already established.

However, some organizations tend to foster creativity in profitable times, but retrench in leaner times, presuming incorrectly that creativity can be turned on and off like a valve. Other companies have managers, or departments, with a high resistance to change, even an aversion to it. This is unfortunate because change is at the forefront of all creativity and innovation, both of which are driven by a dissatisfaction with the status quo.

While most, if not all, company managements profess their desire for more innovation and more creativity in their organizations, they often act in the opposite direction. There are probably two major reasons why: first, any change is usually accompanied by problems; and, second, many executives are evaluated by short-term results, quarter by quarter.

No company accepts frequent failures from an executive. Some tolerate few if any mistakes, and a major error could stunt or terminate one's career. Therefore, the cost of failure is so high that the surest way to avoid it is to avoid change. Don't try anything new; stick to the "tried and true" things that worked in the past.

Justifying New Products

In a company that resists change, the procedures for justifying new products are often quite involved. Various approvals are required, and sometimes re-required in different stages. The vulnerability of a new idea, the difficulties of defending it against critical attack by virtue of its uncertain facts, all lead to problems in justifying new products. If several managers in the company are averse to the problems associated with change, anything new can be blocked or stymied by the need for further justification.

The Price of Past Success

Even in a company that is resistant to change, some managers hit on a big winner, a product they championed that made the original targets, then exceeded them year after year. Will such a manager try to do this a second time and risk failure? Or will he play it safe and not risk his reputation? Some will opt for no change and will not try again. In some company environments, the manager will not chance failure by sticking his neck out, and thus he will avoid a second opportunity for success.

Change and Conflict

Change also means conflict, problems, and more problems, with errors and their correction. As mentioned, some organizations accept change willingly, some reluctantly, whereas still others resist change because of all its associated problems.

When a revised product is being phased in, the true amount of change involved is more than many think. New parts, which sometimes closely resemble the old parts, must now be used. Drawings come out to the shop floor with errors that should have been caught but were not. Written information is frequently difficult to understand or subject to more than one interpretation. Phase-in of the revised product and phase-out of the existing one lead to further difficulty. Old parts are used with the new assembly, requiring rework, and vice versa. About the time when the revised product has been brought on stream, additional inventories of old parts and assemblies turn up unexpectedly. That means more trouble.

These are only a few of the problems that occur in a manufacturing operation when change is introduced. The manufacturing department, the focal point of production activity, tends to be especially resistant to change because of the ongoing difficulties that occur almost every day just in keeping the operation moving. Product changes and improvements add to their worries.

Other levels of a company may be equally resistant to change, not only because change upsets the status quo and causes problems, but because change implies risk. The marketing group may not be fully behind the cost reduction changes because of concern about customer reaction. If the customers view the change as a cheapening of the product, or the salesmen do, there is a serious risk that customer purchases will decline. The loss of business would quickly cause financial problems, perceived to be due to poor marketing.

Similarly, purchasing may resist changing to new vendors—working with new vendors. A new one may not be as reliable in quality and delivery as the old; or may not continue his excellent prices on the second or third repeat order. On the other hand, a purchasing department that is under pressure to keep purchase costs down may bring in many new vendors, and experience poor quality with low price. This, in turn, could trigger a strong resistance to change on the part of quality control or product engineering.

Resistance to change, as described, is not necessarily an emotional issue, based only on feelings and concerns. Quite often it is based on unpleasant past experience that may repeat itself if nothing is done differently to facilitate change. While individuals can have empathy for such reasoned resistance to change, the market does not. Other companies manage to introduce new or improved products and bring them on stream. Over a period of time these product innovations change the customers' buying practices and habits, and an even greater risk is encountered—the risk that the world is passing by and the company is falling behind.

Innovation—whether it be modest or major—can be an effective means of fostering and building a company's image as a progressive leader in its industry. Absence of innovation due to resistance to change can have the opposite effect. A company's image becomes that of a stolid, nonprogressive "buggy whip" type of manufacturer.

Emphasis on the Short Term

Despite clear and obvious reasons why innovation is important and resistance to change is risky, many company managers are ambivalent about innovation. For example, they frequently want higher productivity and improved or new products, but without capital investment. Or they want these innovative achievements without disruption of the existing product flow, profitability, and cost structures.

These requirements may be impossible to meet, particularly without staff increases, capital, or both. As a result, innovations and improvements are incremental, modest, and generally ineffective. The financial restrictions placed on the staff inhibit their ability to bring innovative products through the company system and into the marketplace.

This seemingly irrational behavior of otherwise rational and logical managers is not caused by any quirks in their makeup or specific pressures of the job. Instead it is the effect of a generic problem that grips a huge segment of U.S. industry—overemphasis on immediate, short-term results. The quarterly P&L review, the monthly output figures, the "all-American" emphases on each year being better than before and each quarter showing significant progress toward that goal, are responsible for the general short-term mentality of business leaders. A general adversary position between business and labor, and between business, government, and the media, does not help the situation. However, a major cause is the emphasis on short-term results.

When a manager is expected to produce short-term results in a given three-month period, he must maximize his production output, minimize his input, and thereby achieve the maximum profit possible, consistent with selling products at reasonable price levels with respect to cost. All of these requirements combined are incompatible with change; so it is not surprising that some of the resistance to change occurs in upper management. Change implies short-term production problems, greater time and money inputs, and temporary profit dips; the expectation is for long-term improvements.

After studying U.S. and world politics of the past several hundred years, Professor Jay Forrester, a scientist and educator, observed that the net long-range effect of political decisions made for short-term benefit has been the exact opposite of the intention of the decision makers. Applying that same insight to many corporations, one can speculate that the long-term effect of short-term profit emphasis will be substantially reduced profits and financial problems for these companies.

There is reason to believe this situation can be changed as such insight becomes widely accepted. Companies whose present and future are based on innovation operate in both a long-range and a short-range mode. To a significant degree, they maintain a long-range perspective toward innovation while not losing sight of the reality that short-term profits fund the innovation.

Resistance to change is partly related to people being reluctant to change and partly due to companies being reluctant to change. In both cases, it occurs because change disrupts the status quo, creates problems, is risky, and costs money. Later in this chapter, ways to prepare for change will be discussed and analyzed. Change must occur in any growing organization. For some, in fact, change is the only "constant." And design engineering is usually the place to initiate change, based on the use of the creativity of its staff members.

DEVELOPING CREATIVE POTENTIAL

The basis for all creative action is in the human mind. Every significant contribution to the world—for good or for evil—has come from some person's mind. It may have been shaped and reshaped by another; it may have initially come from a mass of metal, strip of plastic, or beaker of chemicals, to be developed into its final form; but it began with an idea. An idea was the beginning of the wheel, the jet engine, the ballpoint pen, the electric light, nuclear energy, the transistor, solar heating, and a myriad of other ideas that have already made an impact on human life or will in the future.

Ideas come as responses to stimuli, such as requests or demands for new ideas, and sometimes just pop into the head from the subconscious after a period of incubation. Some people have more and better ideas, whereas others do not have any good ideas because no one ever encouraged them to think hard, or gave them clues on how to develop their skills.

Characteristics of Creative Individuals

In the following pages we will be concerned both with determining how to identify the more creative engineer and staff member, and with how to develop a higher level of creative output in the staff one already has. It is important to remember initially that creativity is not restricted to the educated, or to technical staff. Creative ideas can come from any staff member, at any time. A secretary, a floor sweeper, a machine operator may have as good an idea as the highest-paid design engineer in the department.

Creative individuals are present in every organization, whether their creativity is utilized or not. They can be recognized, most of all, by the variety and quantity of ideas they develop.

Table 14-1 covers 30 attributes or characteristics of creative people. No one has all of them, and some may have others not included. The listing indicates that creative people are different; they know it and don't mind being different. They generate a large number of ideas, enjoy or seek out problems, and have a high tolerance for the lack of clarity, the ambiguity, that surrounds problems or new ideas. They are usually high performers, and some get into management, where they may or may not excel. Creative persons accomplish more work than others in less time. They are change seekers and change makers, with, as has been described, both positive and negative consequences for a company.

Importantly, creative people are needed in a design engineering department because of the contributions they make to new and innovative products. Every engineering department and every manager must work to attract some creative talent, and must develop and nurture the creative talent that already exists within the staff.

Enhancing Staff Creativity

The following paragraphs discuss ways in which a technical person can further develop his or her creative potential. These are ideas that an individual can apply, or a manager can teach, in order to enhance the creative abilities of the staff. A person with limited creativity who works at improving his

Table 14-1.   Characteristics of Creative Individuals

1. Curiosity	16. Unconventional ideas
2. Imaginativeness	17. Compulsiveness
3. Self-confidence	18. Unpredictability
4. Open-mindedness	19. Prolific ideation
5. Concentration	20. Unusual sensitivity
6. Persistence	21. Ability to fantasize
7. Adventurousness	22. Problem orientation
8. Sensitivity	23. High level of motivation
9. Skeptical attitude	24. Ability to initiate actions
10. Tolerance of imprecision	25. Flexible responses
11. Resourcefulness	26. Good power of observation
12. Intelligence	27. Introspection
13. Responsiveness	28. Individuality
14. Independence	29. Positive attitude
15. High energy level	30. Acceptance of ambiguity

talent will be much more creative than a more gifted person who does nothing with it. In one sense, creativity can be thought of as a set of unused muscles that can be strengthened by regular exercise.

These ideas are listed in the general sequence in which one would approach a problem situation, but they represent individual areas that can be developed one at a time.

1. Develop a positive, creative attitude. This is a basic requirement. Attitude governs an individual's approach to almost any situation. Creative persons are confident that they can solve any problem that comes their way. So can people who want to become more creative. Most problems become less of a puzzle as they are worked on. The solution often is not clear at the beginning. One must accept as "fact" that he is equal to the task. Otherwise fear and self-doubt can freeze the imagination.

A creative attitude can be built on a foundation of small successes leading to larger ones. A person should regard a failure not as a personal shortcoming but as a mistake, a misunderstanding of the laws of nature, a learning experience, and a stepping-stone to future success. If the reason for failure can be determined, one can profit from it and maintain a positive attitude toward his own creativity.

2. Gather facts, opinions, information. Newsmen, police investigators, problem solvers are all good at this. Creative people are curious about problem situations. They want to know more about them. They ask the who, what, where questions. Often they ask why questions as well.

Rather than look for solutions immediately, they ask more questions about the problem. One asks questions to better understand what's happening. Questions are asked of people close to the problem, people who do not know anything about the problem, and people who do. The more information one has, within limits, the better one can approach a problem. New information, new facts are what is really needed, not just the existing facts and sometimes opinions that are filed in memos and reports.

3. Set problem boundaries. This is sometimes referred to as defining the problem, but setting the boundaries is a more accurate description. A problem usually has many, or *n,* dimensions; but considering a three-dimensional analogy, a problem may start off looking like a huge, amorphous blob of unknowns, and by looking for the limits, one can deduce whether it looks more like a sphere, a cube, or a pyramid—with a blob inside the shape's boundaries.

Boundaries resemble specifications for a problem situation into which one can fit creative solutions. This does not limit creativity any more than a road map with a destination limits the number of highways or sideroads one can take to reach the destination.

Often, one needs to write out the problem and its specification in order to properly understand it, and to make sure others do as well. Dr. Edward De Bono, who teaches lateral thinking, another approach to creativity, says that a properly stated problem almost generates its own solution. Others have stated that a problem well stated is half solved. Both are valuable insights. One boundary, which creative engineers are not particularly fond of, is the time boundary. Yet, as discussed in Chapter 7, time is a precious resource that is lost if not utilized effectively. Putting a time boundary around a problem means that an answer must be reached by a specific date. This will often be a stimulus to a result rather than an unwelcome roadblock. As writers know, working toward a deadline can be stimulating. Often a needed idea or solution will develop simply because it is needed at that point. A time boundary on a problem is just one dimension in the framework for a problem and its creative solution.

4. Use the imagination. Imagination is the faculty in the brain most important to creativity, and unfortunately one that is used too little by too many. When they were children, most adults imagined that brooms were horses, rifles, baseball bats, crutches, and other objects of daydreams. As adults, it was clear to them that a broom is a tool for sweeping a floor not much else.

As children advance in age, their judgment faculty increases as their imagination declines. Thomas Edison said that between the ages of 11 and 15, the average child suffers a "paralysis of curiosity" and a suspension of the powers of observation.

Curiosity and imagination are tightly linked. Children ask why much of the time, to learn new information about the world. Then they take the information and feed it to their imagination. A five-year-old may ask why a spaceship flies into the sky, but he uses the answer to further his imaginative flight from earth to the stars. An older child may ask the same question, why, but uses the answer to help out with a homework problem. Important as education is, school is often a place where there are only right answers and wrong answers. Only one is right, and many are wrong. One soon looks for the right answer and stops imagining the others. That is one reason why imagination is often stifled by increasing education.

It is only by developing an active curiosity that a less creative person can begin to reuse his imagination. The word imagination contains "image," which is a major clue to creative thinking. Images are brief pictures that flash into the mind as one thinks about a problem and searches for solutions.

Imagination should not be restricted to mental pictures. One should try to imagine in all five senses—sight, touch, taste, smell, hearing. A new idea may not normally be associated with hearing, but one engineer came up with a simple way to separate small castings, some of which were inadvertently cast from a lower-quality grade of iron. When he tapped each with a small hammer, one material sounded like a bell, whereas the other had a dull "thunk" sound. This provided a simple, inexpensive method for segregating the castings made from the correct material from those to be scrapped.

Imagination can be stretched by "think-up" games. How many uses are there for a red brick, a wooden pencil, a shoelace? The game of finding 25 uses for a red brick becomes a tough thinking exercise after the first four. Bricks build houses and walls, can be used with boards to make a bookcase, and make roads, but what else? Using the five senses suggests that the rough surface can be used as an abrasive, a tool to texture wet concrete. Its weight holds papers down, or holds down things glued together. Its dimensions may act as a crude measuring stick, its volume and weight as a water saver for a toilet. Using all five senses in a real problem can spark the imagination to come up with many possible answers in "think-up" exercises.

Another imagination stretcher is playing the mental games found in game books. Some involve placing matchsticks in different ways to form numbers or figures; others require extremely careful observation to find a misplaced letter, or a word hidden in a jumble of letters. Still others require interpretations of half cartoons, where the missing portion is hidden by a wall or a fence. These games develop the freewheeling visualization part of imagination. In an actual problem it is important to let the imagination run as freely as it does when playing a game.

In the creative method called synectics, developed by William J. J. Gordon, creativity is enhanced by working on an emotional rather than a logical concept. Members of a group, or individuals, change their perspectives to make the strange familiar and the familiar strange. For example, a material is being processed—molded or extruded—and the process is unsuccessful. Using the synectics method, one tries to imagine how the material "feels" as it is being pushed and squeezed by the forces that shape it. Perhaps it does not "want" to be pushed, or squeezed, but rather to be stretched or folded. If so, why? Why does it "prefer" to be treated differently?

This is a cursory view of synectics, which needs to be studied and used to be appreciated. Several books and articles describe its methodology and application in enhancing the imagination.

The imagination is a person's most important creativity tool (or muscle). Only by frequent use can its abilities be sharpened and strengthened. The foregoing are just a few of the many ways this can be done. Alex Osborne, pioneer in the teaching of creativity, devoted an entire book to the subject, *Applied Imagination,* which was first published 30 years ago. It still is an excellent source of ideas on how to stimulate one's imagination.

5. Be persistent. Many people quit after thinking up three or four ways to use a red brick or a wooden pencil. The first few ideas spring up easily; the remainder come more slowly. One must keep after them, and creative people generally do. Edison tested over 6000 species of vegetable and plant fibers before he found the three bamboo species that provided the first suitable light bulb filaments. German scientist Paul Ehrlich tested 606 compounds before he found the cure for syphilis. That is persistence.

Persistence is a measure of a person's motivation to solve a problem. Many problems will not succumb to the first attack. First ideas tend to be simplistic, the tried and true solutions that have been used before. Many times they no longer work.

Persistence means trying harder, not giving up—searching for another approach. Sometimes one is too practical in the approach. Ideas that are unusual, irrational, may have value: "what if" ideas, where the problem is worked backward from the end to the beginning, to see if a reverse approach will work; or perhaps starting in the middle of a problem and working toward both ends.

Persistence can sometimes be fostered by suspending the laws of nature in one's imagination. Would the problem be solvable if the law of gravity were repealed? In space, there is no gravity; liquids and solids float. What if water could flow uphill? What if plants grew horizontally, or grew without soil? What if heavy objects could be pushed, or slid easily? A number of creative insights can be developed if the imagination replaces the laws of nature with a permissive universe where possible solutions can be developed because the impossible is possible.

Sometimes a problem can be solved by looking at it from the back instead of the front end. A homeowner had a problem with a balky dishwasher. The motor would not drive and appeared to be burned out. However, it turned stiffly by hand, so he lubricated the bearing on the accessible end of the shaft. The motor responded slowly when the power was turned on, and then spun rapidly and operated the diswasher properly. After a few minutes, however, it began to squeal, and finally it stopped turning altogether. Additional lubrication was unsuccessful. But what about the other end of the motor? There was no oil cup to drip lubricant into the closed end, but it was possible to unscrew a cover at this end, thus revealing the end of the shaft, a felt oil reservoir, dry as a bone. The persistent handyman was able to free the shaft with a few squirts of oil, putting the motor (and the dishwasher) back in operation. The moral to this story: be persistent. Sometimes the real problem may not be visible, but is just out of sight, waiting to be resolved.

6. *Brainstorm.* Brainstorming is another way of being persistent, of searching for more and more ways to solve a problem. Alex Osborne started brainstorming as an organized group activity to generate many more ideas than an individual could. The synergism of a small group, one idea inspiring another, one idea building on another, can lead to creative output far

beyond what individual contributors could develop on their own, within the time allowed.

Numerous books and articles have been written about brainstorming, both for and against it. Brainstorming can and does work. Usually a small group of five or less, meeting for a specific length of time—even as short as a half hour—can be effective. As a prelude to such a session the participants should know about the problem in advance so they can think about it. Also, the problem should be reasonably well defined.

The session itself should be relaxed, with a positive orientation. Someone needs to be secretary and record all ideas, preferably on a board or large pad so everyone can see them. The moderator or chairman can be that person. The problem is restated to make sure that everyone understands it, and then brainstorming begins, with someone suggesting an idea as a possible solution.

Brainstorming will be successful or not, depending on how diligently these four simple rules are followed:

1. Avoid negative thinking. No one says "That won't work," or judges any ideas. Sometimes charging a 25-cent fine for a negative comment proves effective.
2. The wilder the idea, the better. Let imagination soar; suspend the rules; be unconventional. A wild idea is frequently the trigger for another thought.
3. Quantity, not quality, is the goal. Stretching the imagination for the 26th use for a red brick can force something novel to surface. If a group comes up with 30 ideas in a hurry, strive for 40. The best idea may be the next one thought up.
4. Try idea combinations. When the thinking slows down, check the list of ideas and have the group look for synergistic, bizarre, fanciful combinations. Sometimes open a dictionary and select a noun at random and try to fit it to the problem. Not surprisingly, this is called a random stimulus.

Once the ideas have been generated and written down, the session is over. No evaluation, no judging of any kind is allowed. Shortly after the meeting, or perhaps the next morning, a participant may have additional ideas to add to the list. These ideas are from the subconscious, which was still "thinking" about the problem.

Brainstorming is used, misused, and underused by many; but as a tactic for a persistent freewheeling attack on a tough problem, it is excellent.

Persistence in generating ideas requires effort; but as Charles F. Kettering, the late, inventive genius of General Motors, said, "Keep on going and the chances are you will stumble onto something. . . . I've never heard of anyone stumbling on something sitting down." Sometimes persistence seems like stumbling, but it does pay off in results.

7. Tolerate ambiguity, loose ends. The creative person does not mind ambiguities, disorder, or loose ends. He may or may not balk at pure chaos. The noncreative person, on the other hand, likes to build an orderly, logical answer and quickly strives to bring order and direction to his efforts.

The creative person obtains data, uses imagination, persistently attacks a problem, but doesn't worry about the ambiguities. He knows that when the problem is solved, everything will make sense. He sees the ambiguities as missing pieces of a jigsaw puzzle, and he works around them, or sets them to one side until some other information turns up to change the ambiguity into an insight.

Similarly, loose ends are analogous to jigsaw puzzle pieces. A loose end can be considered a key part whose identity is well known but whose location in the whole (puzzle) has not been established. (An ambiguity is a loose end that may or may not be part of this puzzle.) A loose end will quickly be tied down once the problem solving has progressed to a point where its location can be determined. The creative person refuses to tie down loose ends too soon, fearing that they may not fit correctly in a force-fit or tie-down location.

The noncreative individual would view disorder as chaos, whereas the creative person views disorder as merely a different structure, a random ordering of things and ideas rather than a logical strucure. Just as RAM's (Random Access Memories) make it possible for information to be retrieved rapidly from a computer, to the creative person disorder is a form of randomness that can be quickly accessed.

Many busy, competent people—creative to some extent—keep disorderly piles of papers, letters, articles, and books on their desks, in what appears to be an unholy mess; but ask for a document, or a fact, about an area they are studying, and in a few moments they will retrieve it from the correct pile.

So, disorder, ambiguity, and loose ends do not cause problems to the creative person, but are something he accepts as the "natural" order of an unsolved problem. In due time all the disorder will be converted into order, loose ends and ambiguity brought into proper perspective and clarified.

8. Playing with ideas. There is a lot of fund involved in playing with ideas as one searches for unique and novel answers to problems. The creative individual usually works harder and has more fun with ideas than the noncreative person. Playing with ideas is sometimes systematic, sometimes random. More often than not, playing with ideas involves association, adaptation, combination, magnification, minifying, rearrangement, and substitution.

This is how the seven concepts can be used when playing with ideas:

- Association: one technology provides an insight for another. Examples: an electronic solution to a mechanics problem, heat transfer to solve a fluid flow problem.
- Combination: two ideas together are better than one. Examples: eraser on a pencil, TV with electronic games, 35 mm slide show with a cassette tape player, ham and eggs.
- Adaptation: a device is used for something other than its original purpose. Examples: airplane seat belts for car safety, floor coverings to decorate walls, dots of glue to replace rivets.
- Magnification: make an item bigger, stronger; put more in a package; provide more pieces, more bulk, larger size everywhere. Examples: earth movers, dumptrucks, jumbo jets.
- Minifying: smaller is better, so reduce weight, or number; reduce time; make an item more compact, easier to handle. Examples: transistor radios, electronic calculators in a wristwatch, miniature batteries, low-calorie dinners.
- Rearrangement: change the order of things or change the sequence of events; make the first last; start in the middle; turn the device upside down, inside out; right becomes left; the cart comes before the horse; do things differently. Examples: front wheel drive cars, side by side refrigerator and freezer.
- Substitution: determine what else can do the job better or at lower cost, or last longer or be more colorful. Examples: metals for plastics, ceramics for metals, plastics for metals, electronic controls instead of mechanical, weathered wood barn planking for room decor.

These ways of playing with ideas are used normally and unconsciously by creative persons. They are quite valuable for the less creative person who wants to develop more creative capability, to realize a greater level of creative output. If used on a regular basis, they can become creative habit patterns after a time.

9. Suspend Judgment. This concept is easy for the creative person because he studies each idea, studies the problem, and tries to make ideas fit in the best way possible. The noncreative person tends to judge each idea as it is generated. Place such a person in a brainstorming session and it ends abruptly, after his second "It won't work" assertion.

The problem here, is that all too often a noncreative person wants to jump to a conclusion, to come up with the right answer quickly. Creative persons take their time. It is true that many experienced persons have a pretty good idea of what will and will not work. But unless judgment is suspended, and new ideas are allowed to surface, creative solutions to problems are too often squelched. Then, the old ideas are tried again and again.

10. Concentrate. When working on a tough problem, the creative person is able to focus on the problem and give it his full attention. Sometimes he sits in his office hunched over his papers, jotting down an occasional thought on a notepad. Or he may leave the office and walk or drive somewhere, to get away from the distractions that break his chain of thought.

Concentrating is not easy; many people have difficulty with it because their internal attention span is relatively short, and self-generated distractions come quickly. However, the creative person becomes absorbed in the problem. He turns it over in his mind, looking at it from different points of view. He reviews the facts and tries to see how they fit into the whole.

Once again the jigsaw puzzle analogy is appropriate. Some persons can become so totally absorbed in a jigsaw puzzle that they look for pieces with hues or patterns or shapes that fit the portion of the puzzle they are working on. An hour, or two or three hours, may pass before the puzzler needs to take a break—to get a drink of water, or stretch his muscles—before going back again. It is exactly that degree of concentration that a creative person brings to the problem at hand.

Whether it be sifting facts, or sorting out the ambiguities and trying to make sense out of contradictory ideas, the creative person brings all of his mental processes to bear. No facet is ignored, no bit of data overlooked. He may spend time considering the various solutions he or others have proposed—how well do they meet some of the criteria?

He may spend hours, even days, concentrating with the full conscious powers of his mind, and yet be unsuccessful. Yet he will not be demotivated or overly disappointed because the creative person knows he has another part of his mind hard at work—the subconscious.

11. Use the subconscious. Creative persons rely on their subconscious mind to continue the process of probing and searching through a problem toward a solution—and they expect the subconscious mind to come up with a high share of ideas. It does.

The mind had been compared to an iceberg. The subconscious mind is the invisible nine-tenths of the human brain, the conscious mind the visible one-tenth. The subconscious quietly processes information, drawing analogies and reviewing data buried deep within the memory. Often a possible solution will float up to the conscious mind disguised as a creative hunch, an intuition. On occasion it will burst forth with the brilliance of a photo flash, insisting that it is recognized!

The subconscious mind helps both the very creative and less creative persons only after the mind has become thoroughly immersed in a problem. One must have spent time thinking up possible solutions, adapting ideas, magnifying them, minifying, combining them into possibilities. And most of these attempts will have been unsuccessful.

When one can do no more, it is time to let go, to let the problem be ignored, to let it incubate in the subconscious. One relaxes and does something else. Some may play a musical instrument (remember Sherlock Holmes?); others will exercise, garden, cut the lawn, listen to music, find and enjoy a TV comedy show. Sometimes daydreaming will unclutter the conscious mind and open it to new inputs or ideas.

A major insight—the flash of illumination, the "Eureka!" or the "Aha!"—comes occasionally, but most often the insight will be just a flicker of light that must be captured and written down. The creative person regularly uses his subconscious; the noncreative person should do so but does not.

These characteristics of creative persons relate to the creative problem-solving process, and so can be used by a manager observing the actions of his subordinates to determine how much natural creative talent his staff members have.

FINDING AND DEVELOPING CREATIVITY IN STAFF MEMBERS

Creative people are both born and made—but they do not necessarily advertise their presence. It is possible to test individuals to find out their level of natural creative ability, but experts are divided on the value of this testing. Thus is it up to the manager to study his people and learn which ones are more creative than others.

Given the sophistication of today's engineers, it is not too difficult for them to fudge the answers in a psychological test and appear to be more creative than they really are. In that event, one of two outcomes is likely: first, the employee's creative inadequacies will surface, showing his deficiencies and ineffectiveness; or, second, he will be motivated to strive mightily to live up to the fictional level he has "claimed," and in doing so will become much more creative than he would have been.

Most of the time, an observant manager can identify his more creative staff members because creative people are exciting to be around. The old saying that "Actions speak louder than words" is appropriate for creative staff members. They can be spotted by the quality and quantity of their suggestions, ideas, and products generated on the job.

Creative people generally are ambitious achievers, who have aspirations to become better and better at what they already do well. They are internally motivated to accomplish results. Sometimes they undertake tough tasks just to prove they can do them. The manager who has one or more creative engineers, technicians, or drafters must learn to manage them effectively, as will be discussed later in this chapter.

If he is so unfortunate as to identify no creative persons on his staff, he should—in the long term—look to hiring new, creative individuals and transferring out some of his uncreative ones. For the near term, he must work with his existing staff to improve their creative output.

Development of Creativity in Staff Members

While the level of creative talent varies among individuals, there is no question that training can substantially improve the output of both the more creative and the less creative people. In a test program run a number of years ago by the electronics division of a company, the objective was to improve the number of ideas generated in response to problem situations.

Before the training program, the individuals were tested to establish them as higher- or lower-creativity engineers. The higher-creativity group developed an average of 106 ideas before training; the lower-creativity group only 60.

After the training course, the higher group improved by 25%, to 133 ideas. The lower group improved their average to 100 ideas, a 67% gain. The lower group showed this increase, according to the instructors, because they had not been aware of how best to utilize the creativity they had, without training.

Training to develop creativity can be done on a company-wide basis internally, through outside seminars, through books, or through departmental instruction or seminars on the creative problem-solving process.

Outside seminars are available through consulting groups that specialize in creativity training, or as university-sponsored seminars, such as those offered by the University of Wisconsin–Extension and other educational institutions. These seminars last two to three days and teach the practicing engineer the "tools of creativity" and how to use them.

If a company does not have a creativity training program, seminars provide an excellent alternative, presuming that the manager has a planned, budgeted program for this training over a period of several years, and sends more than one person at a time. A team of two or three ambitious engineers, regardless of age, can bring back far more enthusiasm and interest than one individual. There is a synergistic effect if associates go to the same or similar programs in succeeding years; the programs act to refresh the information learned and provide all involved with new methods for approaching problems.

Books are another source of ideas on creative thinking and creative problem-solving. There are many excellent books on the market for building a creativity study collection. Osborne's *Applied Imagination* should be the first book purchased. However, books are valuable only to those who take the time to read, study, and learn from them. Many engineers will not take full advantage of books and need a more direct approach.

Engineering Department Creativity Seminar. A series of short talks on the creative problem-solving process can be given by staff members as a do-it-yourself creativity seminar. While this method does not have the scope, polish, or status of a professional seminar, the ideas and methodology demonstrated by enthusiastic staff members can awaken dormant creativity. This is also a demonstration that the manager is interested in improving the creativity of its staff.

Such a program could be handled on a once-a-week basis, for several weeks, for about two hours per session, and should include a reasonable fraction—say, one-fifth or one-fourth—of the staff at a time, or at least four people per session. The topic to be covered is the creative problem-solving process, as outlined in Table 14–2.

Five of the six steps were discussed earlier in this chapter. The sixth, implementation of the solution, is carried out in the same way as implementing a decision, which was discussed in Chapter 7. The information from these two chapters could be used as a basis for the seminar, embellished by outside reading on creativity subjects, using some of the references mentioned in the bibliography provided at the end of this book. In other descriptions of the creative problem-solving process, the words used may be different, but the steps are substantially the same.

An internal program can be substantially improved if the manager is present at the opening session to express his philosophy on the need for creativity—to set the tone—and at the last session to find out if the group has developed any solutions that bear on current problems. He can also use this opportunity to tell the group of any creative solutions that have been found for company problems. This is important. A practical application of suc-

Table 14–2.   The Creative Problem-Solving Process.

1. Formulate the problem to be solved. Set its boundaries; define it.
2. Gather all data relevant to the problem.
3. Organize, study, and digest the data in the context of the problem.
4. Develop a large number of ideas and potential solutions that bear on the problem.
5. Put the problem aside consciously, and wait for the subconscious mind to act upon the information and ideas and report back.
6. Act on the solution ideas received, to implement the solution.

cessful creative problem-solving is much more of a self-motivator than mere words.

By this means, the manager will be able to instill in subordinates an awareness of how important creative ideas are to product development and improvement. His personal backing of the effort reflects the degree of commitment he makes to developing creativity in his staff members. If this support comes from his superiors as well, his task is made much easier.

MANAGING CREATIVE ENGINEERS

Creative engineers are different and must be managed differently from noncreative engineers. Some managers view creative engineers as troublemakers, prima donnas. If managed from that perspective, the creative engineers will become disgruntled, unhappy, and demotivated. One or more may leave the company and go elsewhere.

The number of creative people in a staff may be small. Some authorities believe only one engineer in 50 is creative, others one in ten. To properly manage that small number, the design manager must treat each engineer as an individual and decide to what extent the rules can be bent or broken regarding working hours, styles of dress, financial rewards, and incentives.

In broad terms, the manager must lead his creative persons rather than attempt to direct and control them. This requires clearly defined goals and results. As in many other areas, the manager defines what is needed, but not how it should be accomplished.

These four principles are important in effectively managing creative engineers, and are effective as well in managing all other engineers:

1. Strive to build an environment (climate) that encourages ideas and changes.
2. Develop a positive approach in order to stimulate and encourage each individual's creativity.
3 Be a good listener.
4. Give recognition to good new ideas.

If there is one single factor important in managing creative engineers, it is the manager's personal relationships with his people. Personal relationships include being honest with people, providing straight answers, criticizing frankly and privately, balancing criticism and praise, keeping the staff informed, providing feedback from top management on good work accomplished by individuals, showing concern for personal problems, and providing psychological rewards. None of these suggests that the manager should be a buddy to his creative engineers; he must remain in charge.

This many-faceted personal relationship with each and every individual will enhance the effectiveness of the group and make it possible for creative things to happen. The staff members will appreciate the honest sincerity of the manager, and will respond accordingly.

The manager also must allow some experimentation. Creative ideas sometimes are unconventional and may not appear to be "right" answers; but if none is tried, the creative engineers will soon realize that all of the talk about the need for creativity was only words, that innovative though unconventional ideas are really not wanted, except in theory.

Ascertaining the failure or success of ideas by testing them is more objective and acceptable than making premature judgments. However, this approach may make some managers uneasy. Costly test failures on unusual ideas can be embarrassing and difficult to explain. The task of managing creativity is not necessarily an easy one because there are some risks involved. Nevertheless, the rewards in terms of new and improved products can be substantial and worth all the effort.

A final thought on managing creative engineers: There is pressure on a manager to produce timely results, and this pressure can be a positive influence on creative output. Though some say creativity "does not punch a time clock," a deadline is a strong motivator for results. Several studies on innovation have demonstrated a positive correlation between creative output and fixed deadlines. Necessary ideas can come in on time. The deadline does provide a strong incentive for accomplishment.

Encouraging Creativity in Nonengineering Staff. While the creative engineer is the obvious focal point for further creativity training, it is important to remember that all staff members have creative potential. Technicians, drafters, clerks, secretaries, all are individuals who can make a significant contribution in creative ideas. The wise manager will involve all engineering staff members who wish to be involved in creativity training. This may cause some disgruntled feelings among the more status-conscious engineers, but the total benefit to the department must be considered.

The creative output of these individuals can enhance that of the engineers, since in many projects they are as close to the work and the problems as the engineers. This benefit can be lost if they are not involved in creativity training. It is then easy to develop the attitude that only engineers' ideas are sought after and welcomed. A "turned-off" technician, drafter, clerk, or secretary becomes a talent that droops to a mediocre performance level instead of performing above the average.

MYTHS AND REALITIES ABOUT INNOVATION

One of the myths about innovation is that company managements are innovation-minded and willingly innovate to bring newer products to market. Earlier, the problem of resistance to change was discussed as a barrier to innovations. And it is a major barrier, particularly when it is motivated by an illogical emotional resistance rather than a serious evaluation of the effects of the innovation.

In an objective evaluation of a proposed product or process innovation, a number of issues must be realistically addressed because not all innovations "deserve" to survive just because they are different. Five important issues are represented by the following questions and discussion:

1. Is the product significantly improved by the additional components or the new technology required to make use of the innovation? If the answer is no, or no one is certain, then innovation for its own sake is a waste of resources.

2. Are present business operations displaced or weakened by the new innovation? This is a tough question. If the answer is yes, some difficult decisions must be made. Consider the time when transistors began to replace vacuum tubes. The solid state devices had a much longer life than the tubes, so tube testing and routine replacement were phased out on an industry-wide basis. A segment of replacement business and dealer service opportunities declined. Yet, if that innovation had not been phased in, the business would have perished anyway because competitors would have introduced the transistor and taken over the marketplace. Thus, a difficult business decision is often necessary in which long-term potential gain or loss is not so easy to forecast as the immediate effect.

4. Does the product innovation provide a significant benefit to the user? This question is similar to the previous one in that it helps management make a tough decision a little easier. If the user sees benefits in the long run, he will opt for the innovation, given the choice. Thus whether the long-run effect is to improve business, have no impact, or change the business, value to the user will usually dictate the best course of action compared to any other decision objective.

5. Will the innovative product expand the market? This is an important area because an expanded market means growth. Market-expanding innovations are sought after.

The myth that all companies welcome innovative ideas and will immediately implement them is only a myth. The reality is that innovations must meet some tough and realistic criteria, including those detailed above. Another reality is that resistance to change often hampers innovation. Change must be planned for, in order to be implemented well.

CHANGE—PREPARING FOR IT
AND IMPLEMENTING IT

It seems to be true that the only constant in the world is change. Yet, as said earlier, many managers and staff members resist change because of the risks and uncertainties that accompany it. However, unless change occurs in a business, it will stagnate and be passed by, as competitors change and update their products.

Engineering and marketing departments jointly or separately are the usual initiators of change. Both respond to the need to make improvements in the marketplace through new or existing products. Because of the work involved in developing products and phasing them into production, engineering plays a major role in planning and preparing for change.

Effective planning for change includes involving people early, well in advance of the change. This does not include only the engineering staff working on the project; they were involved from the beginning. It also includes manufacturing and quality control staff and others who will be involved in change that will alter their current activities. Early involvement gives them time to understand and accept the need for change. Then they can adapt to the upcoming situation and more willingly participate in the change.

Involvement means more than being sent a memo describing a future change. It can mean participating in the early design reviews; then the process of change is still anticipated so there is time to adjust to the idea. It may also mean working closely with the design engineer on a day-to-day basis, to understand why the design changes are necessary, and how to better phase them into production.

A written plan is needed to make sure the change has been thought through thoroughly, and that it supports company objectives. The plan must be communicated to those affected, again to give them time to adjust, react, and accept the changes.

The process of change, whenever it involves people, requires their awareness that change is coming, followed by understanding of the need for it. After understanding, acceptance and commitment follow as value to the company can be perceived. As new approaches become standard procedures and the change is accomplished, what was new becomes the norm.

Implementing change should not be so easy that it is done sloppily simply because no one asks a lot of questions. It is far better to have enough resistance to change to ensure careful evaluation of the factors and problems involved.

Most projects can be phased into production without great difficulty if there has been adequate involvement of affected people beforehand. Nevertheless, there are several steps that need to be addressed:

1. The change process must be organized. Change does not "happen" in the right sequence automatically.
2. Change must be implemented when standard operations are ongoing. This can be a difficult task because there must be minimum interruption or disruption to the bread-and-butter, ongoing activities of the area.
3. Using a project task force, or team, as described in an earlier chapter, is one of the most effective ways to phase in change. The size of the task force depends on the scope and importance of the change.

Change is not universally resisted; in many cases some change is welcomed by almost everyone. It keeps the job interesting. People will accept change more readily if they can talk about it and are able to contribute to it. Change is accepted faster in several small packages rather than one large one. Small changes are less threatening, but too many simultaneous changes will slow each other down. In the same vein, the more specialists in an organization, the more likely it is that conflicting changes will be introduced, as each sees the change from his own, narrow perspective. The manager of design engineering must see that his specialists and those of other support groups do not have a free hand in introducing too much change, which can be quite disruptive.

Innovation is the key to the growth development of a company as well as the people who participate in it. Early communication and involvement are important to reducing the resistance to change inherent in innovation, and converting it to commitment. Communications and progress milestones are important during the implementation stage to keep the effort strong. These major factors were discussed in earlier chapters.

SUMMARY

Creativity in staff members, especially in design engineers, is the product of natural talent and of training. The company environment has a great deal to do with fostering creativity, or inhibiting it. Resistance to the changes brought about by creative problem-solving exists partially because of the natural human tendency to retain the status quo and partly because managers are often judged by short-term rather than long-term profit results. This is not always understood, especially by creative people; but it is a reality. The company that truly wants to grow and innovate will create a climate that fosters creativity and enables it to flourish.

Creative people are different, and they know it. Their distinguishing traits are many, and they are oriented to problem solving in unique ways. Not every design staff has naturally creative people, but creative skills can be developed in everyone who is willing to acquire them, by training and practice. Creativity can be managed effectively and in a way only slightly different from the management of engineering staff. Openness, trust, and some freedom of action are necessary; yet deadline pressure and results orientation need to be stressed.

Innovation is the process of introducing creative ideas into the plant via new products or processes. It can be done effectively by planning and preparation and a great deal of communication (spoken words, not memos) throughout the cycle. Small increments of innovation are easier to accept than large ones because they do not threaten familiar activities; yet they do build momentum to implement further changes. Since change is the only constant in a growing viable business, change is a way of life unless a company wants to be passed by its competition and left behind.

15
Patenting Inventions

BACKGROUND

Patents have several different meanings. They have a potential commercial value to a company; have a personal, status, and monetary value to an inventor; and have a mystical value to the general public. This chapter will provide insights into these areas, and will include a section on disclosures of inventions plus some background on the U.S. and foreign patent systems as they exist today.

WHAT IS A PATENT?

Briefly defined, a U.S. patent is a limited grant from the government to an inventor by which he may exclude others from making, using, or selling his invention. It claims the fundamental concept as the original idea of the inventor. A U.S. patent expires 17 years after the date of issue and cannot be renewed. After that time, anyone can use the idea; it is in the public domain.

In exchange for this exclusive right, the inventor (via the patent) teaches the substance of the invention so that anyone competent in the appropriate engineering field can understand and reproduce the device or process. This limited grant for exclusive use is and was intended to encourage invention and innovation, in order to bring new and useful ideas into the commerce of the country. Other countries have similar and different rules regarding patents, as will be reviewed later; but in all cases patent issuance is intended to foster the spread of useful ideas into commerce.

Patent Myths and Fallacies

Before discussing real and perceived values of patents, it is important to review and comment on the perceptions of patents in the outside world. Many perceptions are fallacies and need to be debunked. There are two important ones: first, that the inventor who "builds a better mousetrap" will find the world beats a path to his door; and, similarly, that the easiest road to wealth for an inventor lies in acquiring a patent.

In the first instance, the person making a significant change in technology has a difficult task. It is he who must beat a path to the door of customers, who may be satisfied with the status quo and unwilling to try something new. Examples include Fulton and the steamboat, Morse and the telegraph, Bell and the telephone, the Wright brothers and their airplane. More recent examples include the instant cameras and the xerographic process for making photocopies. Both of these processes have revolutionized their respective technologies and market areas; but, like the older examples, they were not instant sucesses. In every case, the inventor had to overcome the inertia, apathy, and, in some cases, opposition of a disinterested world before being able to commercialize the invention.

Regarding the second fallacy, that wealth comes from patents, the million dollars that is attributed to a patent is largely fictional. Only a very few patents bring in a million dollars or more to the inventor, or to the company to whom they are assigned. Most patents are worth much less. Some are worth less than the money spent for the attorney's services and patent office fees involved.

Why, then, do companies patent inventions? Some patents are drawn to prevent competitors from copying features or variations that are not commercially useful. Sometimes one patent in a particular product area is relatively weak, and the inventor seeks to develop other ideas in the same area, so that the group of patents is relatively strong because they overlap. However, few of the nearly five million U.S. patents that have been issued have been of million-dollar value, although many patents are worthwhile in that they present an obstacle to a competitor wishing to make or sell a direct copy of an invention.

Last, a patent may potentially have a million-dollar value, but only until a competitor is able to design around the patent without infringing it. When this happens, the patent's value may be reduced or nullified.

VALUE OF PATENTS TO A COMPANY

Despite the negative picture suggested by the previous section, patents do have value to a company, particularly one that is growing stronger in its field of technology and wants to remain a leader. The value is largely commercial, although there is also a leadership aspect inherent in a company that has a "portfolio of patents" in an area rather than two or three. Leadership may be in R&D and engineering, which conceived and developed the ideas, and/or in marketing and production, which brought them to the market.

The commercial advantage of a patent to a company occurs for these reasons:

1. Competitors with greater resources cannot move in and take over the market with an identical product.
2. Competitors must then do one of the following:
   (a) Wait 17 years until the patent expires.
   (b) Negotiate a license.
   (c) Make an investment of time and talent to find a noninfringing alternative.
3. If a competitor does infringe the patent, in a product brought to market and sold, he can be sued for damages for past infringement. He can also be enjoined (prevented) by a court from making and selling the infringing product in the future.

The advantages are real, and can propel a company with inventive talent to a position of dominance in an industry. Since the patents can be licensed or even sold, the company can receive monetary value from an invention whether it is produced or not. This can be in the form of a single payment in cash or can be an annual royalty based on a percentage of the licensee's sales.

Patent licensing early in the life of a patent offers an advantage and a risk at the same time. Since most companies' large customers want two sources for a product rather than just one, early licensing can provide this second source. Often, this substantially increases the total market available to both suppliers. On the other hand, there is a risk that the licensee may become a more efficient, lower-cost producer. Then, the licensor would see his market share and profits erode to a much greater extent than the amount of royalties he would receive.

Recognizing there is some risk in licensing, there also are further advantages in licensing a useful patent:

1. Companies can cross-license each other, with an exchange of patent rights, where each has one or more patents to "trade" with equal perceived value.
2. A company may license even though it is reasonably sure its similar product does not infringe the patent—if the license fee is reasonable and it wishes to avoid the possibility of costly litigation.

The total patent picture is more complex than has been presented thus far. The complexity of the patent process, the huge scale of patent operations in some companies, and the vast library of patents already issued have reduced the worth of a patent in many technical fields.

Because of the overlapping of a large number of patents in specific technology areas, companies may be in violation of the patent rights of others, either knowingly or unknowingly. Sorting out the details is a difficult process, requiring large staffs and major efforts to define the technical and legal issues. This is one more reason why companies will seek to cross-license—to arrive at fair agreements in the exchange of patent rights to minimize the possibilities of a court case.

Improvement patents represent another reason for cross-licensing. One company may be issued a patent, and a short time later a second company receives a patent on a major improvement to the product involved. The first company can make the product, but not with the major improvement; the second company cannot make it at all because it does not own the basic patent. In such cases, the two companies will often cross-license each other so both can make the improved product available to customers.

Are Patents Necessary?

Some companies view the patent procedures, the delays, the loss of claims, and the sometimes limited value of the final patent to be enough of a deterrent that they do not bother to seek patents. One entrepreneur in the medical technology field said that his small business was producing equipment well ahead of his competitors. Rather than apply for patents, pay the costs, and experience delays until their issuance, he preferred to apply that effort and funding to develop improvements to his devices and stay two years ahead of his competition.

This strategy can be effective for a small firm in a rapidly growing technology; but it does ignore the possibility of licensing patented ideas, which could be another source of revenue for the future, when the company's product improvement ideas might be less prolific and the competition not so far behind.

There are a few cases in which patents may not be needed:

- The company has the technology to be the lowest-cost manufacturer and produces at a very low cost.
- Manufacturing the product requires a high capital investment, thus reducing the number of potential competitors.
- The product is very complex and difficult to copy.

However, there also is risk involved in not applying for a patent. It is the risk of someone else's patenting the same invention. Inventors in the same technology area are often working on similar ideas but at a different pace. Edison encountered this, as did Alexander Graham Bell. It occurs in today's technologies, as well.

It is entirely conceivable that one company could introduce a product innovation and decide not to apply for a patent, while, at the same time, a second company might complete an identical development effort and apply for the patent. Should it receive a patent and sue the first company for infringement, the second company could win the suit. Under that circumstance, the first company would have to pay to license the patent or possibly be prevented from manufacturing the product it introduced in the first place—an ironic, and complex outcome, but a possible one.

Lawsuits

The commercial value of a patent may be so great that a company decides it must go to court to protect the patented invention from being copied by competitors. Thomas Edison commented, "A patent is a license for a lawsuit." In many cases, this is true.

Legal procedures are very expensive, as well as time-consuming. The pretrial activities disrupt the project work of technical staff members, who must dig out old records, deciphering old drawings and cryptic notes, in order to build a solid case for their attorney's efforts. Typically, costs to a company are over $100,000 in such cases and may go much higher, depending on the case.

It can take several years for a case finally to be concluded. And there is some risk that the patent will be declared invalid. Sometimes the "infringing" competitor has uncovered older or prior patent art, or records of older products that convince the judge that the original patent was not valid. Without a valid patent, there can be no infringement.

The company with a patent that takes the case to court thus has a chance of losing as well as winning. Because of this, in many cases, there will be a concerted effort to license a patent at a reasonable cost to avoid all of the expenses and business disruption associated with litigation, as well as the risk that the patent will be declared invalid.

Patents and New Product Development

The development of a new product is a risky yet rewarding process. Often, the creative efforts that are required to solve design problems are novel, inventive, and the source of a number of patentable ideas. These may be valuable enough to enhance the company's position in its marketplace, particularly in a growth industry.

However, in an industry where the growth rate is slower, the industry more established, new product developments are viewed from a different perspective, that of existing patents and potential ones. Some companies will review the patents already issued to them and their competitors that could impact the potential new product. Then they review the project proposal and make an assessment of what areas may have patentable solutions to the anticipated problem. Only if the potential of additional patent coverage is high enough, will extensive new product development be undertaken.

Patents can be beneficial to the company and often are essential to its growth and market position. Licensing and cross-licensing are methods by which at least two manufacturers are able to offer equivalent products to customers who insist on dual sources. Licensing is also a means of bringing in royalty income for the use of a patent and avoiding the possibility of an expensive lawsuit to stop infringement.

VALUE TO THE INVENTOR/ENGINEER

Patents have several types of value to the inventor and the engineer, including their value as technical reports and as state-of-the-art indicators; their value as motivators; and their monetary value, due to patent awards to inventors.

Patents as Textbooks and State-of-the-Art Technical Reports

A patent issued today has likely been "in process" for two years or more since the date of application. If the patent issues, it has passed the scrutiny of a patent examiner, with some revisions and possibly limitations of claims. What remains is the substance of what the U.S. Patent Office agrees is both novel and inventive. It also reflects the state of the art not much more than two years before, when it was filed.

Textbooks often use material at least five years old or older, especially in technical areas. Technical journal articles are definitely newer than texts, but any detailed information considered proprietary is either omitted or veiled by vague words and inferences that mask its substance. On the other hand, a patent must be so clear that a person skilled in that technical field could understand and build the device himself.

TIPS ON READING A PATENT

There remains the problem of reading a patent. Many engineers avoid reading patents or approach them with great reluctance because of their unfamiliar language and structure. However, patents contain a wealth of information. A patent assumes the reader has little prior knowledge of the patented material (contrasted with technical papers, which assume prior knowledge). Each patent is a reasonably complete discussion of the problem addressed, shortcomings of previous solutions, the inventor's solution, and his ideas about practical use of his findings.

The engineering design manager should become conversant with the patent art his department functions with, and should know how to read and understand patents. Then, he should require that his key engineers and project leaders become familiar with patents in their field, so they can assist him in evaluating the patent art that pertains to their design efforts. The long-range benefits from reading and understanding patents far outweigh the short-term learning process.

With these thoughts in mind, the first tip on reading a patent is to make sure that the patent is worth reading. Some will be; others will not. The first procedure is:

- Read the title. It should describe the invention, but sometimes it may be misleading or simplistic.
- Look at the drawings. The drawings will provide a good idea of what the invention looks like.
- Read the abstract. Many patents have abstracts at the begining. They provide a good idea of the substance of the invention being disclosed.
- Read the first claim. Claims are numbered and appear at the end of the patent. The first claim is usually the broadest and the easiest to read.

Reviewing the Full Patent

After this preview, it will be clear whether the patent should be discarded or reviewed carefully. Review of a patent includes the following:

- Read the background of the invention. This covers both the field of invention and the prior art. Its second portion describes the problems and how prior art has failed to solve them. Usually this section describes the state of the art and provides some technical information.
- The summary of the invention, which includes the objectives the inventor tries to meet, should be scanned carefully; but do not spend too much time on it, or on the brief description of the drawings.
- Thoroughly review the description of the preferred embodiment. Patent law requires that the invention be described in full, using clear and concise terms so that anyone skilled in the art can make and use the invention. In this section, the invention is described in detail, with examples.

    When each component of the invention is mentioned for the first time, it will be followed by a reference number, corresponding to that component in the drawings. It is a good idea to unstaple the drawings from the patent to study these parts, or to make a photocopy of the drawings and mark them in color. If there are examples of test data, they should be studied carefully if they are of interest.

- Read the claims. Everything described in the claims should have been described earlier in the description. The first claim is the most general and is a good invention summary. It is worth reading more than once. The succeeding claims should be reviewed, as they are generally more detailed and restricted than the first. Groups of succeeding claims are often dependent on an earlier one.

- If the patent is of serious interest in an area being worked on in a product development progam, it is prudent to look up the patents cited as references. These patents were evaluated by the examiner as being closest to the invention. They may impose some limitations on the patent that may not be obvious from reading only the patent and its claims.

 In a situation where the patent relates to development work being done by engineering, the manager should consider ordering the patent's file history from the patent office. These documents provide a record of all the official correspondence between the patent office and the inventor, or his attorney.

 In this area, the advice of the company's patent attorney can be valuable. The important relationship of the engineering manager and the patent attorney regarding patent disclosures and other patent matters will be discussed later in this chapter.

- Summarize the patent. After the patent is evaluated, a brief one- or two-paragraph summary of its key points should be written so that they can be recalled in the future.

A Caution on Patent Review Summaries.   One note of caution is in order. If the engineer reviewing the patent believes that a product being designed by him or by others infringes a patent he has reviewed, he should report his views to the engineering manager and/or to the patent attorney. He should not formalize this view in writing, however, If he does, an undesirable problem may be created by his written statements.

The following hypothetical example could occur in real life: After reviewing a patent, an engineer sent his manager a letter, indicating that their new design probably infringed that patent. But in the opinion of the patent attorney the company's product did not infringe, and production began shortly. A year later, the company was charged with infringement of the patent mentioned. After unsuccessful negotiation efforts, a lawsuit was started.

Pre-trial procedures include "discovery." Here, each side must provide the opposing attorney with copies of all relevant correspondence and documents. As our example trial progressed, the letter surfaced. The judge asked the company's general manager why the company proceeded into production without obtaining a license, since one of its professional employees, an engineer, was certain the product infringed the patent in question.

The startled manager replied that he had relied on his patent attorney's opinion rather than the engineer's. The judge then asked him if he respected the engineer's technical and professional knowledge. When he responded yes, the judge said that the manager should have equally considered his engineer's opinion on the patent question. The company eventually lost the case and had to pay substantial additional damages in settlement. This was partly due to the negative influence of the engineer's written viewpoint, which affected the judge's decision.

In describing this hypothetical case, a patent attorney pointed out that the engineer should not have written a letter in any case. Had his letter stated that their product did not infringe the patent in question, that view would not have positively influenced the judge's viewpoint. He would have said it was not a legal opinion.

The implication is clear. Anything written can be used against one's case in court, but not necessarily for it. Written viewpoints of engineers, if they relate negatively to a potential infringement, can thus be damaging. They should not be written. An oral communication to the attorney will provide a strong enough statement that the infringement possibility needs to be thoroughly studied and that a legal opinion is required, not an engineer's.

Patents as Reference Material

A study of patents is invaluable for reviewing the current and past technology in a given area where the company has an interest. Besides studying the cited references mentioned earlier, one can find additional patents of possible interest by a search of the patent classification.

Each patent is categorized by a class and a subclass, clearly noted on the patent. Cited reference patents also have class and subclass identification, which may be the same as, or slightly different from, that of the initial patent. Using a list of these classes and subclasses, a search for other related patents can be made, either by the engineer, by another employee, or with the help of a patent attorney.

The U.S. Patent Office publication *Official Gazette,* issued weekly, lists new patents in a variety of ways, including class and subclass. A perusal of each issue would keep one current. Review of a year's worth (50 issues) would be a minor chore, but review of the last 10 or 15 years would prove very laborious unless one used the annual summary. A search of that type can be done at relatively low cost through a computer data retrieval service if the company uses one, or through a patent attorney's office.

The list of patent numbers thus obtained will be a good starting point. All of the patents can be ordered at a nominal cost. (At this time, the cost is one dollar each.) If the list is formidable and the company is located near a major city, the actual patents can be studied by a visit to a Patent Deposit Library (PDL). There are 30 of these libraries in 20 states, in addition to the Public Search Room of the Patent and Trademark Office located in Washington, D.C. The company patent attorney would have a list of locations. Some PDL collections have only the last few years of patents, but most of them have files of patents that go back 100 years or more. It is a good idea to find out the collection size of a nearby PDL before arranging a trip there.

PATENT AWARDS AND REWARDS

Probably because of the myths of world acclaim and wealth that still surround the idea of patents, there is a psychological boost when an engineer receives a patent. Besides any financial award or commendation, there is the satisfaction of knowing that one of his ideas was good enough for the company to invest time and money in applying for a patent. Furthermore, it survived an adverse examination in the U.S. Patent Office, whose job it is to ascertain whether an idea is novel and patentable, or not.

Besides that, there are the friendly jokes and congratulations from coworkers, who recognize the intangible value of a patent as an achievement. It is an achievement that has value because it provides a feeling of accomplishment, which others sense and perhaps envy a little. The inventor/engineer whose idea became a patent now joins an elite group, those who have achieved their first patent. It is a good feeling, which is repeated each time another patent is received.

With receipt of more than one patent, the engineer begins to be considered an inventor as well as an engineer. With several patents to his credit, he begins to be recognized as an idea man, a problem solver, a creative individual. The image is not a fictional one generated in a general way, but factual, attested to by recognition of the patent office. The image is deserved, and can be a valuable reminder of performance at salary review times, or when times are tough and layoffs must be considered. The engineer whose ideas have been patented has distinguished himself as a valuable employee.

PATENT POLICIES—ASSIGNMENTS AND AWARDS

Most companies require that engineering and other professional staff members sign a pre-employment agreement that gives the company the rights to the employee's future inventions. Whether this is intrinsically right or wrong is a subject for prolonged debate. A substantial investment in product development, tooling, and marketing is necessary to take an inventive idea to the marketplace, with the risk that the idea will be a commercial failure. If it is successful in the market, the inventor would like to have a share in that success. This is reasonable, but he is not the only individual who provides substantial effort to make the idea successful. So, rewards should not necessarily go only to the inventor.

An inventive employee who believes that although his inventions have brought the company substantial profits, he is not being compensated adequately, is in a paradoxical situation. He may be difficult to manage because of his dissatisfaction, and may be somewhat resentful. Yet, his inventive contributions are part of the growth of the company and must be encouraged.

There are no easy answers to this. The design manager may have an ongoing problem if his company has inadequate patent award policies. He can alleviate some of the problem by organizing patent recognition activities. Recognition within the company, within the engineering department, could be one type of award for a company to use when it does not provide a monetary award for patents based on dollar value to the company.

Since its commercial value is often uncertain, when a patent issues, an award based on value is difficult to determine. To avoid this complication and make a strong effort to encourage and stimulate ideas, some companies establish an equal award for every patent. This recognizes that each has potential value to the company. The amount of the award is somewhat nominal. In many cases it is $100 to $200 per patent, split equally between inventors of record. These monetary awards, presenting along with a plaque at an appropriate dinner, meeting, or ceremony, provide recognition that the patent has value for the company, and that the inventor is a person of value.

Patent awards of greater value may be issued if a company has derived substantial profits from the commercialization of a particular patented idea. For example, IBM Corporation distributed awards totaling $225,000 among the nine inventors of the computer programming language FORTRAN. Aerospace companies such as Lockheed reward patents with amounts between $1,000 and $20,000.

Whether a patent award should be modest or large, and whether it should reflect monetary benefit to the company or recognition of company-supported creativity, are issues that will be discussed and argued for a long time. There is much to be said for any approach that recognizes the creative thinking of the inventor and rewards his contributions in some way.

Since the basis for a company's growth is the development of new ideas, one particular company encourages the flow of creative ideas by beginning with modest awards for disclosed invention ideas. The procedure is to award shares of company stock equal to about $50 when an idea is disclosed and assessed to have technical merit making it worth further study. A second award of equal value is paid when a patent application is filed on the idea. A third award equal to the sum of the first two is paid when the patent issues. Thus, the inventor's contribution is reidentified with each milestone toward patenting.

This system recognizes and encourages the submission of many ideas because it rewards almost all of them. It recognizes that not all ideas are significant enough to be filed on, and that not all filed applications become patents; but it recognizes that from many good ideas, a few excellent ones emerge.

DISCLOSURE PROCEDURES

Disclosures are the first step in the transition from idea to patent, as the inventor takes a concept that he has been working on, converts it to written form, and submits it to others for review, comment, and action.

A system of disclosures that encourages submission of ideas is important, rather than one that, by its complexity, discourages them. A good test of that factor is a survey of how many staff members have submitted only one disclosure. If the number is more than 50%, and if their comments about why there has been only one tend to be standardized and center about disclosures being "too much trouble," the disclosure procedure needs to be reviewed and revised.

A good disclosure system should be easy for the inventor to use, cover the essential items, and be a legally acceptable document. In that context it is important that the company management seek the advice of its patent attorney early in the process of developing or writing a disclosure procedure. It should be correct from the beginning.

A good disclosure system will assure that potentially valuable patent rights will be properly protected, and engineers will be encouraged to disclose their ideas.

Characteristics of a Poor Disclosure System

Before discussing a good disclosure system, it will be helpful to consider the characteristics of a poor one, and how an engineer/inventor might try to use it. A typical disclosure system will require that the disclosure be filled out according to a specific format. Of course, it must be understandable and familiar to the engineer/inventor's colleagues who will read it. A number of computations may be necessary to prove that the concept fits a known theory. The original design concept may require modification, since a model should be built and tested to make sure it works before one applies for a patent.

As the disclosure form is being prepared, it will be necessary for the inventor to answer a number of key questions:

- When was the invention conceived?
- When was it included in a product?
- When and to whom was the invention disclosed?

This is pretty straightforward. Next, when the forms, drawings, and test records are readied, the inventor must approach colleagues to "witness and understand" the invention. This can be difficult—particularly so if one prospective witness, an older engineer who is skilled in the technology, thinks the idea is old, and signs it somewhat reluctantly. Then a second one may sign it with the comment that the patent committee will stifle it because "they don't accept anything."

In just such a situation, somewhat disappointed by his colleagues' lack of enthusiasm and failure to recognize talent, one engineer nevertheless placed the completed and witnesses disclosure forms in an interoffice envelope addressed to the company patent department.

Then he waited, and waited. His work on other projects was important, but did not seem half as important as the disclosure. Eventually some mail arrived from the patent department. It contained no check, no promise of an award—just a Xeroxed form that tersely assigned a number to the invention disclosure and asked that it be used on any correspondence regarding the disclosure.

With that bit of reassurance, the engineer went back to his assignments—his invention was on its way. So he want back to work, and waited for another response, but wrote no more disclosures.

In due course—which seemed an eternity—another response came from the patent department. This time it was a printed form, with a number of possible reasons why the disclosure would not be filed. The one checked was "Insufficient novelty over prior art." The disclosure was dropped.

And, for that promising engineer who might have become an inventor, that's all there was. Probably he never will submit another disclosure.

The disclosure procedure was complex, difficult, time-consuming, and somewhat demeaning. The disclosure was treated as if it had little potential value, and its perusal seemed an act of condescension by an unseen patent department.

Part of the problem was a poor system, and part of it the engineer's lack of knowledge of the workings of the system.

The process of submitting disclosures usually follows this sequence:

1. Disclosures of inventions are made on forms with detailed accompanying instructions.
2. Disclosures must be "witnessed and understood" by two (or more) competent persons.
3. Disclosures are forwarded to a patent department or a review committee where they are evaluated.
4. Sometimes the disclosure is submitted to the employee's manager, who forwards it to the committee with his comments.
5. Numbering is the first step in a disclosure evaluation. It involves many people and is designed to create a clear legal position for the conception date, novelty, and ownership of the invention.

Because of this impersonal sequence, the disclosure appears to be in an unfriendly environment from its initiation until a decision is made to file or drop it. These procedures will deter many, although not the determined engineer who wants to patent so strongly that he will try and try again regardless of the system.

Improving a Disclosure System

One guiding principle for improving a disclosure system is "Make it easy for the inventor." If that can be done, the primary objective of the disclosure system can be better achieved—more inventive ideas will be generated.

In his book *The Management of Inventive People,* T. Clark suggests this uncomplicated approach to making it easy to submit disclosures.*

1. Disclosures should be on simple forms, headed only by the words "What I have invented is . . . ." The inventor proceeds to describe his invention. Inventors will become used to providing sketches that are labeled consistently and referenced in the text. It gets easier to do this with more practice in describing how devices work.
2. Eliminate the signature by witnesses. The inventor's signature, the date, and verification that it has been received by another party, such as a date stamp plus a prompt letter of acknowledgment, will be enough to complete the disclosure. Anything more adds to its complexity and demotivates too many inventors.
3. The procedure should cover "inventive disclosures," not "patent disclosures." Many valuable inventions are not patentable, or not

---

* Quotations from *The Management of Inventive People.* Copyright 1980 by Trevor Clark. Permission of Cove Press.

economically so. Others are patentable but are more valuable as product applications. So, a system that provides an award or reward for a good invention disclosure, whether patentable or not, will encourage more ideas.

A second principle is "Keep the inventor informed." Lack of knowledge of how a disclosure is doing discourages an inventor because he thinks his disclosure is being ignored, is not of value, and that any other ideas disclosed will similarly be ignored. A disclosure system is relatively slow, in terms of all of the elapsed time that occurs between idea generation and issuance of a patent, but most inventors do not know this. They also do not realize that even though the process is slow, this delay gives them more time to perfect the idea, or think of other, alternative ways to broaden the patent.

Both the patent attorney and the engineering design manager can help to improve this situation. The attorney can occasionally advise the engineer/inventor of the status of his idea, and how long the next procedural step may take. The manager, also, can keep the engineer informed about recent actions taken, and make sure the engineer has been thinking about possible ways to broaden his idea. He also can have the engineer review, and comment on how to circumvent, any patents that may have been received as part of a patent search. More information on these areas will be developed in the next section of this chapter.

Encouragement of invention disclosures leads to new and innovative answers to problems, and, beyond that, to new product possibilities that may not have been considered. The dual approach of simpler disclosures and information feedback to the engineer/inventor has much merit. No one can predict who will have those better ideas; so if the submission of many ideas is encouraged, a small number of good, new inventions will be uncovered. This approach can bring innovative benefits to a company, if used.

FILING PROCEDURES FOR A U.S. PATENT

It now takes about two years or longer for a U.S. patent to issue, from the date of its application. The following section, which covers the major steps involved from invention disclosure to a patent, explains why the whole procedure is lengthy. These steps include the first patentability review, file search, application preparation, application filing, office actions, responses, amendments, and, finally, patent issuance.

Patentability Review

When an invention disclosure is reviewed, whether by one person or by a committee, several questions should be asked and answered:

1. What is the potential value of the new product (product improvement) created by this invention?
2. What might our competitors do with this invention?
3. What other inventions do we have that could be strengthened by this one?
4. What are the novel aspects of this invention?
5. Does it resemble any patent art we are familiar with?
6. Would the inventor's concept be of more value to the company if it were treated as a trade secret rather than a patentable idea?
7. How likely is the idea to achieve its commercial potential if a patent issues?

After these questions have been answered and discussed, the patent review committee will make a decision on whether the disclosure should be considered for filing or dropped. Members of the committee will usually include technical and marketing management executives. The company's patent attorney is a key member of the review committee and provides input to it. If he has spent a number of years with the firm, he is quite familiar with the company's patent position, competitors' patents in the company's technology field, and possible problems in obtaining a patent on a particular idea.

The inventor's manager may or may not be a member of the patent review committee, but he is usually asked to provide a recommendation addressing the first three questions on possible use of the invention. If he has experience in the company's patent art, he may comment on several of the other questions as well.

If the idea is to be dropped, the reasons for this should be clear, and the inventor should be told why. This should be done by a letter from the patent attorney, and followed up by positive verbal reinforcement from the engineering manager. The manager's input is important because the disappointment of a disclosed idea's being dropped will be tempered by awareness of the manager's interest and encouragement to generate more ideas.

File Search

If the decision is to file for a patent, a patent application can be prepared, or a search for prior art patents of significance can be made first. While a search may cost a few hundred dollars, it is preferable to filing without one. The purpose of the search is to uncover similar patents that may have already issued and to determine how to avoid their restrictions.

Omitting the search saves some money, but if the patent office uncovers the similar patents referred to, it may reject all of the claims or place major restrictions on the claims submitted. This could render any patent that issues much less valuable than it would have been.

The file search usually covers patent classification numbers appropriate to the device invented, and should include not only patents issued to competitors, but those issued to non-competitors who may have patents in the area.

After receiving the search, the attorney will review the patents involved to get an understanding of the prior art. He may elect to broaden the search if he believes it will be beneficial to do so, or he may not. If not, he contacts the inventor and his manager, briefs them on the search results, and sends the patents to the inventor for review and comment.

His objective is twofold: to have the inventor determine where his invention differs from prior art and to determine what limitations prior art may impose on a patent filed on the invention. Sometimes the inventor is shocked, other times pleased, at the quantity or the scope of other patents in the field. For a first-time inventor, the extent of patent art can be discouraging, but the more experienced inventor looks for shortcomings in prior art, and ways in which the major patents can be avoided by his invention.

After a joint review by the attorney, the inventor, and his manager, a much better idea of the novelty of the invention and the limitations of prior art can be presented to the patent review committee. The decision may be to close the disclosure and not file for a patent because prior art has anticipated the invention, and the likelihood of obtaining a patent is low. On the other hand, if the disclosure should be filed as a patent application, it can be done with the additional input from the inventor showing improvements over prior art and perhaps covering other areas not previously considered.

Preparing the Application

After the decision to file, the patent attorney begins drafting the patent application, using the format described earlier in the chapter. He uses the formal, somewhat stilted language of patents to make sure the invention is understood by the legal minds in the patent office who will evaluate it. He also writes it with a view to the competition's patent attorneys, who later will dissect the patent and try to find weaknesses and loopholes in it.

He will have the patent drawings prepared by a patent drafter who is skilled in the necessary representations of shapes, circuit schematics, or chemical formulas necessary to illustrate the concepts filed. When the draft is completed, he will ask the inventor and his manager to review it looking for errors and omissions, and will ask for their concurrence that the proposed claims cover the broadest possible interpretations.

Generally, the claims will be written in three groups. The first group will strive for the broadest possible coverage. Several independent claims may be written or, sometimes, only one. The second group of claims is less general and somewhat more specific. The third group is extremely specific and covers the exact structure illustrated, to prevent a direct copy. Claims are written this way for two reasons: first, to obtain the broadest possible coverage of the concept if all the claims are allowed; and, second, to achieve, as a minimum, protection of the actual invention if the other groups of claims are rejected.

After the draft has been reviewed and edited as necessary, the application is ready for final typing and filing.

Filing a U.S. Patent Application

Prior to mailing, the patent application is retyped, via a word processor, in the required columnar structure, with numbered lines. This is to enable the examiner and any other subsequent patent reviewer to refer to a specific location, such as "column 3, lines 12 to 14," without any confusion as to location of a specific part of the text.

The inventor performs a final review of a copy of the patent application. He signs two documents: an oath that declares he is the inventor and it was his original idea, and a second that he grants power of attorney to the patent attorney for any correspondence with the patent office relative to that application. The first document is notarized to assure that the inventor signed it. He may, instead, sign a declaration, in lieu of oath, that does not require notarization. The documents assign the invention to the inventor's company. With the required fees included, the application is sent by the attorney to the U.S. Patent Office, in Washington, D.C., to await its turn for office action by a patent examiner.

As a point of interest, the foregoing sections considered was only one inventor. In some cases there are two, and occasionally more than two. The patent attorney during the course of his prefiling activity tries to discern who contributed what to the invention. If it appears that all are co-inventors, then each of them is involved in the evaluation of the patent drafts and in signing the appropriate forms.

Office Actions

After the patent application has been filed, a notification of receipt is sent by the patent office. It is then necessary to wait for the first office action. At the time of this writing time the waiting period before the first office action occurs is often longer than one year.

Most inventors are elated when the application is filed; they naturally expect something will happen soon. So, it is prudent to advise the inventor in advance that the patent office does not move quickly, and that at least one year will pass before any information is fed back. The response time is a function of the number of patent examiners and their backload of patent cases. There does not appear to be an easy solution to this time problem. Whether one is impatient or patient, does or does not follow up, will not affect the outcome significantly.

The patent office response, the office action, will in many cases be a rejection of some or all of the claims on the basis of prior art, which will be cited. The examiner may find a foreign patent and a U.S. patent (or several) that, in combination or singly, negate the novelty of the claims, or render them obvious.

As disheartening as this appears, the attorney and the inventor need to take a close look at the examiner's comments and the rejected claims to determine where they agree, and where they disagree with the examiner's findings. Often one can find an argument to the contrary, or at least a strong difference of opinion as to the extent to which the references cited relate to and negate the invention as claimed. One patent attorney commented that he expects patent office rejections on at least some of the claims in applications he has written, for if there were none, he would not have written the claims broadly enough.

The attorney's reply to an office action within a specific time, usually three months, states counter arguments as to why the claims are not anticipated by the cited patents. In many cases, however, claims may have to be amended to meet the examiner's valid points. This may reduce some of the potential scope of the patent when it issues—in other words, may weaken it.

This is one of the reasons a pre-application patent search is important. If the prior art is thoroughly studied before the patent application is made, there will be fewer surprises in the office actions. The claims will be written with greater awareness of the prior art and will result in a stronger patent. If at least some of the prior art is known, the application can be written to allow easier future amendments if they are required.

After the office action hs been responded to, the return response by the patent examiner will occur within a few months. It could be an allowance of all claims as argued or amended, or a complete rejection of all claims as a "final" rejection. The patent attorney can appeal the examiner's decision if he believes it to be in error. This appeal is sent to the Board of Appeals of the U.S. Patent Office, which could either completely support the examiner's position or overturn his decision and allow all of the claims, depending on how it evaluates the case.

As a last resort, the attorney could have the patent case tried before the Court of Customs and Patent Appeals in Washington, D.C. Since all of these procedures are time-consuming and expensive, the patent's potential value to the company because of the inventiveness of the idea is the major reason to pursue a case to higher levels of appeal.

The Patent Issues

Assuming the patent application survives the rigors of the examiner's evaluations and office actions, the attorney will receive a notification from the patent office. It will state that the most recent set of claims has been allowed and a patent will issue in due course. That means that the final fees need to be paid, and the patent probably will issue in six months or so. It must wait its turn for final processing and listing in the OG (Official Gazette).

The actual issuance of a patent is sometimes anticlimactic to the attorney because of all of the arguments and modifications that were necessary during the prosecution of the case. However, the engineer/inventor has now entered the ranks of those who are the proud possessors of a U.S. patent in their own name, and the company will benefit from the protection this patent will offer to the new product line it covers.

Statutory Bars to Patents

One patent concern that needs to be addressed is the statutory bar. This is the legal limitation of a one-year time period between public use or sale of the invented device and the date of patent application. The statute is brief and to the point. It states that a patent may not be issued if:

the invention was patented or described in a printed publication in this or a foreign country or in public use or on sale in this country, more than one year prior to the date of the application for patent in the United States.

This one-year period allows the inventor a little time to develop his invention commercially before rushing to the patent office with an application. The definitions of publication and public use are rather precise and legalistic. The interested reader is advised to consult a patent attorney for a practical interpretation of these terms. The same advice applies to public sale, which will be discussed in the next paragraph. In this regard, the information contained is believed accurate and has been checked by a practicing patent attorney but is given only for information purposes. It should not be construed as legal advice.

The term "on sale" generally means an actual sale or an offering for sale. The sale can be for a product including the invention or products that were made using the invented process. The statutory bar on filing an application can sometimes exist without delivery of the product. This area is extremely complex, and the best approach to avoid a statutory bar problem is to make sure that the invention is completed and a patent applied for, prior to any publication, use, or potential offer of sale.

FOREIGN PATENTS

Not everyone realizes that a U.S. patent provides protection for an invention only in the United States, its territories and possessions, and nowhere else. An invention patented in the United States can be copied and sold in foreign countries without legal recourse for the inventor. However, if such an infringing product is sold in the United States, the importer, the distributor, and perhaps the foreign manufacturer can be sued under the law, and it is possible to prohibit the entry of the products into the United States.

In order to obtain protection in most high-technology foreign countries, it is necessary, in almost every case, to file for a patent in the foreign country. There are a few countries in which a U.S. patent can be registered for a measure of protection. However, the registration is often contingent on the patent owner's agreement to use the patent's technology in that country.

Generally a patent must be filed in a foreign country within one year of the U.S. patent filing. If this is done, maximum patentability protection is afforded, and a foreign filing date retroactive to the U.S. filing date will be granted. This early filing has both advantages and disadvantages. With the earlier effective filing date, it is possible to avoid concern about an "absolute novelty" requirement. For example, a magazine article about the new device or a description in a trade brochure would prevent subsequent patent filing in France because such minor descriptions violate its absolute novelty rule. The use of the device in the U.S. allowed by U.S. law (one year to file) would also prohibit obtaining a patent under French law.

Since most U.S. office actions are not received for over a year, the European patent may be filed without any knowledge of how the U.S. application is perceived by the U.S. Patent Office. This can be a serious disadvantage in considering foreign filings, but there is no way around this shortcoming at present. The disadvantage, of course, is that a foreign patent must be applied for prior to any knowledge as to how the U.S. application will fare in the patent office. If all claims are rejected, and the application is eventually abandoned, all the money spent on foreign filings may have been wasted.

Filing in foreign countries can be important if the company plans to market the products in specific countries, to manufacture the products overseas, or to license others to manufacture. A foreign patent will easily cost $2,000 or more to obtain, and after it issues, an annual fee, or annuity, is required to keep the patent in force. In some countries this fee increases each year; in others it is constant. The life of the patent in a foreign country is up to 20 years from the application filing date, depending on the particular country.

Group Filing of Foreign Patents

Before filing in one or more European countries, it is worthwhile to consider the EPC (European Patent Convention) approach, which was initiated in 1978. A patent filing must be submitted to the EPC within the normal one year of U.S. filing. A single, common format is specified regardless of how many countries will later be involved.

The application is sent to a single office to establish that it complies with the required format. Then it is submitted to a searching office (the European Patent Office in Munich) where a search of foreign (including U.S.) art is made. The applicant is provided with a copy of the search report, which lists the most pertinent prior art patents that could be located.

If the applicant wishes to pursue this parent application further, he designates the member countries of the EPC (see list) where patent protection is sought and pays the required fee. If the European Patent Office determines the application is patentable, the application is transferred to the patent office in each designated country. Each country issues a specific patent. Protection of the patent and enforcement are then governed by the laws of each country.

These are the 11 member countries of the EPC:

Austria	Netherlands
Belgium	Sweden
France	Switzerland
Italy	United Kingdom
Leichtenstein	West Germany
Luxembourg	

The advantage of the EPC approach is its single-patent format and one series of examinations. If he receives an unfavorable search report initially, the inventor and his company have the option not to pursue the application further and may save extra fees.

A disadvantage or an advantage, depending on one's viewpoint, is cost. The fees are high enough that unless filings are planned in four countries or more, it is less costly to apply in separate countries; but if more than four countries will be filed in, the reduced cost per country and the unified approach to evaluating the application make the EPC filing approach attractive.

When filing is done in countries other than the EPC group, the PCT group may be of interest as well. This group signed a Patent Cooperation Treaty (PCT), initiated in 1970, that operates in a manner generally similar to EPC. The countries participating in the PCT are listed below:

Australia	Hungary	Romania
Austria	Japan	Senegal
Belgium	Leichtenstein	Sri Lanka
Cameroon	Luxembourg	Sweden
Central African	Madagascar	Switzerland
Republic	Malawi	Tchad
Congo	Mauritania	Togo
Denmark	Monaco	U.S.S.R.
Finland	Netherlands	United Kingdom
France	North Korea	United States
Gabon	Norway	West Germany

For filing done in other foreign countries, the requirements of each country must be met in terms of fees, translation charges, and other matters. Almost all countries charge substantial fees for the process of obtaining a patent and maintaining it after issue. Clearly, foreign filings should not be undertaken casually because of the high costs involved, but international business can be facilitated if patents are in place in countries where sales and/or manufacturing agreements are anticipated. Some countries are party to no international treaties, and still others have no patent laws. The counsel of the company's international marketing group, blended with that of the patent attorney, is clearly needed. Together, they can provide a basis for present and future patent protection in world countries where the company will do significant business. As international trade expands, so does the need for foreign patents.

SUMMARY

Obtaining patents on innovative products is important to enhance the sales and profits of the company and protect the inventions from being copied. Patentable ideas generally come from inventors within the company, and the process of bringing these ideas from disclosure to patent application should be made as simple and convenient to the inventor as possible, in order to encourage the flow of a large number of ideas.

The process of obtaining a patent is a legal one and requires the services of competent patent counsel, well versed in the company's area of patent art. All must be aware that the process is relatively slow, and be patient. Office actions must be handled effectively with a view toward furthering the company's interest in obtaining a patent.

Foreign filing for a patent of a U.S. invention must be done within one year after the U.S. filing, to avoid complications in several countries that have very stringent rules about "public use" or "absolute novelty" of the idea. Because of the initial costs and the maintenance costs associated with foreign patents, countries must be selected with care. Anticipation of business trends and directions is important because a patent can rarely if ever be obtained in a foreign country for a number of years after it has issued in the United States.

Patents are an important form of technical literature that is valuable for understanding the current levels of technical processes and products. This reality is often missed by those who are "turned off" by their legalistic wording and unfamiliar structure. As one becomes familiar with patents, they become less formidable and much more informative.

Many other areas about patents could have been covered here, but were not. These include disclosures from inventors outside the company, patent agents, and patent interferences, for example. These and other areas are important and can best be approached with the aid of a patent attorney, who will be a valuable ally to the design engineering manager interested in improving the invention output of his department.

16
Products Liability
and Engineering Design Management

OVERVIEW OF PRODUCT LIABILITY ISSUES

In recent years, there has been a spectacular increase in the number of product liability claims and lawsuits. Because of a more demanding interpretation of manufacturers' responsibilities under the law, more of these claims have been successful, and the cost of settling them in or out of court has become extremely expensive.

It is no longer adequate to design a product that meets its specifications and works adequately within the intended use limits. It is necessary also to determine how the product could be misused by a person who either does or does not fully understand the instructions and warnings. The manufacturer, and therefore the engineering manager and his project engineers, must foresee unusual or unexpected uses of the product. Using a screwdriver to open a paint can, for example, might be an unexpected use. However, it would not usually be dangerous.

A product can be considered defective in a court of law if using the product—without attention to the instructions or warnings—would expose the user or bystanders to unreasonable risk of injury or harm.

To reduce the likelihood of product liability suits, engineering design must adhere to principles of conservative design to ensure not only the design, but also the manufacture of reasonably safe products. The term "reasonably safe" will be explained later.

Not all products are equally likely to be involved in a product liability suit, but none is immune. In a 1980 survey, the nine "top" product categories involved in liability suits were: prescription drugs, valves, automobiles and trucks, campers and mobile homes, miscellaneous machinery, electrical appliances, nonindustrial chemicals, cloth and fabric, and asbestos. However, the list changes in time as accidents or industrial illnesses occur and claims are made. The cyanide lacing of nonprescription pain relievers with resulting lawsuits and the asbestos-related lawsuits leading a major asbestos manufacturer to declare bankruptcy are two recent examples. It appears that consumer-usage-related products will dominate the products liability litigation scene.

A product liability case can develop whether the product has a design defect (in which every product can potentially cause problems) or a manufacturing defect (where only an occasional product is defective). Thus, it behooves the manager of engineering design to become aware of existing liability laws, their application in the courts, and how to reduce the probabilities that any of his products will become the target of a lawsuit.

This chapter deals with those aspects of product liability. It is not intended to replace competent legal counsel as a do-it-yourself guide. Rather it should make possible more meaningful dialogue with an attorney.

When questions arise concerning product liability, a lawyer specializing in that field should be consulted. Even if the situation appears clear-cut, an outside opinion can be extremely valuable. It is important to remember what will be restated later in this chapter: the term "defect" is defined by the courts and not by a manufacturer. If these defects can cause harm to a user, misuser, or bystander, a lawsuit can develop should any harm or injury occur. The ancient principle "caveat emptor" (let the buyer beware) is being replaced by a newer one: "let the manufacturer beware."

BASIC CONCEPTS OF PRODUCTS LIABILITY LAW

"Products liability" is a legal term that describes an action where an injured party seeks to recover damages for personal injury or loss of property from a seller, when it is alleged that the injuries came from a defective product. The injured party is the plaintiff, the seller is often the manufacturer and is the defendant, and personal injury is the predominant cause of today's liability lawsuits.

The product defect can be a production/manufacturing defect where a product that did not meet the manufacturer's own standards was sold to a customer, or it can be a design defect where the product does meet the manufacturer's standards; but its use, in either case, injures the plaintiff.

Most product liability cases are founded on one of these three legal principles:

- Negligence
- Strict or implied warranty
- Express warranty and misrepresentation

Negligence is conduct that "involves an unreasonable risk of causing damage," or conduct that "falls below the standard established by law for the protection of others against unreasonable risk of harm." Negligence does not mean that the defendant intended harm or that he acted recklessly in bringing about harm.

The test for negligence focuses on what the manufacturer should have known at the time the product was manufactured. It is an evaluation of the technology and information available to the industry. Whether the manufacturer conformed to industry standards may or may not be adequate. A court will decide if he acted reasonably or not.

For example, in a case involving a steam boiler explosion and injury to the plaintiff, the manufacturer was found negligent because a hydrostatic test was used to establish boiler integrity. The court determined that the manufacturer should have used another test, developed by a University of Wisconsin professor, that had a higher probability of discovering flaws in the metal and was a feasible alternative.

The difference between negligence and strict liability is subtle. Negligence relates to the reasonableness of the manufacturer's conduct, that is, the standards he applied to the product before it left his hands. Strict liability relates to the reasonableness of the product in the environment of its use. Under strict liability, any of the sellers—retailer, wholesaler, distributor—as well as the manufacturer can be held liable. A component part supplier is potentially liable, also. In other words, any member of the distribution chain could be liable because strict liability focuses on the condition of the product in its environment, not on the reasonableness of the conduct of the sellers.

Types of Warranties

Implied warranty assumes that a product offered for sale will be reasonably safe for its intended use. The legal codes covering implied warranty are not as clearly defined, and an implied warranty does not have as strong a legal position as strict liability.

Express warranty exists when a seller warrants or represents that a product has certain characteristics or will perform in a certain way, and the product fails to meet these criteria. If the buyer can then prove that his injury resulted from the failure of the product to meet the warranty, liability is established. There frequently is a question of a salesman's exaggeration, called "puffing." If a used car salesman talks about the car's "A-1" condition, that may not mean much if it is eight years old. This type of inflated sales talk cannot be treated as an express warranty when it represents a commendation of the seller that he wants the buyer to believe. To say that a product is "the best on the market" is to provide an opinion and not an express warranty.

Misrepresentation is not the same. It involves the willful or unwillful statement that a product will meet the use criteria, when, in fact, it will not. It is very close to violation of an express warranty.

Regardless of which liability principle is involved in a lawsuit, the plaintiff must establish that:

- The product was defective.
- The defect existed when the product left the seller.
- The defect caused the harm.

On the other hand, in some states, the plaintiff's claim can be defeated if it is proved that he has been negligent in using the product and contributed to the problem by his conduct. In these situations, either the suit will be defeated, or a percentage of negligence will be determined and any awards reduced by the amount of negligence assigned to the plaintiff.

Litigation Trends

There is no general rule that can be applied to litigation situations. Since product liability laws vary from one state to another in both their detail and application, nothing can be taken for granted, especially in the area of contributory negligence. Lawsuits for physical harm are becoming more frequent, and damage claims are escalating, regardless of the company's ability to pay.

It was mentioned earlier that nine categories of businesses have had the majority of product liability lawsuits. However, no industry or business is immune from the possibility of a lawsuit. While a company can do relatively little with existing products in use, short of a product recall—as automotive manufacturers do—it can begin to change its internal procedures so as to significantly reduce the probability of shipping a defective product that could cause harm and be the cause of a lawsuit.

The following sections deal with approaches that the engineering design manager and his staff can use toward that end.

PRODUCT/ENGINEERING/MANAGEMENT VULNERABILITY

Engineering management and company management are vulnerable in four areas of preparing a product for the market:

- Design
- Manufacturing and materials

- Packaging, installation, and application
- Warnings and labels

Design

Design is, of course, the heart of a product. From the standpoint of product liability, the design of the product must be examined adversely. There is an ever-present possibility that a concealed danger has been created by the design. The danger could exist for several reasons:

1. The materials selected had inadequate strength or failed to comply with accepted industry or user standards.
2. Necessary safety devices were not included in the design. When these devices are offered as extra cost options, courts have often acted unfavorably toward the manufacturer.
3. The designer, working toward normal uses, failed to consider possible unsafe conditions due to use or misuse of the product that were readily foreseeable by him. Whether or not he actually perceived this misuse is not the concern of the court.

Manufacturing and Materials

As mentioned earlier, a defect can occur in design or in manufacturing. Whatever the reason, if the defect caused the accident leading to personal injury, liability is involved. "A defect can emerge from the mind of the designer as well as the hand of the workman," one judge stated.

Manufacturing and materials take into account the production processes as well as the materials that are processed. If tooling is worn and stamped parts are weakened, defects can occur. Manufacturing defects may occur in areas where quality control inspection is on a sampling basis, or there may be 100% inspection and one or more defective parts get through. Or, a process requiring heat to cure a product could cause defects. If a heater burned out and it was undetected, a part could be undercured and have inadequate strength; whereas under other circumstances a part could become overcured from too much heat, and be brittle.

Material themselves change with environment and time. They age. In foreseeing use and abuse over time, one must evaluate the adverse changes that occur as the materials used in a product age.

Packaging and Installation Instructions

Packaging.   At the very minimum, packaging must protect the product from the time it leaves the factory until the user unpacks it for use. It must keep out moisture if the product would be damaged by excessive moisture or humidity. It must protect the product not only from breakage, but, more important, from developing internal damage, or cracks, which would be invisible but would cause the product to have a defect.

One unusual case of a packaging defect is worth noting. A manufacturer of wooden doors that had glass windows in the top made an overseas shipment of his product. Because of concern about glass breakage, with the doors stacked flat, the glass was packed separately. Doors were packed in groups, each group stacked approximately 3½ feet high. A cardboard cover was placed around the doors for protection, and two steel bands secured the stack. The ends were not covered and gave the appearance of solid wood to the casual observer. The openings in the doors were in line with one another, and this void was covered with the cardboard sheeting. Markings were minimal, besides the words "fine doors."

A longshoreman carrying a 100-pound sack of flour walked across the doors and was injured as he fell through the void area. In the trial, the manufacturer claimed this occurred because of product abuse, but the court decided the injury was caused because of the manufacturer's packaging method. He should have known it is customary for longshoremen to walk on material already loaded. Thus, the failure to foresee how the product would be "used" in transit ended with the manufacturer making a substantial payment for damages.

Installation Instructions.   Installation instructions can be the source of a future problem if they are not clear, but are ambiguous and subject to more than one interpretation. They must not only be written clearly, but also must be capable of being followed by someone relatively unfamiliar with the product, who is installing and or assembling it for the first time.

The best way to check out installation instructions is to have someone unfamiliar with the product use the instructions to install or assemble it. If this person has any trouble following the instructions or interpreting their meaning, the difficulties must be cleared up because they will cause similar or worse problems in the field.

Warnings and Labels

Warnings, usually in the form of labels as well as printed statements in user manuals, are intended to alert the user to potential dangers in the product's use. To have any effect, they must warn of the dangers inherent in the failure to follow instructions or warnings.

Since foreseeable dangers and misuses of the product play an important role in litigation, the manufacturer must try to foresee extraordinary uses of his products. A product is considered defective if its use without warnings or instructions would expose the user or bystander to unreasonable risk.

However, warnings are not a solution to design defects. When risks could be reduced significantly by redesign, a court may decide the manufacturer is liable for damages because he decided to try to cut the risk by warnings instead of redesign.

On the other hand, warning labels are a way of dealing with hazards and risks where redesign of the product cannot be accomplished without increasing the costs to a point where it would be an unsalable product, too expensive for the marketplace. This is known as the cost-utility factor.

Advertising

One other area where product liability exposure exists is in advertising. While the engineering design manager is not usually involved in advertising development or layout, he may be able to influence or approve the choice of words or phrases that are important. These are some recommendations to consider, many of which are in common use:

1. Truth: All claims should be true, not false, deceptive, or capable of being misleading or misunderstood.
2. Guarantees: Avoid the concept of guarantees in ads.
3. Safety: Unless the device is a safety mechanism, do not use the word.
4. Superlatives: In general, they can be troublesome. Words such as "safest," "foolproof" should be avoided. "Bigger," "simple to operate" are better choices.
5. Photos and illustrations: These should not show the product being used incorrectly or unsafely. Do not show it with safety guards or gates removed.
6. Accuracy: Consider legal and engineering review of proposed ads to provide technical accuracy and catch any exaggerated performance statements. The "risk" of a dull ad must be prevented by creative advertising approaches within the concerns of product liability considerations.

In all of these liability areas, there is a substantive risk of a lawsuit should some type of defective product be sold in the marketplace. The amount of risk is impossible to forecast because the defect may or may not be a flaw in the design, material, manufacture, warnings, instructions, or advertising. To repeat, a defect in product liability law is defined by the courts, not the manufacturer. However, most defects will relate to a flaw in one of these areas, often stemming from the design, materials, or use of the product. The design engineering manager can reduce these liability risks in the ways described in the following section.

REDUCING PRODUCT LIABILITY RISKS BY DESIGN

The design process must provide for a broad analysis of the overall design, including fail-safe design and redundancy, hazards analysis, and product safety audits, to reduce the risk of a product liability lawsuit. Greater attention must be paid to codes and standards, failure analysis, comprehensive design reviews. Some of these methods were mentioned in earlier chapters.

The manager's responsibilities are broadened to include these extra efforts and their documentation, in his work toward both liability prevention and a safer product. Life cycle design concepts must be used because of the aging effects of materials in various normal and abnormal usage environments.

To bring more structure to the concept of product safety by design, these seven concepts can be useful:

1. Define and describe the scope of the product's intended use.
2. Identify all of the environments where the product will be used.
3. Describe the user population, and their familiarity with this or a similar product.
4. Consider all of the possible hazards, including estimates of the probability of their occurrence and the seriousness of any resulting harm.
5. Determine alternative approaches—design features, production techniques, material changes, design simplicities or revisions, warnings—that could significantly reduce or eliminate the hazards.
6. Evaluate these and other alternatives relative to the expected performance of the product, including:
   (a) What hazards may be introduced by the alternatives.
   (b) The effect of these alternatives on the usefulness of the product.
   (c) Their effect on the cost of the product.
   (d) How the revised product compares to others being sold.
7. From this review, make decisions about which features are to be included in the design.

In all these considerations, the effects on cost must be weighed. However, cost includes more than materials, labor, burden, and selling expense. A broader view including the cost of possible injuries, with resulting settlements or lawsuits, must be taken. These are part of the product's true cost and should affect decisions about safety features or fail-safe design. However, estimating these costs can be somewhat subjective rather than objective, and these considerations may even influence a company's decision to enter the product into the marketplace. These final decisions are never easy, but with adequate attention to product safety factors as well as costs, some of the subjectivity can be eliminated.

Documentation

In the process of developing a safer design, a certain amount of documentation is involved, so that management can assess what is being done, what decisions are being made, and why. Admittedly, documentation is a two-edged sword. In the event of litigation, the other side will have access to information about why various trade-offs were made and for what reason. But, on the positive side, since the manufacturer is held to a standard of what he "should have known" about the product, it makes sense that he have a wide knowledge of it—both its strengths and its weaknesses.

The documentation areas important to new product development include the following:

1. Hazard and risk data, taken from historical files, field experience, and laboratory tests.
2. Design safety analyses—from design reviews and including fault trees, failure modes and effects analyses (FMEA), and hazard analyses.
3. Warning and instruction developments—how they were developed, what information was included, what excluded.
4. Standards—use of industry standards, in-house standards, mandated design or performance standards, testing beyond standards.
5. Quality assurance program—methods for selecting the procedures used, limitations of the methods, production records on defects noted, disposition of lots containing defective parts.
6. Product performance data. This covers the reporting procedures, complaint file and follow-up data, field returns and analysis of returns (failures and nonfailures), product recall (extent and reasons), and records of all product modifications.
7. Decision making—who made which decisions for what reasons, and what the results were.

Documentation in these areas, in the long run, will reduce product liability suits and damage claims—not because a judge or jury will be more understanding when they review the company's file, but because the company and its staff will know more about its products, where safety improvements are needed. Retaining this documentation will foster better, safer designs of new products early in the design process, when changes and improvements are least expensive to incorporate. The design reviews, emphasizing safety as well as performance, will also review packaging, labeling, instructions, warnings, advertising materials—since all of these are part of the total expected performance of the product.

Much of this decision making resides with design engineering management, but the decision process will be integrated into the total management philosophy of more reliable products.

Guidelines for Design Engineering Management

The following is a checklist of 19 guideline items for the design engineering manager. It amplifies many statements made earlier and provides specifics for review.

1. Consider safety to the user throughout the entire product life cycle.
2. Design to meet or exceed a nationally recognized standard.
3. Select materials and components of appropriate quality and with small deviation of key properties.
4. Apply accepted analysis techniques to determine whether all electrical, mechanical, and thermal stresses are within published limits.
5. Test the device using accelerated aging methods from a recognized standard if one is available.
6. Make failure and hazards analyses of the product for various stages of its life.
7. Make a worst-case analysis to make sure the product is not hazardous then.
8. Provide sufficient, clear notes on drawings and specifications to eliminate hazards.
9. Maintain a permanent record of the product development history from the original needs analysis to completion of pilot run testing and approval. The record needs to be complete enough that others can explain key decisions later.
10. Conduct design reviews, including in them persons knowledgeable about the product's transportation, installation, and end use.
11. Work with advertising to avoid overstatement of product performance.
12. Use warning labels when appropriate. Make sure they are understandable, adequate, complete, durable, and visible when in place.
13. Have all products inspected after manufacture for conformance to key requirements. However, make sure any test levels selected are not too severe and do not cause damage to the product.
14. Submit the product to an independent testing laboratory, for their evaluation and their approval, if possible.
15. Installation and use instructions must be unambiguous, and need to be checked out by someone unfamiliar with the product.
16. Determine what service or maintenance may be necessary to keep the

product safe and operating to specifications. Provide this information to the user with the installation instructions.

17. Encourage sales and service personnel to report any complaints related to injury or economic loss.

18. Run accelerated life tests on samples selected at random from a production run. Compare them against design tests, and take any remedial action indicated.

19. Document the trade-offs made in the design where the improvement in safety needed would make the product unsalable. These studies may be necessary in a product defense.

These guidelines, used in a climate of safety-conscious design, will reduce the risks of designing and manufacturing unsafe products, and will enable product safety to be one of the priorities in product design.

HAZARDS, RISKS, DANGERS

These three terms are often used in safety and product liability discussions, sometimes without proper definition.

- *Hazard* is a condition that has the potential to cause harm or injury to people, animals, or property.
- *Risk* is the probability of an accident occurring when a hazard is present.
- *Danger* is the combination of hazard and risk.

Products can possess inherent hazards of their own and contribute to other hazards. For example, with a bicycle chain and sprocket, the "pinch point" is such an inherent hazard. A compilation of major hazards is shown in Table 16-1, which lists hazard sources associated with equipment, and those associated with human performance and behavior. The evaluation of hazards from product failures is carried on by the design engineer as part of a failure modes and effects analysis (FMEA), described in Chapter 4.

Hazards can sometimes be avoided or reduced by fail-safe design, backup redundancy, and interlocks. While it is not possible to eliminate all hazards from a product, the effort by a design engineer to do so accomplishes two things—it reduces the probabilities of an injury causing a lawsuit and reduces the human misery associated with a serious injury.

It follows, then, that potential design hazards should be eliminated early. The product needs to be designed to stand up to misuse and abuse, in addition to the expected, normal use.

Table 16-1.   Inherent Hazard Sources.

A. Associated with Equipment:

Chemical	*Mechanical*	*Miscellaneous*
Corrosion	Weight	Noise
Toxicity	Acceleration	Light intensity
Flammability	Stability	Stroboscopic effect
Pyrophoricity	Vibration	Temperature effect
Explosives	Rotation	Pressure, suction
Oxidizing agents	Translation	Emissions
Photoreactivity	Reciprocation	Ventilation
Hydroreactivity	Pinch or nip points	Ignition sources
Carcinogens	Punching, shearing	Decomposition
Shock sensitivity	Sharp edges	Slipperiness
	Cam action	Moisture
Electrical	Stored energy	Aging
Shock	Entrapment	
Short circuit	Impact	
Sparking	Cutting actions	
Arcing		
Explosion	*Radiation*	
Radiation	Alpha, beta, gamma	
Overheating	X-rays	
Insulation failure	Infrared, ultraviolet	
	Radio and microwaves	

B. Human Hazards:

Personal	*Human Errors*	*Environmental*
Ignorance	Failure to perform	Weather
Boredom, Loafing	Incorrect performance	Noise
Negligence	Incorrect supervision	Temperature
Carelessness	Incorrect training	Light
Horseplay	Overqualification	Floor texture
Smoking	Poor judgment	Ventilation
Alcohol or drug use		Complexity
Sickness		Comfort on the job
Exhaustion		Warnings
Cultural background		Social factors
Stress		

PRODUCT LIABILITY SUITS

Only a small percentage of all product liability suits ever get to trial. A pretrial investigation by both sides often results in a decision by the plaintiff to settle out of court, or to drop the case.

If the plaintiff's attorney initiates a suit, one of the early actions is called the process of discovery. In discovery, the opposing sides are required to prepare copies of all pertinent documents and correspondence relative to the product. By this means, each side has a better chance to assess the strength or weakness of the opponent's position.

A review of the documentation will show if the defendant did or did not design as safe a product as possible, and what trade-offs occurred in the design. The files may or may not contain information detrimental to the defendant's case.

Should be product be put on trial, the key issue is not its design and documentation, but whether the product contained a flaw or defect—in the environment of its use—that caused an accident resulting in injury or property damage.

A product liability case contains these elements:

1. Description of the product, the accident, and the defect.
2. Description of the unreasonably dangerous nature of the product.
3. Establishment of the cause–effect relationship between the defect and the resulting injury.
4. Explanation of how the immediate cause of the accident is related to items 2 and 3.

The flow of evidence may not be as distinct as these four items. The difficulties occur during a trial because a proper description of the product is not made, and the four areas become muddled and confused. A potentially good case can be severely weakened by the attorney's lack of attention to sequence in explanation, as the jury will become confused. The flow diagram, Fig. 16–1, indicates schematically the flow of issues in a product liability case.

Since the question of technical causation (is the product failure responsible for the injury?) can short-circuit the lawsuit at the beginning, it deserves the attention of attorneys and experts early in the litigation.

MANAGER ON THE WITNESS STAND

It is unlikely that a manager will be called on to testify for his product, or defend his design on a witness stand. Lawyers familiar with product liability cases agree that the chances of this occurring are close to zero. These technical areas, for both the plaintiff and the defense, are covered by technical experts who are skilled in their expertise and able to evaluate probable scenarios based on the evidence of the accident and the product.

Though the manager will probably never sit in the witness box, he is indirectly on trial, should one of his products be involved in a product liability suit. Thus, if he can prevent a lawsuit by a safe improved design, he should

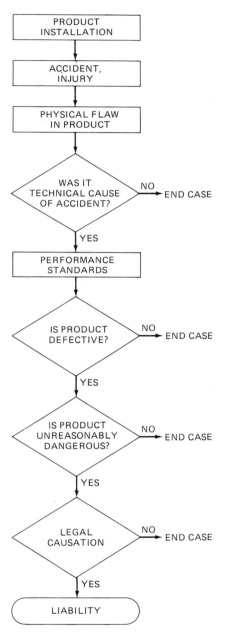

Fig. 16-1. Flow chart of legal and technical issues in product liability litigation.

do so. In his own evaluation of a design, he should keep these seven questions in mind, when deciding if one of his products has a defective and unreasonably dangerous nature:

1. How useful and desirable is the product?
2. What is the availability of other, safer products that meet the same needs?
3. What is the likelihood of serious injury for someone using or misusing the product?
4. How obvious is the danger?
5. What is the level of common knowledge and normal problem expectation of the danger?
6. To what extent can injury be avoided by care in the use of the product or by heeding a warning?
7. What are the probabilities of eliminating the danger without seriously impairing the usefulness of the product or making it unduly expensive?

While some of the questions are subjective and others reasonably objective, the final decision about the product's potential hazards is a blend of all seven answers and thus is somewhat subjective, depending to some extent on the product's environment and end use.

SUMMARY

Products liability lawsuits are an expensive fact of industrial life, encompassing a wide variety of products. The number of lawsuits tripled from 1975 to 1982, filling court dockets, increasing insurance costs, raising prices, forcing bankruptcies. It shows no signs of slowing down.

A large part of the problem stems from the absence of national, uniform guidelines for products liability law. Only 28 states have product liability statutes, no two of them identical. As a result, it is nearly impossible for a manufacturer to know beforehand whether a product will be safe enough to be free from a lawsuit in every state where it is sold. With these kinds of uncertainties, the need for safer products is clear.

Bibliography

Chapter 1

Bronikowski, R. "Managing the Engineering Interface." *Machine Design,* 8/23/79.
Levitt, T. "Innovative Imitation." *Harvard Business Review,* Sept./Oct. 1966.
Levitt, T. *Marketing for Business Growth.* McGraw-Hill, Inc., 1974.
Ryckman, P. "Write the Ad before You Design a Product." *Advertising Age,* 12/17/79.
Shaffer, R. "New Home Computer That Can Emulate Others Threatens to Shake Up the Industry." *The Wall Street Journal,* 1/12/82.

Chapter 2

Bronikowski, R. "Will That New Product Pay Off?" *Machine Design,* 7/6/78.
Humphrey, G. *Turning Uncertainty to Advantage.* McGraw-Hill, Inc., 1979.
Mathey, C. "The Management of Product Planning." Paper No. 75-WA/Mgt-6 presented at the Winter Meeting of the American Society of Mechanical Engineers (ASME), 1975.

Chapter 3

Bronikowski, R. "How to Fail Your Way to Design Success." *Machine Design,* 1/7/82.
Davis, S. and Lawrence, P. "Problems of Matrix Organizations." *Harvard Business Review,* May/June 1978.
Ryckman, P. "Write the Ad before You Design a Product." *Advertising Age,* 12/17/79.

Chapter 4

Eastman, R. "Engineering Information Released Prior to Final Design Freeze." *IEEE Engineering Management Review,* May 1980.

Chapter 5

Howell, R. "Multi-Project Control." *Harvard Business Review,* Mar./Apr. 1968.
McLeod, T. *Management of Research Development and Design in Industry.* Gower Publishing Company Ltd., 1971.

Chapter 6

Coates, J. J. "Engineering Techniques—What do they do, why use them, how are they managed?" *Chemical Engineering,* 6/28/80.
Haynes, M. "How Do You Score as a Delegator?" *Chemical Engineering,* 9/12/77.
Kerzner, H. *Project Management.* Van Nostrand Reinhold Co., 1979.
Mackenzie, R. *The Time Trap.* AMACOM, 1972.
Mercer, J. L. "Organizing for the 80's—What about Matrix Management?" *Business,* July/Aug. 1981.
Moe, D., "CAD/CAM means Higher Productivity." *Wisconsin Business Journal,* Mar. 1982.
Pyhrr, P. *Zero Base Budgeting.* John Wiley & Sons, Inc. 1973.
Raudsepp, E. "Why Managers Don't Delegate." *Chemical Engineering,* 9/25/78.
Rosenbaum, B. "Creating the Behavior You Want." *Chemical Engineering,* 4/20/81.

Chapter 7

Ashkenas, R. and Schaffer, R. "Manager Can Avoid Wasting Time." *Harvard Business Review,* May/June 1982.
Bronikowski, R. "Mastering the Subtleties of Decision Making." *Machine Design,* 8/21/80.
Bronikowski, R. "Down to Earth Advice for the New Manager." *Machine Design,* 5/10/79.
Druckers, P. *The Effective Executive.* Harper and Row, 1966.
Hines, G. "A Short Course for Supervisors of All Levels." *Industrial Management Journal,* Feb. 1968.

Chapter 8

Holzmann, R. "To Stop or Not—The Big Research Decision." *IEEE Engineering Management Review,* June 1974.
Murphy, T. "Design and Analysis of Industrial Experiments." *Chemical Engineering,* 6/6/79.

Chapter 9

Bittel, L. *The Nine Master Keys of Management.* McGraw-Hill, Inc., 1972.
Burgess, J. "Making the Most of Design Reviews," *Machine Design,* 7/4/68.

Dailey, C. *Entrepreneurial Management*. McGraw-Hill, Inc., 1971.

Martin, F. "The Importance of Engineering Checking," *Consulting Engineer*, 2/74

Miles, L. *Techniques of Value Analysis and Engineering*. McGraw-Hill, Inc., 1961.

Pitt, H. "Specifications: Laws or Guidelines?" *Quality Progress*, July 1981.

"Compendium on Reliability," *IEEE Spectrum*, Oct. 1981.

Staff writer, "Formal Design Review." *Machine Design*, 3/30/77.

Chapter 10

Howard, G. "A Lesson in Listening." *Machine Design*, 12/3/64.

Karger, D. and Murdick, R. "A Management Information System for Engineering and Research." *IEEE Engineering Management Review*, May 1977.

Raudsepp, E. "How to Sharpen Your Listening Skills." *Machine Design*, 5/9/68.

Tichy, H. *Effective Writing for Engineers–Managers–Scientists*. John Wiley & Sons, Inc., 1967.

"Are They Telling You the Truth?" from the Learning Curve. *Electrical World Magazine*, 7/1/80.

Chapter 11

Auger, B. "Fine Tuning Your Business Meetings." *Chemical Engineering*, 6/16/80.

Phillips, L. "Reading Between the Lines of Communications." *IEEE Engineering Management Review*, June 1981.

Raudsepp, E. "We've Got to Stop Meeting Like This." *Machine Design*, 2/11/82.

Sigband, H. "Engineering and Conducting the Successful Meeting." *Chemical Engineering*, 8/19/74.

"How to Make a Business Meeting Pay" from the Learning Curve. *Electrical World Magazine*, 2/15/78.

Chapter 12

Dunn, D. (ed.) "The Serious Business of Using Jokes in Public Speaking." *Business Week*, 9/6/83.

Quinn, J. "A Short Course in Writing." *Industrial Research and Development*, Nov. 1968.

Thomas, E. *How to Write Clearly*. International Paper Company, 1981, 1982.

Chapter 13

Berra, R. "Getting Results through People." *Machine Design*, 5/12/66.

Bhasin, R. "Replacing Poor Performers with Good New Ones." *Chemical Engineering*, 9/7/81.

Cohen, H. *You Can Negotiate Anything*. Lyle Stuart, 1980.

Emery, D. *The Compleat Manager*. McGraw-Hill, Inc., 1970.

Goldberg, E. and Cohen, E. "A Primer on Playing Politics within Corporations." *IEEE Engineering Management Review,* Dec. 1978.

Harrison, J. "How to Stay on Top of the Job." From *Paths towards Personal Progress. Harvard Business Review,* 1980.

Jacobs, J. "Four Keys to Motivating Engineers." *Chemical Engineering,* 8/14/78.

Ramo, S. *The Management of Innovative Technological Corporations.* John Wiley & Sons, Inc., 1980.

Raudsepp, E. "Getting Extra Miles from Performance Appraisals." *Machine Design,* 4/8/82.

Raudsepp, E. "Interviewing Candidates for Staffing Openings." *Machine Design,* 11/9/76.

Raudsepp, E. "Reducing Engineer Turnover." *Machine Design,* 9/9/82.

Sterner, F. "Managing and Motivating Engineers." *Professional Engineer,* July 1969.

Chapter 14

Bronikowski, R. "Seven Steps to Power Your Creativity." *Chemical Engineering,* 7/31/78.

Buggie, F. *New Product Development Strategies.* AMACOM, 1981.

Comella, T. "How to Manage Creativity without Killing It." *Machine Design,* 3/6/79.

Irwin, P. and Langham, F. "The Change Seekers." *Harvard Business Review,* Jan./Feb. 1966.

Levitt, T. "Innovative Imitation." *Harvard Business Review,* Sept./Oct. 1966.

Sinnet, G. "Challenge of Professional Development." *Industrial Research and Development,* Feb. 1970.

Steiner, G. *The Creative Organization.* The University of Chicago Press, 1971.

"Planned Creativity Pays Off" from *Nation's Business,* Jan. 1957. Copyright © 1957 by Chamber of Commerce of the United States.

"Creativity" reprint of Aug. 1964 from *Product Engineering.*

Chapter 15

Clark, T. *The Management of Inventive People.* Cove Press, 1980.

Karger, D. "Patents: Who needs them?" *Machine Design,* 1/22/76.

Kimball, A. "Patenting US Inventions Abroad." *IEEE Transactions on Professional Communications,* June 1979. (patents)

Schall, E. "Hints on Reading a Patent." *Chemical Engineering,* 11/7/77.

Stipp, D. "Lab Legacy." *The Wall Street Journal,* 9/9/82.

Terragno, P. "Patents as Technical Literature." *IEEE Transactions on Professional Communications.* June 1979, (patents)

Treplow, L. "Patent Fallacies," chapter in *Patent Background for Engineers.* Published by Allis-Chalmers Electrical Review, 1951.

Wolber, W. "The Business Value of Patents." *IEEE Transactions on Professional Communications,* June 1979. (patents)

General Information Concerning Patents from the U.S. Department of Commerce, Commissioner of Patents and Trademarks, 1979.

Chapter 16

Irving, R. "Is Liability Insurance Still Costly and Scarce?" *Iron Age,* Aug. 1980.

Lloyd, D. and Lipow, M. *Reliability: Management, Methods and Mathematics.* Lloyd & Lipow Publishers, 2nd ed., 1977.

Thorpe, J. and Middendorf, W. *What Every Engineer Should Know about Product Liability.* Marcel Dekker Inc., 1979.

Walton, O. "Effects of Advertising on Product Liability." *IEEE Engineering Management Review,* Dec. 1977.

Weinstein, A. and others. *Product Liability and the Reasonably Safe Product.* John Wiley & Sons, Inc., 1978.

Index

Index